Air Pollution from Motor Vehicles

Standards and Technologies for Controlling Emissions

Air Pollution from Motor Vehicles

Standards and Technologies for Controlling Emissions

Asif Faiz
Christopher S. Weaver
Michael P. Walsh

With contributions by

Surhid P. Gautam
Lit-Mian Chan

The World Bank
Washington, D.C.

© 1996 The International Bank
for Reconstruction and Development/The World Bank
1818 H Street, N.W., Washington, D.C. 20433, U.S.A.

The findings, interpretations, and conclusions expressed in this publication are those of the authors and do not necessarily represent the views and policies of the World Bank or its Board of Executive Directors or the countries they represent. Some sources cited in this paper may be informal documents that are not readily available.

The complete backlist of publications from the World Bank is shown in the annual *Index of Publications*, which contains an alphabetical title list (with full ordering information) and indexes of subjects, authors, and countries and regions. The latest edition is available free of charge from Distribution Unit, Office of the Publisher, The World Bank, 1818 H Street, N.W., Washington, D.C. 20433, U.S.A., or from Publications, The World Bank, 66, avenue d'Iéna, 75116 Paris, France.

Cover photos: Asif Faiz

Asif Faiz is currently chief of the Infrastructure and Urban Development Operations Division of the World Bank's Latin America and the Caribbean Country Department I. Christopher S. Weaver and Michael P. Walsh coauthored this book as consultants to the World Bank.

Library of Congress Cataloging-in-Publication Data

Faiz, Asif.
        Air pollution from motor vehicles : standards and technologies
    for controlling emissions/ Asif Faiz, Christopher S. Weaver, Michael P.
    Walsh, with contributions by Surhid Gautam and Lit-Mian Chan.
            p.   cm.
        Includes bibliographical references (p.   ).
        ISBN 0-8213-3444-1
        1. Motor vehicles—Pollution control devices. 2. Automobiles—
    Motors—Exhaust gas—Law and legislation—United States.
    I. Weaver, Christopher S.   II. Walsh, Michael P.   III. Title.
    TL214.P6F35   1996
    363.73'1—dc20   95-37837
                                                             CIP

Contents

Boxes

Figures

Tables

Preface

Because of their versatility, flexibility, and low initial cost, motorized road vehicles overwhelmingly dominate the markets for passenger and freight transport throughout the developing world. In all but the poorest developing countries, economic growth has triggered a boom in the number and use of motor vehicles. Although much more can and should be done to encourage a balanced mix of transport modes—including nonmotorized transport in small-scale applications and rail in high-volume corridors—motorized road vehicles will retain their overwhelming dominance of the transport sector for the foreseeable future.

Owing to their rapidly increasing numbers and very limited use of emission control technologies, motor vehicles are emerging as the largest source of urban air pollution in the developing world. Other adverse impacts of motor vehicle use include accidents, noise, congestion, increased energy consumption and greenhouse gas emissions. Without timely and effective measures to mitigate the adverse impacts of motor vehicle use, the living environment in the cities of the developing world will continue to deteriorate and become increasingly unbearable.

This handbook presents a state-of-the-art review of vehicle emission standards and testing procedures and attempts to synthesize worldwide experience with vehicle emission control technologies and their applications in both industrialized and developing countries. It is one in a series of publications on vehicle-related pollution and control measures prepared by the World Bank in collaboration with the United Nations Environment Programme to underpin the Bank's overall objective of promoting transport development that is environmentally sustainable and least damaging to human health and welfare.

Air Pollution in the Developing World

Air pollution is an important public health problem in most cities of the developing world. Pollution levels in megacities such as Bangkok, Cairo, Delhi and Mexico City exceed those in any city in the industrialized countries. Epidemiological studies show that air pollution in developing countries accounts for tens of thousands of excess deaths and billions of dollars in medical costs and lost productivity every year. These losses, and the associated degradation in quality of life, impose a significant burden on people in all sectors of society, but especially the poor.

Common air pollutants in urban cities in developing countries include:

- Respirable particulate matter from smoky diesel vehicles, two-stroke motorcycles and 3-wheelers, burning of waste and firewood, entrained road dust, and stationary industrial sources.
- Lead aerosol from combustion of leaded gasoline.
- Carbon monoxide from gasoline vehicles and burning of waste and firewood.
- Photochemical smog (ozone) produced by the reaction of volatile organic compounds and nitrogen oxides in the presence of sunlight; motor vehicle emissions are a major source of nitrogen oxides and volatile organic compounds.
- Sulfur oxides from combustion of sulfur-containing fuels and industrial processes.
- Secondary particulate matter formed in the atmosphere by reactions involving ozone, sulfur and nitrogen oxides and volatile organic compounds.
- Known or suspected carcinogens such as benzene, 1,3 butadiene, aldehydes, and polynuclear aromatic

hydrocarbons from motor vehicle exhaust and other sources.

In most cities gasoline vehicles are the main source of lead aerosol and carbon monoxide, while diesel vehicles are a major source of respirable particulate matter. In Asia and parts of Latin America and Africa two-stroke motorcycles and 3-wheelers are also major contributors to emissions of respirable particulate matter. Gasoline vehicles and their fuel supply system are the main sources of volatile organic compound emissions in nearly every city. Both gasoline and diesel vehicles contribute significantly to emissions of oxides of nitrogen. Gasoline and diesel vehicles are also among the main sources of toxic air contaminants in most cities and are probably the most important source of public exposure to such contaminants.

Studies in a number of cities (Bangkok, Cairo, Jakarta, Santiago and Tehran, to name five) have assigned priority to controlling lead and particulate matter concentrations, which present the greatest hazard to human health. Where photochemical ozone is a problem (as it is, for instance, in Mexico City, Santiago, and São Paulo), control of ozone precursors (nitrogen oxides and volatile organic compounds) is also important both because of the damaging effects of ozone itself and because of the secondary particulate matter formation resulting from atmospheric reactions with ozone. Carbon monoxide and toxic air contaminants have been assigned lower priority for control at the present time, but measures to reduce volatile organic compounds exhaust emissions will generally reduce carbon monoxide and toxic substances as well.

Mitigating the Impacts of Vehicular Air Pollution

Stopping the growth in motor vehicle use is neither feasible nor desirable, given the economic and other benefits of increased mobility. The challenge, then, is to manage the growth of motorized transport so as to maximize its benefits while minimizing its adverse impacts on the environment and on society. Such a management strategy will generally require economic and technical measures to limit environmental impacts, together with public and private investments in vehicles and transport infrastructure. The main components of an integrated environmental strategy for the urban transport sector will generally include most or all of the following:

- *Technical measures involving vehicles and fuels.* These measures, the subject of this handbook, can dramatically reduce air pollution, noise, and other adverse environmental impacts of road transport.

- *Transport demand management and market incentives.* Technical and economic measures to discourage the use of private cars and motorcycles and to encourage the use of public transport and non-motorized transport modes are essential for reducing traffic congestion and controlling urban sprawl. Included in these measures are market incentives to promote the use of cleaner vehicle and fuel technologies. As an essential complement to transport demand management, public transport must be made faster, safer, more comfortable, and more convenient.

- *Infrastructure and public transport improvements.* Appropriate design of roads, intersections, and traffic control systems can eliminate bottlenecks, accommodate public transport, and smooth traffic flow at moderate cost. New roads, carefully targeted to relieve bottlenecks and accommodate public transport, are essential, but should be supported only as part of an integrated plan to reduce traffic congestion, alleviate urban air pollution, and improve traffic safety. In parallel, land use planning, well-functioning urban land markets, and appropriate zoning policies are needed to encourage urban development that minimizes the need to travel, reduces urban sprawl, and allows for the provision of efficient public transport infrastructure and services.

An integrated program, incorporating all of these elements, will generally be required to achieve an acceptable outcome with respect to urban air quality. Focus on only one or a few of these elements could conceivably make the situation worse. For example, building new roads, in the absence of measures to limit transport demand and improve traffic flow, will simply result in more roads full of traffic jams. Similarly, strengthening public transport will be ineffective without transport demand management to discourage car and motorcycle use and traffic engineering to give priority to public transport vehicles and non-motorized transport (bicycles and walking).

Technical Measures to Limit Vehicular Air Pollution

This handbook focuses on technical measures for controlling and reducing emissions from motor vehicles. Changes in engine technology can achieve very large reductions in pollutant emissions—often at modest cost. Such changes are most effective and cost-effective when incorporated in new vehicles. The most common approach to incorporating such changes has been through the establishment of *vehicle emission standards*.

Chapter 1 surveys the vehicle emission standards that have been adopted in various countries, with emphasis on the two principal international systems of standards, those of North America and Europe. Chapter 2 discusses the test procedures used to quantify vehicle emissions, both to verify compliance with standards and to estimate emissions in actual use. This chapter also includes a review of vehicle emission factors (grams of pollutant per kilometer traveled) based on investigations carried out in developing and industrial countries.

Chapter 3 describes the engine and aftertreatment technologies that have been developed to enable new vehicles to comply with emission standards, as well as the costs and other impacts of these technologies. An important conclusion of this chapter is that major reductions in vehicle pollutant emissions are possible at relatively low cost and, in many cases, with a net savings in life-cycle cost as a result of better fuel efficiency and reduced maintenance requirements. Although the focus of debate in the industrial world is on advanced (and expensive) technologies to take emission control levels from the present 90 to 95 percent control to 99 or 100 percent, technologies to achieve the first 50 to 90 percent of emission reductions are more likely to be of relevance to developing countries.

Hydrocarbon, carbon dioxide, and nitrogen oxide emissions from gasoline fueled cars can be reduced by 50 percent or more from uncontrolled levels through engine modifications, at a cost of about U.S.$130 per car. Further reductions to the 80 to 90 percent level are possible with three-way catalysts and electronic engine control systems at a cost of about U.S.$600 - $800 per car. Excessive hydrocarbon and particulate emissions from two-stroke motorcycles and three-wheelers can be lowered by 50 to 90 percent through engine modifications at a cost of U.S.$60 - $80 per vehicle. For diesel engines, nitrogen oxide and hydrocarbon emissions can be reduced by 30 to 60 percent and particulate matter emissions by 70 to 80 percent at a cost less than U.S.$1,500 per heavy-duty engine. After-treatment systems can provide further reductions in diesel vehicle emissions although at somewhat higher cost.

Measures to control emissions from in-use vehicles are an essential complement to emission standards for new vehicles and are the subject of chapter 4. Appropriately-designed and well-run in-use vehicle inspection and maintenance programs, combined with remote-sensing technology for roadside screening of tailpipe emissions, provide a highly cost-effective means of reducing fleet-wide emissions. Retrofitting engines and emission control devices may reduce emissions from some vehicles. Policies that accelerate the retirement or relocation of uncontrolled or excessively polluting vehicles can also be of value in developing countries where the high cost of vehicle renewal and the low cost of repairs result in a very slow turnover of the vehicle fleet, with large numbers of older polluting vehicles remaining in service for long periods of time.

The role of fuels in reducing vehicle emissions is reviewed in chapter 5, which discusses both the benefits achievable through reformulation of conventional gasoline and diesel fuels and the potential benefits of alternative cleaner fuels such as natural gas, petroleum gas, alcohols, and methyl/ethyl esters derived from vegetable oils. Changes in fuel composition (for example, removal of lead from gasoline and of sulfur from diesel) are necessary for some emission control technologies to be effective and can also help to reduce emissions from existing vehicles. The potential reduction in pollutant emissions from reformulated fuels ranges from 10 to 30 percent. Fuel modifications take effect quickly and begin to reduce pollutant emissions immediately; in addition, they can be targeted geographically (to highly polluted areas) or seasonally (during periods of elevated pollution levels). Fuel regulations are simple and easy to enforce because fuel refining and distribution systems are highly centralized. The use of cleaner alternative fuels such as natural gas, where they are economical, can dramatically reduce pollutant emissions when combined with appropriate emission control technology. Hydrogen and electric power (in the form of batteries and fuel cells) could provide the cleanest power sources for running motor vehicles with ultra-low or zero emissions. Alternative fuel vehicles (including electric vehicles) comprise less than 2 percent of the global vehicle fleet, but they provide a practical solution to urban pollution problems without imposing restrictrions on personal mobility.

Technical emission control measures such as those described in this handbook do not, by themselves, constitute an emission control strategy, nor are they sufficient to guarantee environmentally acceptable outcomes over the long run. Such measures can, however, reduce pollutant emissions per vehicle-kilometer traveled by 90 percent or more, compared with in-use uncontrolled vehicles. Thus a substantial improvement in environmental conditions is feasible, despite continuing increases in national vehicle fleets and their utilization. Although technical measures alone are insufficient to ensure the desired reduction of urban air pollution, they are an indispensable component of any cost-effective strategy for limiting vehicle emissions. Employed as part of an integrated transport and environmental program, these measures can buy the time necessary to bring about the needed behavioral changes in transport demand and the development of environmentally sustainable transport systems.

Acknowledgments

This handbook is a product of an informal collaboration between the World Bank and the United Nations Environment Programme, Industry and Environment (UNEP IE), initiated in 1990. The scope and contents of the handbook were discussed at a workshop on Automotive Air Pollution—Issues and Options for Developing Countries, organized by UNEP IE in Paris in January 1991. The advice and guidance provided by the workshop participants, who are listed on the next page, is gratefully acknowledged.

It took nearly five years to bring this work to completion, and in the process the handbook was revised four times to keep up with the fast-breaking developments in this field. The final revision was completed in June 1996. This process of updating was greatly helped by the contributions of C. Cucchi (Association des Constructeurs Europeans d'Automobiles, Brussels); Juan Escudero (University of Chile, Santiago); Barry Gore (London Buses Ltd., United Kingdom); P. Gargava (Central Pollution Control Board, New Delhi, India); A.K. Gupta (Central Road Research Institute, New Delhi, India); Robert Joumard (Institute National de Recherche sur les Transports et leur Sécurité, Bron, France); Ricardo Katz (University of Chile, Santiago); Clarisse Lula (Resource Decision Consultants, San Francisco); A.P.G. Menon (Public Works Department, Singapore); Laurie Michaelis (Organization for Economic Co-operation and Development/International Energy Agency, Paris); Peter Moulton (Global Resources Institute, Kathmandu); Akram Piracha (Pakistan Refinery Limited, Karachi); Zissis Samaras (Aristotle University, Thessaloniki, Greece); A. Szwarc (Companhia de Tecnologia de Saneamento Ambiental, São Paulo, Brazil); and Valerie Thomas (Princeton University, New Jersey, USA). We are specially grateful to our many reviewers, particularly the three anonymous reviewers whose erudite and compelling comments induced us to undertake a major updating and revision of the handbook. We hope that we have not disappointed them. Written reviews prepared by Emaad Burki (Louis Berger International, Washington, D.C., USA); David Cooper (University of Central Florida, Orlando); John Lemlin (International Petroleum Industry Environmental Conservation Association, London); Setty Pendakur (University of British Columbia, Canada); Kumares Sinha (Purdue University, Indiana, USA); Donald Stedman (University of Colorado, Denver); and by Antonio Estache, Karl Heinz Mumme, Adhemar Byl, and Gunnar Eskeland (World Bank) proved invaluable in the preparation of this work. In addition, we made generous use of the literature on this subject published by the Oil Companies' European Organization for Environmental and Health Protection (CONCAWE) and the Organization for Economic Cooperation and Development (OECD).

We owe very special thanks to José Carbajo, John Flora, and Anttie Talvitie at the World Bank, who kept faith with us and believed that we had a useful contribution to make. We gratefully acknowledge the support and encouragement received from Gobind Nankani to bring this work to a satisfactory conclusion. Our two collaborators, Surhid P. Gautam and Lit-Mian Chan spent endless hours keeping track of a vast array of background information, compiling the data presented in the book, and preparing several appendices. Our debt to them is great.

We would like to acknowledge the support of Jeffrey Gutman, Anthony Pellegrini, Louis Pouliquen, Richard Scurfield and Zmarak Shalizi at the World Bank, who kept afloat the funding for this work despite the delays and our repeated claims that the book required yet another revision. Jacqueline Aloisi de Larderel, Helene Genot, and Claude Lamure at UNEP IE organized and financed the 1991 Paris workshop and encouraged us to complete the work despite the delays. We would like to record the personal interest that Ibrahim Al Assaf, until recently the Executive Director for Saudi Arabia at the World Bank, took in the conduct of the work and the encouragement he offered us .

Paul Holtz provided editorial assistance and advice. Jonathan Miller, Bennet Akpa, Jennifer Sterling, Beatrice Sito, and Catherine Ann Kocak, were responsible for artwork and production of the handbook.

In closing we are grateful for the patience and support our families have shown us while we toiled to finish this book. Many weekends were consumed by this work and numerous family outings were canceled so that we could keep our self-imposed deadlines. Without their understanding, this would still be an unfinished manuscript. Very special thanks to our wives, Surraya Faiz, Carolyn Weaver, and Evelyn Walsh.

Asif Faiz
Christopher S. Weaver
Michael P. Walsh

November 1996

Participants at the UNEP Workshop

The workshop on Automotive Air Pollution — Issues and Options for Developing Countries, sponsored by the United Nations Environment Programme, Industry and Environment (UNEP IE), was held in Paris, January 30-31, 1991. The titles of the particpants reflect the positions held at the time of the workshop.

Marcel Bidault
Chief, Directorate of Studies and Research
Renault Industrial Vehicles, *France*

David Britton
International Petroleum Industry
Environmental Conservation Association
IPIECA, *United Kingdom*

Asif Faiz
Highways Adviser
Infrastructure and Urban Development Division
The World Bank, *U.S.A.*

Hélène Genot
Senior Consultant
UNEP IE, *France*

Barry Gore
Vehicle Engineer
London Buses Ltd., *United Kingdom*

M. Hublin
President, Expert Group on Emissions and Energy
European Automobile Manufacturers Association, *France*

Claude Lamure
Director
National Institute for Transport and Safety Research (INRETS), *France*

Jaqueline Aloisi de Larderel
Director
UNEP IE, *France*

Tamas Meretei
Professor, Institute of Transportation Sciences, *Hungary*

Juan Escudero Ortuzar
Executive Secretary
Special Commission for the Decontamination of the Santiago Metropolitan Region, *Chile*

Peter Peterson
Director, Monitoring Assessment and Research Centre (MARC)
UNEP/GEMS, *United Kingdom*

John Phelps
Technical Manager, European Automobile Manufacturers Association, *France*

Claire van Ruymbeker
Staff Scientist, Administration for Air Quality, *Mexico*

Zissis C. Samaras
Associate Professor
Aristotle University, Thessaloniki, *Greece*

Kumares C. Sinha
Professor of Transport Engineering, Purdue University, Indiana, *U.S.A.*

Michael P. Walsh
International Consultant
Arlington, Virginia, *U.S.A.*

1

Emission Standards and Regulations

Motor vehicle emissions can be controlled most effectively by designing vehicles to have low emissions from the beginning. Advanced emission controls can reduce hydrocarbon and carbon monoxide emissions by more than 95 percent and emissions of nitrogen oxides by 80 percent or more compared with uncontrolled emission levels. Because these controls increase the cost and complexity of design, vehicle manufacturers require inducements to introduce them. These inducements may involve mandatory standards, economic incentives, or a combination of the two. Although mandatory standards have certain theoretical disadvantages compared with economic incentives, most jurisdictions have chosen them as the basis for their vehicle emissions control programs. Vehicle emission standards, now in effect in all industrialized countries, have also been adopted in many developing countries, especially those where rapid economic growth has led to increased vehicular traffic and air pollution, as in Brazil, Chile, Mexico, the Republic of Korea, and Thailand.

Because compliance with stricter emission standards usually involves higher initial costs, and sometimes higher operating costs, the optimal level of emission standards can vary among countries. Unfortunately, the data required to determine optimal levels are often unavailable. Furthermore, economies of scale, the lead-time required and the cost to automakers of developing unique emission control systems, and the cost to governments of establishing and enforcing unique standards all argue for adopting one of the set of international emission standards and test procedures already in wide use.

The main international systems of vehicle emission standards and test procedures are those of North America and Europe. North American emission standards and test procedures were originally adopted by the United States, which was the first country to set emission standards for vehicles. Under the North American Free Trade Agreement (NAFTA), these standards have also been adopted by Canada and Mexico. Other countries and jurisdictions that have adopted U.S. standards, test procedures or both include Brazil, Chile, Hong Kong, Taiwan (China), several Western European countries, the Republic of Korea (South Korea), and Singapore (for motorcycles only). The generally less-stringent standards and test procedures established by the United Nations Economic Commission for Europe (ECE) are used in the European Union, in a number of former Eastern bloc countries, and in some Asian countries. Japan has also established a set of emission standards and testing procedures that have been adopted by some other East Asian countries as supplementary standards.

In setting limits on vehicle emissions, it is important to distinguish between *technology-forcing* and *technology-following* emission standards. Technology-forcing standards are at a level that, though technologically feasible, has not yet been demonstrated in practice. Manufacturers must research, develop, and commercialize new technologies to meet these standards. Technology-following standards involve emission levels that can be met with demonstrated technology. The technical and financial risks involved in meeting technology-following standards are therefore much lower than those of technology-forcing standards. In the absence of effective market incentives to reduce pollution, vehicle manufacturers have little incentive to pursue reductions in pollutant emissions on their own. For this reason, technology-forcing emission standards have provided the impetus for nearly all the technological advances in the field.

The United States has often set technology-forcing standards, advancing emissions control technology worldwide. Europe, in contrast, has generally adopted technology-following standards that require new emission control technologies only after they have been proven in the U.S. market.

Incorporating emission control technologies and new-vehicle emission standards into vehicle production is a necessary but not a sufficient condition for achieving low emissions. Measures are also required to ensure the durability and reliability of emission controls throughout the vehicle's lifetime. Low vehicle emissions at the time of production do little good if low emissions are not maintained in service. To ensure that vehicle emission control systems are durable and reliable, countries such as the United States have programs to test vehicles in service and recall those that do not meet emission standards. Vehicle emission warranty requirements have also been adopted to protect consumers.

International Standards

Vehicle emission control efforts have a thirty-year history. Legislation on motor vehicle emissions first addressed visible smoke, then carbon monoxide, and later on hydrocarbons and oxides of nitrogen. Reduction of lead in gasoline and sulfur in diesel fuel received increasing attention. In addition, limits on emissions of respirable particulate matter from diesel-fueled vehicles were gradually tightened. Carcinogens like benzene and formaldehyde are now coming under control. For light-duty vehicles, crankcase hydrocarbon controls were developed in the early 1960s, and exhaust carbon monoxide and hydrocarbon standards were introduced later in that decade. By the mid-1970s most industrialized countries had implemented some form of vehicle emission control program.

Advanced technologies were introduced in new U.S. and Japanese cars in the mid- to late 1970s. These technologies include catalytic converters and evaporative emission controls. As these developments spread and the adverse effects of motor vehicle pollution were recognized, worldwide demand for emission control systems increased. In the mid-1980s, Austria, the Federal Republic of Germany and the Netherlands introduced economic incentives to encourage use of low-pollution vehicles. Australia, Denmark, Finland, Norway, Sweden, and Switzerland adopted mandatory vehicle standards and regulations. A number of rapidly industrializing countries such as Brazil, Chile, Hong Kong, Mexico, the Republic of Korea, Singapore, and Taiwan (China) also adopted emission regulations.

In 1990, the European Council of Environmental Ministers ruled that all new, light-duty vehicles sold in the EU in 1993 meet emission standards equivalent to 1987 U.S. levels. They also proposed future reductions to reflect technological progress. While Europe moved toward U.S. standards, the United States, particularly California, moved to implement even more stringent legislation. Also, in 1990, the U.S. Congress adopted amendments to the Clean Air Act that doubled the durability requirement for light-duty vehicle emission control systems, tightened emission standards further,

mandated cleaner fuels, and added cold temperature standards. The California Air Resources Board (CARB) established even more stringent regulations under its Low-Emission Vehicle (LEV) program.

Efforts are now being made to attain global harmonization of emission standards. Emissions legislation is being tightened in many member countries of the Organization for Economic Co-operation and Development (OECD). Harmonization of emission standards among countries can reduce the costs of compliance by avoiding duplication of effort. Development of a new emission control configuration typically costs vehicle manufacturers tens of millions of dollars per vehicle model, and takes from two to five years. By eliminating the need to develop separate emission control configurations for different countries, harmonization of emission standards can save billions of dollars in development costs. Such harmonization would greatly facilitate international exchange of experience with respect to standards development and enforcement activities, particularly between industrialized and developing countries.

The independent standards development and enforcement activities of the California Air Resources Board require a staff of more than 100 engineers, scientists, and skilled technicians, along with laboratory operating costs in the millions of dollars per year. The total state budget for California's Mobile Source Program is U.S.$65 million a year. This figure substantially exceeds the entire environmental monitoring and regulatory budget of most developing nations.

Harmonization of emission standards in North America was an important aspect of the NAFTA involving Canada, Mexico, and the United States. The ECE and the EU have established common emission regulations for much of Europe. The United Nations Industrial Development Organization (UNIDO) is supporting work to harmonize emission regulations in southeast Asia. A proposal submitted by the United States would expand the ECE's functions by creating an umbrella agreement under which any country could register its emission standards, testing procedures, and other aspects of its vehicle emission regulations as international standards. A mechanism would also work toward regulatory compatibility and the eventual development of consensus regulations. Agreement has already been reached on harmonized emission requirements for some engines used in off-highway mobile equipment.

U.S. Standards

California was the first U.S. state to develop motor vehicle emission standards and, because of the severe air quality problems in Los Angeles, remains the only state with the authority to establish its own emission standards. In the past several decades California has often established vehicle emission requirements that were later adopted at the U.S. federal level. The national effort to control motor vehicle pollution can be traced to the 1970 Clean Air Act, which required a 90 percent reduction in emissions of carbon monoxide, hydrocarbons, and nitrogen oxides from automobiles. The Act was adjusted in 1977 to delay and relax some standards, impose similar requirements on trucks, and mandate vehicle inspection and maintenance programs in areas with severe air pollution. Further amendments to the Act, passed in 1990, further tightened vehicle emission requirements.

Because of the size of the U.S. auto market, vehicles meeting U.S. emission standards are available from most international manufacturers. For this reason, and because U.S. standards are generally considered the most innovative, many other countries have adopted U.S. standards.[1]

Light-duty vehicles. The U.S. emission standards for passenger cars and light trucks that took effect in 1981 were later adopted by several countries including Austria, Brazil, Canada, Chile, Finland, Mexico, Sweden, and Switzerland. Compliance with these standards usually required a three-way catalytic converter with closed-loop control of the air-fuel ratio, and it provided the impetus for major advances in automotive technology worldwide. The 1990 Clean Air Act amendments mandated even stricter standards for light-duty and heavy-duty vehicles, and also brought emissions from nonroad vehicles and mobile equipment under regulatory control for the first time.

The evolution of U.S. exhaust emission standards for light-duty, gasoline-fueled vehicles is traced in table 1.1. In addition to exhaust emission standards, U.S. regulations address many other emission-related issues, including control of evaporative emissions, fuel vapor emissions from vehicle refueling, emissions durability requirements, emissions warranty, in-use surveillance of emissions performance, and recall of vehicles found not to be in compliance. Regulations that require onboard diagnostic systems that detect and identify malfunctioning emission systems or equipment are also being implemented.

The 1990 Clean Air Act amendments mandated implementation of federal emission standards identical to 1993 California standards for light-duty vehicles. These Tier 1 emission standards (to be phased in between 1994 and 1996) require light-duty vehicle emissions of volatile organic compounds to be 30 percent less and

---

1. As U.S. standards are used by many other countries and are considered a benchmark for national standards around the world, they are treated as de-facto international standards.

Table 1.1 Progression of U.S. Exhaust Emission Standards for Light-Duty Gasoline-Fueled Vehicles
(grams per mile)

Model year	Carbon monoxide	Hydrocarbons	Nitrogen oxides
Pre-1968 (uncontrolled)	90.0	15.0	6.2
1970	34.0	4.1	—
1972	28.0	3.0	—
1973–74	28.0	3.0	3.1
1975–76	15.0	1.5	3.1
1977	15.0	1.5	2.0
1980	7.0	0.41	2.0
1981	3.4	0.41	1.0
1994–96 (Tier 1)	3.4 (4.2)	0.25[a] (0.31)	0.4 (0.6)
2004 (Tier 2)[b]	1.7 (1.7)	0.125[a] (0.125)	0.2 (0.2)

— Not applicable
Note: Standards are applicable over the "useful life" of the vehicle, which is defined as 50,000 miles or five years for automobiles. The durability of the emissions control device must be demonstrated over this distance within allowed deterioration factors. Figures in parenthesis apply to a useful life of 100,000 mile, or ten years beyond the first 50,000 miles.
a. Non-methane hydrocarbons.
b. The U.S. Environmental Protection Agency (EPA) could delay implementation of tier 2 standards until 2006.
Source: CONCAWE 1994

emissions of nitrogen oxides to be 60 percent less than the U.S. federal standards applied in 1993. Useful-life requirements are extended from 80,000 to 160,000 kilometers to further reduce in-service emissions. Requirements for low-temperature testing of carbon monoxide emissions and for on-board diagnosis of emission control malfunctions should also help reduce in-service emissions.

In response to the severe air pollution problems in Los Angeles and other California cities, CARB in 1989 established stringent, technology-forcing vehicle emission standards to be phased in between 1994 and 2003. These rules defined a set of categories for low-emission vehicles, including *transitional low-emission vehicles (TLEV)*, *low-emission vehicles (LEV)*, *ultra low-emission vehicles (ULEV)*, and *zero-emission vehicles (ZEV)*. These last two categories are considered as favoring natural gas and electric vehicles, respectively. Table 1.2 summarizes the emission limits for passenger cars and light-duty vehicles corresponding to these low-emission categories.

In addition to being far more stringent than any previous emission standards, the new California standards are distinguished by having been designed specifically to accommodate alternative fuels. Instead of hydrocarbons, the new standards specify limits for organic emissions in the form of non-methane organic gas (NMOG) which is defined as the sum of non-methane hydrocar-

Table 1.2   U.S. Exhaust Emission Standards for Passenger Cars and Light-Duty Vehicles Weighing Less than 3,750 Pounds Test Weight
(grams per mile)

Standard	Year implemented	50,000 miles or five years			100,000 miles or ten years		
		Carbon monoxide 75°/20°F	Hydrocarbons	Nitrogen oxides	Carbon monoxide 75°F	Hydrocarbons	Nitrogen oxides
Passenger car[a] (Tier 0)	1981	3.4/—	0.41	1.0	—	—	—
Light-duty truck[a] (Tier 0)	1981	10/—	0.80	1.7	—	—	—
Tier 1[b]	1994–6	3.4/10.0	0.25 NMHC	0.4	4.2	0.31 NMHC	0.6
Tier 2	2004	1.7/3.4	0.125 NMHC	0.2	—	—	—
California Low-Emission Vehicle/Federal Clean-fuel Fleet programs							
Transitional low-emission vehicle (TLEV)	1994[c]	3.4/10	0.125 NMOG	0.4	4.2	0.156 NMOG	0.6
Low-emission vehicle (LEV)	1997[c]	3.4/10	0.075 NMOG	0.2	4.2	0.090 NMOG	0.3
Ultra low-emission vehicle (ULEV)	1997[c]	1.7/10	0.040 NMOG	0.2	2.1	0.055 NMOG	0.3
Zero-emission vehicle (ZEV)	1998[c]	0	0	0	0	0	0

— Not applicable
NMHC = non-methane hydrocarbons
NMOG = non-methane organic gases
Note: The federal Tier 1 standards also specify a particulate matter limit of 0.08 gram per mile at 50,000 miles and 0.10 gram per mile at 100,000 miles. The California standards also specify a maximum of 0.015 gram per mile for formaldehyde emissions for 1993 standard, transitional low-emission, and low-emission vehicles, and 0.008 grams per mile for ultra low-emission vehicles. Likewise, for benzene, a limit of 0.002 gram per mile is specified for low-emission and ultra low-emission vehicles. For diesel vehicles, a particulate matter limit of 0.08 gram per mile is specified for 1993 standard, transitional low-emission, and low-emission vehicles, and 0.04 gram per mile for ultra low-emission vehicles at 100,000 miles.
a. Except for California.
b. Equivalent to California 1993 model year standard.
c. To be phased in over a ten-year period; expected year of phase-in.
Source: CONCAWE 1994, Chan and Weaver 1994

bons, aldehydes, and alcohol emissions, and thus accounts for the ozone-forming properties of aldehydes and alcohols tests that are not measured by standard hydrocarbon tests. The new standards also provide for the non-methane organic gas limit to be adjusted with reactivity adjustment factors. These factors account for the differences in ozone-forming reactivity of the NMOG emissions produced by alternative fuels, compared with those produced by conventional gasoline. This provision gives an advantage to clean fuels such as natural gas, methanol, and liquified petroleum gas, which produce less reactive organic emissions.

The 1990 Clean Air Act amendments also clarified the rights of other states to adopt and enforce the more stringent California vehicle emission standards in place of federal standards. New York and Massachusetts have done so. In addition, the other states comprising the "Ozone Transport Region" along the northeastern seaboard of the United States (from Maine to Virginia) have agreed to pursue the adoption of the California standards in unison. This has prompted the auto industry to develop a counter-offer, which is to implement California's LEV standard throughout the U.S. The auto industry offer would not include California's more-restrictive ULEV and ZEV standard, which are required under Massachusetts and New York law.

Motorcycles. Current U.S. and California emission standards for motorcycles are summarized in table 1.3. Unlike other vehicles, motorcycles used in the U.S. can meet these emission standards without a catalytic converter. The most important effect of the U.S. federal emission standards has been the elimination of two-stroke motorcycles, which emit large volumes of hydrocarbons and particulate matter. California standards, though more stringent than the federal ones, can still be met without a catalytic converter. Motorcycle standards in the United States are lenient compared with standards for other vehicles because the number of motorcycles in use is small, and their emissions are considered insignificant compared with other mobile emission sources.

Medium-duty vehicles. In 1989, CARB adopted regulations that redefined vehicles with gross vehicle weight ratings between 6,000 and 14,000 pounds as medium-

Table 1.3   U.S. Federal and California Motorcycle Exhaust Emission Standards
(grams per kilometer)

Standard	Engine type/size (cubic centimeters)	Carbon monoxide	Hydrocarbons
U.S. Federal			
1978	50–170	17.0	5.0
	170–750	17.0	5+0.0155 (D-170)[a]
	More than 750	17.0	14.0
1980 to present	All models	12.0	5.0
California			
1978–79	50–169	17.0	5.0
	170–750	17.0	5+0.0155 (D-170)[a]
	More than 750	17.0	14.0
1980–81	All models	12.0	5.0
1982–February 1985	50–279	12.0	1.0
	More than 280	12.0	2.5
March 1985–1987	50–279	12.0	1.0
	More than 280	12.0	1.4
1988 to present	50–279	12.0	1.0
	280–699	12.0	1.0
	More than 700	12.0	1.4

a. D is the engine displacement in cubic centimeters.
Source: Chan and Weaver 1994

duty vehicles. Previously, vehicles under 8,500 pounds gross vehicle weight were defined by both the CARB and by the U.S. Environmental Protection Agency (EPA) as light duty, while those weighing more than 8,500 pounds gross vehicle weight were defined as heavy duty and subject to emission standards based on an engine dynamometer test. The U.S. EPA still classifies vehicles according to the old system, though vehicles weighing between 6,000 and 8,500 pounds are subject to somewhat less stringent standards (table 1.4).

CARB recognized that large pickup trucks, vans, and chassis have more in common with light-duty trucks than with true heavy-duty vehicles. Light-duty trucks are subject to more rigorous emission control requirements than larger vehicles. Medium-duty gasoline- and alternative-fueled vehicles are tested using the same procedure as light-duty vehicles, but with heavier simulated weight settings. Medium-duty vehicles that have diesel engines or that are sold as incomplete chassis have the option of certifying under the heavy-duty engine testing procedures instead. CARB has also established LEV and ULEV emission standards for these engines. Presently, the only engines capable of meeting the ultra low emission vehicle standards use natural gas or methanol as fuel.

Heavy-duty vehicles. Limits on pollutants from heavy-duty engines were adopted by the United States in 1970. The current transient test procedure was introduced in 1983. Current U.S. and California emission reg-

ulations for heavy-duty vehicle engines are summarized in table 1.5. The 1991 and 1994 emission standards were established by regulations adopted in 1985. Engines meeting the 1994 standards are now being sold. The 1990 Clean Air Act amendments established still more stringent particulate levels for urban buses, and a new standard of 4.0 g/bhp-hr for nitrogen oxides will take effect in 1998. The U.S. EPA has also adopted low-emission vehicle and ultra low emission vehicle standards for heavy-duty vehicles covered under the Clean-Fuel Fleet program. In July 1995 the U.S. EPA and the Engine Manufacturers Association agreed that the limits on nitrogen oxides and hydrocarbons equivalent to the ultra low emission vehicle standard would become mandatory for all engines in 2004. At present, the only heavy-duty engines capable of meeting these standards use methanol or natural gas as fuel. Engine manufacturers expect to be able to meet the standards using diesel engines with exhaust gas recirculation by 2004.

Evaporative emissions. Evaporative emission limits apply to vehicles fueled by gasoline or alcohol fuels. Both CARB and the EPA limit evaporative hydrocarbon emissions from light-duty vehicles to 2.0 grams per test, which is considered effectively equivalent to zero (a small allowance is needed for other, non-fuel related organic emissions from new cars, such as residual paint solvent). California also applies this limit to motorcycles, but the U.S. EPA does not regulate motorcycle evaporative emissions. New, more stringent evaporative

Table 1.4　U.S. Federal and California Exhaust Emission Standards for Medium-Duty Vehicles
(grams per mile)

Standard (FTP-75)	Year implemented	50,000 miles or five years			120,000 miles or eleven years		
		Carbon monoxide	Hydrocarbons	Nitrogen oxides	Carbon monoxide	Hydrocarbons	Nitrogen oxides
U.S. federal	1983	10.0	0.80	1.7	—	—	—
California/U.S. Tier 1	1995/1996[a]						
0–3,750 pounds		3.4	0.25 NMHC	0.4	5.0	0.36 NMHC	0.55
3,751–5,750 pounds		4.4	0.32 NMHC	0.7	6.4	0.46 NMHC	0.98
5,751–8,500 pounds		5.0	0.39 NMHC	1.1	7.3	0.56 NMHC	1.53
8,500–10,000 pounds[b]		5.5	0.46 NMHC	1.3	8.1	0.66 NMHC	1.81
10–14,000 pounds[b]		7.0	0.60 NMHC	2.0	10.3	0.86 NMHC	2.77
California Low-Emission Vehicle/Federal Clean-Fuel Fleet programs							
Low-emission vehicle (LEV)	1998[a]						
0–3,750 pounds		3.4	0.125 NMOG	0.4	5.0	0.180 NMOG	0.6
3,751–5,750 pounds		4.4	0.160 NMOG	0.7	6.4	0.230 NMOG	1.0
5,751–8,500 pounds		5.0	0.195 NMOG	1.1	7.3	0.280 NMOG	1.5
8,501–10,000 pounds[b]		5.5	0.230 NMOG	1.3	8.1	0.330 NMOG	1.8
10–14,000 pounds[b]		7.0	0.300 NMOG	2.0	10.3	0.430 NMOG	2.8
Ultra low-emission vehicle (ULEV)	1998[a]						
0–3,750 pounds		1.7	0.075 NMOG	0.2	2.5	0.107 NMOG	0.3
3,751–5,750 pounds		2.2	0.100 NMOG	0.4	3.2	0.143 NMOG	0.5
5,751–8,500 pounds		2.5	0.117 NMOG	0.6	3.7	0.167 NMOG	0.8
8,501–10,000 pounds[b]		2.8	0.138 NMOG	0.7	4.1	0.197 NMOG	0.9
10–14,000 pounds[b]		3.5	0.180 NMOG	1.0	5.2	0.257 NMOG	1.4

— Not applicable

Note:　NMHC–Non-methane hydrocarbons, NMOG–Non-methane organic gas. Emission standards for medium-duty vehicles also include limits for particulate matter and aldehyde emissions.

a.　Expected year of phase-in.

b.　California non-diesel vehicles only. All U.S. and California diesel-fueled vehicles weighing more than 8,500 pounds are subject to heavy-duty testing procedures and standards.

Source:　CONCAWE 1994

test procedures scheduled to take effect during the mid-1990s will have the same limit of 2.0 grams per test, but running-loss emissions will be limited to 0.05 grams per mile. The new standard is nominally the same as the old one, but more severe test conditions under the new test procedures will impose much greater compliance requirements on manufacturers.

Evaporative and refueling emissions have become a more significant fraction of total emissions as a consequence of the steady decline in exhaust hydrocarbon emissions. To address this problem, and to encourage the introduction of vehicles using cleaner fuels, the U.S. EPA has defined a special category of vehicles called inherently low-emission vehicles (ILEVs). These vehicles must meet the ultra low-emission standard for emissions of nitrogen oxides and the low-emission vehicle standards for carbon monoxide and non-methane organic gas. They must also exhibit inherently low evaporative emissions by passing an evaporative test with the evaporative control system disabled. Gasoline-fueled vehicles cannot meet this standard. Inherently low-emission vehicles are eligible for certain regulatory benefits, including exemption from "no-drive" days and other time-based

transportation control measures. As of mid-1995, just two vehicle models were certified as inherently low-emission vehicles, and both were fueled by compressed natural gas (CNG).

U.N. Economic Commission for Europe (ECE) and European Union (EU) Standards

The vehicle emission standards established by the ECE and incorporated into the legislation of the EU (formerly the European Community) are not directly comparable to those in the United States because of differences in the testing procedure.[2] The relative emissions measured using the two procedures vary with the vehicle's

2.　Besides the member states of the EU, China, the Czech Republic, Hong Kong, Hungary, India, Israel, Poland, Romania, Saudi Arabia, Singapore, Thailand, the Slovak Republic, and countries in the former U.S.S.R. and the former Yugoslavia require compliance with ECE regulations. Austria, Denmark, Finland, Norway, Sweden, and Switzerland have adopted U.S. standards. Following their admission into the European Union in 1995, Austria, Finland, and Sweden must comply with EU regulations; a four-year transitional period has been agreed after which national emission standards must either be harmonized with EU regulations or renegotiated.

Table 1.5 U.S. Federal and California Emission Standards for Heavy-Duty and Medium-Duty Engines

	Exhaust emissions (g/bhp-hr)						Smoke opacity[a]
	Total hydrocarbons	Hydrocarbons (non-methane)	Nitrogen oxides	Carbon monoxide	Particulate matter	Formaldehyde	
U.S. Federal heavy-duty regulation							
1991 HDV diesel	1.3	—	5.0	15.5	0.25	—	20/15/50
1991 LHDV gasoline	1.1	—	5.0	14.4	—	—	—
1991 MHDV gasoline	1.9	1.2[b]	5.0	37.1	—	—	—
1994 HDV diesel	1.3	0.9[b]	5.0	15.5	0.10	—	20/15/50
1994 LHDV otto[c]	1.1	1.7[b]	5.0	14.4	—	—	—
1994 MHDV otto[c]	1.9	1.2[b]	5.0	37.1	—	—	—
1994 transit bus	1.3	1.2[b]	5.0	15.5	0.07	—	20/15/50
1996 transit bus	1.3	1.2[b]	5.0	15.5	0.05	—	20/15/50
1998 HDV diesel	1.3	1.2[b]	4.0	15.5	0.10	—	20/15/50
1998 transit bus	1.3	1.2[b]	4.0	15.5	0.05	—	20/15/50
2004 (proposed) HDV Opt.A	1.3	2.4[d]		15.5	0.10	—	20/15/50
2004 (proposed) HDV Opt. B	1.3	2.5[d, e]		15.5	0.10	—	20/15/50
Federal clean fuel fleet standards regulation							
LEV - Federal fuel	NR	3.8[d]		14.4	0.10	—	20/15/50
LEV - California fuel	NR	3.5[d]		14.4	0.10	—	20/15/50
ILEV	NR	2.5[d]		14.4	0.10	0.050	20/15/50
ULEV	NR	2.5[d]		7.2	0.05	0.025	20/15/50
California heavy-duty regulation							
1991 HDV diesel	1.3	1.2[f]	5.0	15.5	0.25	0.10[g]	20/15/50
1991 LHDV otto[c,h]	1.1	0.9	5.0	14.4	—	0.10[g]	—
1991 MHDV otto[h]	1.9	1.7[f]	5.0	37.1	—	0.10[g]	—
1994 HDV diesel	1.3	1.2[f]	5.0	15.5	0.10	0.10[g]	20/15/50
1994 urban bus	1.3	1.2[f]	5.0	15.5	0.07	0.10[g]	20/15/50
Optional bus std. 1994[i]	1.3	1.2[f]	0.5-3.5	15.5	0.07	0.10[g]	20/15/50
1996 urban bus	1.3	1.2[f]	4.0	15.5	0.05	0.05[g]	20/15/50
Optional bus std. 1996[i]	1.3	1.2[f]	0.5-2.5	15.5	0.05	0.05[g]	20/15/50
California medium-duty regulation[j]							
Tier 1	NR	3.9[d]		14.4	0.10	—	20/15/50
LEV 1992-2001	NR	3.5[d]		14.4	0.10	0.05	20/15/50
2002-2003	NR	3.0[d]		14.4	0.10	0.05	20/15/50
ULEV 1992-2003	NR	2.5[d]		14.4	0.10	0.05	20/15/50
2004+Opt.A.	NR	2.5[d, e]		14.4	0.10	0.05	20/15/50
2004+Opt. B	NR	2.4[d]		14.4	0.10	0.05	20/15/50
SULEV	NR	2.0[d]		7.2	0.05	0.05	20/15/50

— Not applicable

NR = Not regulated; HDV= Heavy-duty vehicle; LHDV= Light heavy-duty vehicle (<14,000 lb. GVW); MHDV= Medium heavy-duty vehicle (>14,000 lb. GVW); ILEV= Inherently low-emission vehicle; LEV= Low-emission vehicle; ULEV= Ultra low-emission vehicle; SULEV= Super ultra low-emission vehicle.
a Acceleration/lug/peak smoke opacity.
b Non-methane hydrocarbon (NMHC) standard applies instead of total hydrocarbion (THC) for natural gas engines only.
c Replaced by "medium duty" vehicle classification beginning 1995.
d These standards (NMHC+NO_x) limit the sum of NMHC and NO_x emissions.
e NMHC limited to 0.5 g/bhp-hr.
f Use of NMHC instead of THC standard is optional for diesel, LPG, and natural gas engines.
g Methanol-fueled engines only. From 1993-95, limited to 0.10 g/bhp-hr, subsequently to 0.05 g/bhp-hr.
h Includes spark-ignition gasoline and alternative fuel engines, except those derived from heavy-duty diesels.
i Optional standards. Engines certified to these standards may earn emission credits.
j Optional standards for diesel and diesel-derived engines and engines sold in incomplete medium-duty vehicle chassis.
Source: CONCAWE 1995

emission control technology, but test results in grams per kilometer are generally of the same order.

Until the mid-1980s, motor vehicle emission regulations in Europe were developed by the ECE for adoption and enforcement by individual member countries. It had been a common practice for the EU to adopt standards and regulations almost identical to those issued by the ECE. In terms of stringency (i.e. level of emission control technology required for compliance) the European standards have lagged considerably behind the U.S. standards. Much of this lag has been caused by the complex, consensus-based approach to standard setting used by the ECE and by the difficulty of obtaining agreement between so many individual countries, each with its own interests and concerns. With the recent shift to decision procedures requiring less-than-unanimous agreement within the European Union, it has been possible to adopt more stringent emission standards. The stringency of the most recent EU emission standards is now closer to that of the U.S. standards. For all practical purposes the ECE no longer promulgates standards that have not been agreed first by the EU.

Unlike the U.S. standards, the ECE emission standards apply to vehicles only during type approval and when the vehicle is produced (conformity of production). Once the vehicle leaves the factory and enters service, the manufacturer has no liability for its continued compliance with emission limits. Surveillance testing, recall campaigns, and other features of U.S. emissions regulation are not incorporated in the European regulatory structure. As a result, manufacturers of such vehicles have little incentive to ensure that the emission control systems are durable enough to provide good control throughout the vehicle's lifetime.

Light-duty vehicles. These vehicles were the first to be regulated, beginning in 1970, to conform to the original ECE Regulation 15. The regulation was amended four times for type approval (ECE 15-01, implemented in 1974, ECE 15-02 in 1977, ECE 15-03 in 1979, and ECE 15-04 in 1984) and twice for conformity of production (1981 and 1986). Regulation ECE 15-04 was applied to both gasoline and diesel-fueled light-duty vehicles, whereas earlier regulations applied only to gasoline-fueled vehicles. The emission limits included in these regulations were based on the ECE 15 driving cycle (van Ruymbeke and others 1992).

The ECE did not adopt emission standards requiring three-way catalytic converters until 1988 (ECE regulation 83), and then only for vehicles with engine displacement of 2.0 liters or more. Less stringent standards were specified for smaller vehicles, in order to encourage the use of lean-burn engines. Although ECE 83 was also adopted as European Community Directive 88/76/

EEC, this regulation was not implemented in national legislation by any European country, in anticipation of the adoption of the Consolidated Emissions Directive, 91/441/EEC. This latter directive was adopted by the Council of Ministers of the European Community in June 1991. Under the Consolidated Emission Directives, exhaust emission standards for passenger cars (including diesel cars) are certified on the basis of the new combined ECE-15 (urban) cycle and extra-urban driving cycle (EUDC). In contrast to previous directives, a common set of exhaust emission standards (including durability testing) were applied to all private passenger cars (both gasoline and diesel-engined) irrespective of engine capacity. The standard also covers vehicle evaporative emissions. Limit values for passenger car emissions are shown in table 1.6. These limits became effective July 1, 1992 for new models, and on December 31, 1992 for all production.

In March 1994, the Council of Ministers of the European Community adopted Directive 94/12/EC which provides for more stringent emission limits for passenger cars from 1996 onwards (table 1.6). These standards again differentiate between gasoline and diesel vehicles, but require significant emission reductions from both fuel types. These standards make separate provisions for direct-injection (DI) diesel engines to meet less-stringent standards for hydrocarbons plus oxides of nitrogen and for particulate matter, until September 30, 1999.

In contrast to previous directives, production vehicles must comply with the type approval limits. There is also a durability requirement for vehicles fitted with pollution control devices. Implementation of these emission standards by EU member States is mandatory and unlike previous directives, not left to the discretion of individual national governments. Directive 94/12/EC also required that new proposals must be prepared before June 30, 1996 to implement further reductions in exhaust emissions by June 1, 2000[3] (CONCAWE 1995).

Limit values for emissions of gaseous pollutants from light-duty trucks and commercial vehicles were also established in the Consolidated Emissions Directive, but the actual limits were identical to the limits established in ECE 15.04, and did not include a limitation on

---

3. In June 1996, the European Commission proposed to adopt the following exhaust emission limits for cars to become effective in years 2000 and 2005 (Walsh 1996a; Plaskett 1996).

		2000	2005
• Gasoline-fueled (g/km)			
	CO	2.30	1.00
	HC	0.20	0.10
	NO_x	0.15	0.08
• Diesel-fueled (g/km)			
	CO	0.64	0.50
	$HC+NO_x$	0.56	0.30
	NO_x	0.50	0.25
	PM	0.05	0.025

Table 1.6 European Union Emission Standards for Passenger Cars with up to 6 Seats
(ECE15+EUDC test procedure, grams per kilometer)

	91/441/EEC[a]		94/12/EC[b]	
	Type approval	Conformity of production	Gasoline	Diesel
CO	2.72	3.16	2.2	1.0
HC + NO$_x$	0.97[c]	1.13	0.5	0.7[d]
PM	0.14[c]	0.18	—	0.08[d]
Evap. Emissions (g/test)	2.0	2.0	2.0	—

— Not applicable

Note: Directive 94/12/EC applies to both type approval and conformity of production.

a. Effective dates:
   (i) All light-duty vehicles except direct-ignition (DI) diesels; new models July 1, 1992, all models Dec. 31, 1992.
   (ii) DI diesels, July 1, 1994.

b. Effective dates:
   (i) Gasoline and IDI diesels; new models Jan 1, 1996, all models Jan 1 1997.
   (ii) DI diesels Oct 1, 1999.

c. DI diesel limits until June 30, 1994 were 1.36 g/km (HC+NO$_x$) and 0.19 g/km (PM).

d. Less stringent standards apply to DI diesel until Sept 30, 1999: 0.9 g/km (HC+NO$_x$) and 0.10 g/km (PM).

Source: CONCAWE 1995

particulate matter emissions. Ministerial Directive 93/59/EEC did finally modify the emission limits for light trucks and commercial vehicles. Table 1.7 shows the emission standards established for light trucks and commercial vehicles by this directive. These standards became effective with the 1994 model year. For light trucks with reference mass less than 1,250 kg, the standards are equivalent to those established for passenger cars by the Consolidated Emissions Directive; heavier vehicles are allowed somewhat higher emissions. Light truck standards comparable in strictness to the passenger car standards of 93/59/EEC have not yet been proposed.

The present European emissions standards for passenger cars and light commercial vehicles are comparable to the U.S. standards adopted in the early 1980s. The emission control technologies required to meet these standards are similar. The main emission control requirements for gasoline vehicles include three-way catalytic converters with feedback control of the air-fuel ratio through an exhaust gas oxygen sensor. The 1996 European emissions standards for passenger cars are more stringent, following the example set by the U.S. Tier 1 standards.

Motorcycles. Although the ECE has issued emission standards for motorcycles (ECE regulation 40.01) and mopeds (ECE Regulation 47), these regulations are only now being adopted in the EU (table 1.8). In addition, Austria and Switzerland have established their own technology-forcing emission standards for motorcycles and mopeds. The moped standards are sufficiently strict to require catalytic converters, at least on two-stroke engines.

Heavy-duty engines. European regulation of heavy-duty vehicle engines has lagged behind U.S. standards for the same reasons as that for light-duty engines. ECE regulation 49.01, for gaseous emissions and ECE regulation 24.03 for black smoke emissions (table 1.9), in effect until July 1992, was comparable in stringency to U.S. regulations from the 1970s, and could be met with little or no effort by diesel-engine manufacturers. The Clean Lorry Directive (91/542/EEC), compulsory throughout the EU, reduces particulate and gaseous emissions for heavy-duty vehicles in two stages. The first-stage standards (Euro 1), which took effect in July 1992, are comparable in stringency to 1988 U.S. standards, while the second-stage standards (Euro 2) are comparable to 1991 U.S. levels (table 1.10). An even more stringent third-stage standard is under discussion, as is a change from the current steady-state emissions testing procedure to a transient cycle similar to the one used in the United States (Baines 1994).

Country and Other Standards

This section summarizes the vehicle emission standards adopted by a number of individual countries, as of early 1995. Because emission standards often change, readers who require precise information are advised to obtain up-to-date information from the legal authorities of the country involved. The Oil Companies' European Organization for Environmental Protection and Health

Table 1.7   European Union 1994 Exhaust Emission Standards for Light-Duty Commercial Vehicles (Ministerial Directive 93/59/EEC)

(grams per kilometer)

Vehicle category	Reference mass (kg)[a]		Carbon monoxide	Hydrocarbons + nitrogen oxides	Particulate matter[b]
			Exhaust emissions		
Light trucks[c]	RM ≤ 1,250	Type-approval	2.72	0.97	0.14
		Conformity of production	3.16	1.13	0.18
	1,250 ≤ RM ≤ 1700	Type-approval	5.17	1.4	0.19
		Conformity of production	6.0	1.6	0.22
	RM > 1,700	Type-approval	6.9	1.7	0.25
		Conformity of production	8.0	2.0	0.29

a. Reference mass (RM) means the mass of the vehicle in running order less the uniform mass of a driver of 75 kg and increased by a uniform mass of 100 kg.
b. Diesel vehicles only.
c. Includes passenger vehicles with seating capacity greater than six persons or reference mass greater than 2,500 kg.
Source: CONCAWE 1995

Table 1.8   ECE and Other European Exhaust Emission Standards for Motorcycles and Mopeds

(grams per kilometer)

Regulation, engine type	Carbon monoxide	Hydrocarbons	Nitrogen oxides	Testing procedure
ECE 40				
Two-stroke, less than 100 kilograms	16.0	10.0	—	ECE cycle
Two-stroke, more than 300 kilograms	40.0	15.0	—	ECE cycle
Four-stroke, less than 100 kilograms	25.0	7.0	—	ECE cycle
Four-stroke, more than 300 kilograms	50.0	10.0	—	ECE cycle
ECE 40.01				
Two-stroke, less than 100 kilograms	12.8	8.0	—	ECE cycle
Two-stroke, more than 300 kilograms	32.0	12.0	—	ECE cycle
Four-stroke, less than 100 kilograms	17.5	4.2	—	ECE cycle
Four-stroke, more than 300 kilograms	35.0	6.0	—	ECE cycle
ECE 47 for mopeds				
Two-wheel	8.0	5.0	—	ECE cycle
Three-wheel	15.0	10.0	—	ECE cycle
Switzerland				
Two-stroke	8.0	3.0	0.10	ECE 40
Four-stroke	13.0	3.0	0.30	ECE 40
Moped	0.5	0.5	0.10	ECE 40
Austria				
Motorcycles (<50 cc and >40 km/h)				
Two stroke (before Oct. 1, 1991)	13.0	6.5	2.0	ECE 40
Two stroke (from Oct. 1, 1991)	8.0	7.5	0.1	ECE 40
Four stroke (before Oct. 1, 1991)	18.0	6.5	1.0	ECE 40
Four stroke (from Oct. 1, 1991)	13.0	3.0	0.3	ECE 40
Motorcycles (<50 cc)				
Two stroke (before Oct. 1, 1990)	12.0-32.0	8.0-12.0	1.0	ECE 40
Two stroke (from Oct. 1, 1990)	8.0	7.5	0.1	ECE 40
Four stroke (before Oct. 1, 1990)	17.5-35.0	4.2-6.0	0.8	ECE 40
Four stroke (from Oct. 1, 1990)	13.0	3.0	0.3	ECE 40
Mopeds (<50 cc and <40 km/h) From Oct. 1, 1988	1.2	1.0	0.2	ECE 40

— Not applicable
Note: This table does not show ECE40 and ECE40.1 limits for Reference Weight, R (motocycle weight+75 kg) of more than 100 kg and less than 300 kg. Furthermore only limits for type approval are shown in this table. See CONCAWE (1995) for additional information and applicable limits for conformity of production.
Source: CONCAWE 1992; 1994; 1995

Table 1.9   Smoke Limits Specified in ECE Regulation 24.03 and EU Directive 72/306/EEC
(smoke emission limits under steady state conditions)

Nominal flow (liters/second)	Absorption coefficient (m^{-1})
42	2.26
100	1.495
200	1.065

Note: Intermediate values are also specified. Opacity under free acceleration should not exceed the approved level by more than 0.5 m^{-1}

Although the free acceleration test was intended as a means of checking vehicles in service it has not proved entirely successful. A number of different methods have been proposed by various countries, but there is no generally accepted alternative method of in-service checking.

Source: CONCAWE 1994

Table 1.10   European Exhaust Emission Standards for Heavy-Duty Vehicles for Type Approval
(grams per kilowatt hour)

Regulation	Effective date		Carbon monoxide	Nitrogen oxides	Hydrocarbons	Particulate matter
	New models	All production				
ECE 49 (13-mode)			14.0	18.0	3.5	a
ECE 49.01 (88/77/EEC)	April 1988	October 1990	11.2 (13.2)	14.4 (15.8)	2.4 (2.6)[b]	a
Clean lorry directive (91/542/EEC)						
Stage 1 (Euro 1)	July 1992	October 1993	4.5 (4.9)	8.0 (9.0)	1.1 (1.23)	0.36–0.61[c] (0.40–0.68)
Stage 2 (Euro 2)	October 1995	October 1996	4.0 (4.0)	7.0 (7.0)	1.1 (1.10)	0.15–0.25[c] (0.15–0.25)
Stage 3 (Euro 3)	1999 (tentative)	n.a.	2.5	5.0	0.7	less than 0.12

n.a. = Not available
a. Smoke according to ECE Regulation 24.03, EU Directive 72/306/EEC.
b. Figures in parentheses are emission limits for conformity of production.
c. Depending on engine rating.
Source: Havenith and others 1993; CONCAWE 1994

(CONCAWE) in Brussels, has also published a series of reports summarizing vehicle emission and fuel standards worldwide. The most recent such report (CONCAWE 1994) is a comprehensive source of information on motor vehicle emission regulations and fuel specifications worldwide.

Argentina

Decree 875/94 of National Law 2254/92 issued in 1994 establishes national emission standards for new and used motor vehicles (table 1.11). The Decree also assigns the Secretaría de Recursos Naturales y Ambiente Humano as the responsible agency for enforcing and updating these standards. These emission limits were reinforced by the Joint Resolutions 96/94 and 58/94 issued by the Secretaries of Transport and Industry in March 1994. Emission limits are established with a different compliance schedule for trucks and urban passenger transport vehicles. In addition, emission limits for particulate matter are being established for the years 1996 and 2000. The exhaust emission standards for new light-duty gasoline-fueled vehicles, new heavy-duty vehicles, diesel-fueled vehicles,

and for all used vehicles appear to be based on ECE regulations. Another regulation required the retirement of buses older than 10 years (about 3,500 buses) in 1995. Municipal emission standards in the Capital Federal are embodied in Ordinance No. 39,025 and appear to be tighter than national emission limits. It is understood that Argentine standards conform closely to Brazilian standards although implementation is delayed because of the current limited availability of unleaded gasoline.

Australia

In Australia's federal system of government, the power to introduce motor vehicle legislation, including emission regulations, lies with state governments. This is the opposite of the situation in the United States. The Australian Transport Advisory Council (ATAC) is composed of federal and state transport ministers who meet twice a year. The Council can agree to adopt emission and safety standards which, while not binding on the states, are usually adopted in state legislation. States have acted unilaterally when agreement is not reached within the Council, however.

Table 1.11 Exhaust Emission Standards (Decree 875/94), Argentina

Model year	*Emission limits for new light-duty gasoline or diesel vehicles*				
	Carbon monoxide (g/km)	Hydrocarbons (g/km)	Nitrogen oxides (g/km)	Carbon monoxide[a] in low gear (% v)	Hydrocarbons[a] in low gear (ppm)
1994	24.0	2.1	2.0	3.0	660
1995	12.0	1.2	1.4	2.5	400
1997	2.0	0.3	0.6	0.5	250
1999	2.0	0.3	0.6	0.5	250
Model Year	*Emission limits for new heavy-duty gasoline or diesel vehicles*				
	Carbon monoxide (g/kWh)	Hydrocarbons (g/kWh)	Nitrogen oxides (g/kWh)	Carbon monoxide[a] in low gear (% v)	Hydrocarbons[a] in low gear (ppm)
1995	11.2	2.4	14.4	3.0	660
1997	11.2	2.4	14.4	2.5	400

a. For gasoline vehicles.
Source: Boletin Oficial 1994

The Council is advised on vehicle emission matters by a hierarchy of committees: the Motor Transport Groups (comprising senior federal and state public servants), the Advisory Committee on Vehicle Emissions and Noise (ACVEN) comprising lower-level federal and state public servants, and the ACVEN Emissions Sub-Committee, which includes public servants, representatives from the automotive and petroleum industries as well as consumers. The Committee also provides advice to the Australian Environment Council, which has some emissions responsibilities.

Before 1986, passenger car emission standards were based on the 1973–74 U.S. requirements (ADR27). Since January 1986, manufacturers are required to meet a standard equivalent to 1975 U.S. requirements (ADR37). Current requirements for commercial gasoline-fueled vehicles are based on regulations from New South Wales and Victoria (table 1.12). Proposals are being considered to introduce stricter emission standards for passenger cars (equivalent to 1980 U.S. standards) beginning in January 1996. Smoke opacity limits (ADR 30 and ADR 55) apply to diesel-fueled vehicles.

Brazil

The Brazilian emissions control program, PROCONVE, was introduced by the national environmental board CONAMA in May 1986. The first emission standards for light-duty vehicles took effect in 1987, but these standards were lenient enough to be met by engine modifications alone. More stringent emission standards, comparable to those adopted by the United States in 1975, took effect in 1992. Compliance with these standards usually requires an open-loop catalytic converter, electronic fuel injection, or both. The Brazilian Congress has also enacted new legislation (No. 8723) effective October 1, 1993, setting strict emission standards for passenger vehicles for the rest of the decade. Exhaust emission standards equivalent to those adopted in the United States in 1981 are scheduled to take effect in 1997; compliance with these standards usually requires a three-way catalytic converter and electronic fuel injection with feedback control of the air–fuel ratio. More-stringent limit values will be introduced by 2000 and will match the U.S. standards. Crankcase emissions have been prohibited since 1977; evaporative emissions are limited to 6 grams per test. Brazilian regulations, like U. S. regulations, require an emissions warranty of 80,000 kilometers for light-duty vehicles and 160,000 kilometers for heavy-duty vehicles. Alternatively, emissions must be 10 percent below the set limits. The Brazilian fuel program also promotes the use of ethanol, both in pure form and as an additive for gasoline. Ethanol, although considered a cleaner-burning fuel than gasoline, can result in excessive emissions of aldehydes, especially acetaldehyde. For this reason, the 1992 and 1997 standards limit aldehyde emissions as well as emissions of hydrocarbons, carbon monoxide, and nitrogen oxides (table 1.13).

Control of heavy-duty diesel emissions has lagged behind that of light-duty vehicles. Limits on smoke emissions took effect in 1987 for buses and in 1989 for trucks. These limits follow European standards and test procedures. The first limits on gaseous emissions from diesel engines, also based on European practice, took effect in 1993. More-stringent standards, based on the Clean Lorry legislation of the European Union, were recently adopted. These provided for the first-stage standards in 80 percent of new buses in 1994 and 80 percent of all new heavy-duty vehicles by 1996. The second-stage limits (comparable in stringency to current U.S. standards for heavy-duty engines) are to be phased in between 1998 and 2002 (table 1.14). A system of prototype and production certification, based on the U.S. procedure has been established. Certification takes about 180 days. All manufacturers must submit statements specifying emissions of

Table 1.12 Exhaust Emission Standards for Motor Vehicles, Australia
(grams per kilometer)

Regulation	Effective date	Carbon monoxide	Hydrocarbons	Nitrogen oxides	Particulate matter	Test procedure	Evaporative emissions (grams per test)
Passenger cars							
ADR 27A/B/C	July 1976	24.2	2.1	1.9	—	FTP 75	2.0 (Canister)
	January 1982	22.0	1.91	1.73	—	FTP 75	6.0 (SHED)[b]
ADR 37[a]	January 1986	9.3	0.93	1.93	—	FTP 75	2.0 (SHED)
		(8.45)	(0.85)	(1.75)	—	(FTP 75)	(1.9 SHED)
Proposed standards	January 1996	4.34	0.26	1.24	2.0	FTP 75	
	January 2000	2.11	0.26	0.63	2.0	FTP 75	
Commercial vehicles (gasoline) NSW (Clean Air Act) and							
Victoria (statutory rules)	1992	9.3	0.93	1.93	—	FTP 75	2.0 (SHED)

— Not applicable

Note: Figures in parentheses apply to certification vehicles.

a. The higher figures apply to production vehicles, which must meet the limits from 150 kilometers to 80,000 kilometers or for five years, whichever occurs first.

b. SHED = Sealed Housing for Evaporative Determinations.

Source: CONCAWE 1994

Table 1.13 Exhaust Emission Standards for Light-Duty Vehicles (FTP-75 Test Cycle), Brazil
(grams per kilometer)

Year effective	Carbon monoxide	Hydrocarbons	Nitrogen oxides	Aldehydes[f]	Particulate Matter[g]	Carbon monoxide at idle (% v)[b]	Evaporative (grams per test)[i]
1988[a]	24	2.1	2.0	—	—	3.0	—
1989[b]	24	2.1	2.0	—	—	3.0	—
1990[c]	24	2.1	2.0	—	—	3.0	6.0
1992[d]	24	2.1	2.0	—	—	3.0	6.0
1992[e]	12	1.2	1.4	0.15	0.05	2.5	6.0
1994	12	1.2	1.4	0.15	0.05	2.5	6.0
1997	2	0.3	0.6	0.03	0.05	0.5	6.0
2000	Limits in line with U.S. standards						

— Not applicable

Notes: Effective January 1, 1988, no crankcase emissions permitted. Effective January 1, 1990, evaporative emissions limited to 6 grams per test (SHED). Exemptions possible for manufacturers with production less than 2,000 vehicles per year.

a. New cars only.

b. For certain specified models.

c. For all models, except those not derived from light-duty vehicles.

d. For models not derived from light-duty vehicles.

e. For models not covered in footnote d.

f. For alcohol-fueled vehicles only.

g. For diesel-fueled vehicles only.

h. Idle CO for alcohol in gasohol-fueled vehicles.

i. Evaporative emissions expressed as propane for gasohol-fueled and ethanol for alcohol-fueled vehicles.

Source: CETESB 1994; CONCAWE 1994

all models. Manufacturers of light-duty trucks (over 2,000 kg GVW) have the option to choose either the LDV or HDV test procedures for certification.

Canada

New standards for cars, light-duty trucks, and heavy duty trucks were introduced in 1987, bringing Canadian standards in line with then current U.S. limits (table 1.15). Limits for heavy-duty trucks are expected to be tightened in line with U.S. standards between 1994 and 1996. The Canadian federal government has announced plans to bring passenger car emission standards in line with the limits established in the U.S. Clean Air Act, pursuant to the provisions of NAFTA. Standards of 0.25 grams per mile for hydrocarbons, 3.4 grams per mile for carbon monoxide, and 0.4 grams per mile for nitrogen

Table 1.14 Exhaust Emission Standards for Heavy-Duty Vehicles (ECE R49 Test Cycle), Brazil
(grams per kilowatt hour)

Vehicle class	Effective date	Percent vehicles	Carbon monoxide	Hydrocarbons	Nitrogen oxides	Particulate matter	Smoke K
All vehicles	January 1, 1989	—	—	—	—	—	2.5[b]
	January 1, 1996	20	11.2	2.45	14.4		
		80	4.9	1.2	9.0	0.7/0.4[a]	
	January 1, 2000	20	4.9	1.2	9.0	0.7/0.4[a]	
		80	4.0	1.1	7.0	0.15	
	January 1, 2002	100	4.0	1.1	7.0	0.15	
All imports	January 1, 1994	100	4.9	1.2	9.0	0.7/0.4[a]	2.5[b]
	January 1, 1996	100	4.9	1.2	9.0	0.7/0.4[a]	
	January 1, 1998	100	4.0	1.1	7.0	0.15	
Urban buses	January 10, 1987	—	—	—	—	—	2.5[b]
	January 3, 1994	20	11.2	2.45	14.4	—	2.5[b]
		80	4.9	1.2	9.0	—	
	January 1, 1996	20	11.2	2.4	14.4	—	
	January 1, 1998	80	4.9	1.2	9.0	0.7/0.4[a]	
		20	4.9	1.2	9.0	0.7/0.4[a]	
	January 1, 2002	80	4.0	1.1	7.0	0.15	
		100	4.0	1.1	7.0	0.15	

— Not applicable
Note: * k=soot (g/m^3) *x gas flow (l/sec), applies to all vehicles.
a. Particulate emissions (PM) 0.7 g/kWh for engines up to 85 kWh; 0.4 g/kWh for engines above 85 kWh. Crankcase emissions must be nil, except for some turbocharged diesel engines if there is a technical justification.
b. Applies from this date onwards.
Source: CETESB 1994; CONCAWE 1994

Table 1.15 Exhaust Emission Standards for Light- and Heavy-Duty Vehicles, Canada

Vehicle type	Year effective	Carbon monoxide	Hydrocarbons	Nitrogen oxides	Diesel particulates	Testing procedure
Light-duty vehicles (grams per kilometer)						
Cars and light-duty trucks	1975–87	25.00	2.00	3.10	—	FTP 75
Cars	1988	2.11	0.25	0.62	0.12	FTP 75
Trucks less than 1,700 kilograms	1988	6.20	0.20	0.75	0.16	FTP 75
Trucks more than 1,700 kilograms	1988	6.20	0.50	1.10	0.16	FTP 75
Heavy-duty vehicles (grams per brake horsepower-hour)						
Less than 6,350 kilograms	1988	14.4	1.1	6.0	—	U.S. transient
More than 6,350 kilograms	1988	37.1	1.9	6.0	—	U.S. transient
Diesels	1988	15.5	1.3	6.0	0.6	U.S. transient
	1994	15.5	1.3	5.0	0.1	U.S. transient

— Not applicable
Source: CONCAWE 1994

oxides will probably be required as of 1996. The Federal Transport Minister and representatives of the automotive industry have agreed that cars sold in Canada from 1994 to 1996 will meet the same emissions standards as those sold in the United States. The U.S. manufacturers have committed themselves to market 1991 and subsequent model heavy-duty engines meeting U.S. standards in Canada in the absence of specific regulations.

Air quality legislation enacted by British Columbia in 1994 gives the province the authority to set standards for vehicle emissions and fuels. Based on the California model, the provincial standards applicable to the densely populated southern regions of the province are the most stringent in Canada.

Chile

Chilean authorities have adopted regulations requiring all new light-duty vehicles to meet 1983 U.S. emission limits. These regulations have been applied to the Santiago metropolitan area and surrounding regions since

September 1992, and were extended nationally in September 1994. Regulations have also been adopted requiring new heavy-duty trucks and buses to be equipped with engines meeting U.S. or European emission standards. U.S. 1991 or Euro 1 standards were required for buses in Santiago beginning September 1993, and for all heavy-duty vehicles in September 1994. U.S. 1994 or Euro 2 standards will be required for Santiago buses in 1996, and nationwide in 1998. An emissions test facility for certification and enforcement purposes is under development.

China

Regulation of motor vehicle emission in China has been guided by legislation enacted by the Standing Committee of the National People's Congress in 1979 and 1987 and the State Council in 1991: the "Environmental Protection Law of the People's Republic of China" (1979), the "Law of the People's Republic of China on the Prevention and Control of Air Pollution" (1987), and the "Detailed Rules and Regulations for the Law of the People's Republic of China on the Prevention and Control of Air Pollution" (1991). Based on these laws, the National Environmental Protection Agency, which is responsible for formulating emission standards and testing procedures, has issued 11 motor vehicle emission control standards and formulated a Management Procedure and Technical Policy to control emissions.

Standards for light-duty vehicles, adopted in 1979, are equivalent to ECE regulation 15-03 and include testing procedures for type approval and conformity of production. The standard test procedure lasts 13 minutes and has four cycles with no intermission. Each cycle covers 15 working phases (idling, acceleration, deceleration, steady speed, and so on). These standards were developed by the Changchun Automotive Research Institute and submitted to the State Environmental Protection Administration by the Chinese Automotive Industry Federation (CSEPA 1989). The Environmental Protection Administration adopted performance targets for motorcycles in 1985 (GB 5366-85), and exhaust emission standards for light-duty vehicles in 1989 (GB11641-89). The light-duty vehicle standards apply to cars, passenger vans, and light-duty freight vehicles (reference mass 3,500 kilograms or less) operating at a minimum speed of 50 kilometers an hour.

Exhaust emission standards for heavy-duty vehicles, adopted in 1983 (Regulations No. GB 3842-83, 3843-83, and 3844-83) consist only of carbon monoxide and hydrocarbon limits determined at idle and apply both to new and in-use vehicles (table 1.16). China is considering legislation for heavy-duty gasoline-engine vehicles based on two runs of the U.S. 9-mode cycle used by U.S. EPA during 1970-1983. Proposed limits are given in table 1.17. Idle emission standards have also

Table 1.16 Exhaust Emission Limits for Gasoline-Powered Heavy-Duty Vehicles (1983), China

	Idle	
	Carbon monoxide	Hydrocarbons
Vehicles	*(% v)*	*(ppm)*
New	5.0	2500
In-use	6.0	3000
Imports	4.5	1000

Source: CONCAWE 1994

been adopted for heavy-duty gasoline engines, and consideration is being given to establishing mass emission limits for these engines in grams per kilowatt hour. Revised or new emission standards and testing procedures that came into force in 1994 are listed in table 1.18.

Chinese motor vehicle regulations require that all domestically produced vehicle models must be listed in the "Index of Enterprises Producing Motor Vehicles and their Products" issued annually by China National Automotive Industry Corporation (CNAIC) and the Ministry of Public Security. Before a vehicle model is listed in the Index, the vehicle should pass an approval test carried out by "The Type Approval Organization for New Motor Vehicle Products." Recently, the responsibility for issuing the index and type approvals has been transferred to the Auto Industry Bureau of the Ministry of Machinery. Included in the approval tests are idling emissions tests for gasoline-fueled vehicles and free acceleration smoke tests for diesel-fueled vehicles. In addition, a full load smoke test is required for diesel engines. In case of imported motor vehicles, the type approval tests, are conducted by authorized laboratories of the State Administration for Import and Export Commodity Inspection.

Colombia

New emission standards for gasoline- and diesel-fueled vehicles were established in 1996 by Resolution 5 of the Ministry of the Environment and Transport. These standards establish carbon monoxide and hydrocarbons emission limits for two mean-sea-level ranges of 0 to 1,500 meters, and 1,501 to 3,000 meters (table 1.19).

Additional emission standards have been adopted by the Ministry of the Environment for light, medium, and heavy-duty gasoline- and diesel-fueled vehicles to come into effect with model year 1997 (table 1.20).

Eastern European Countries and the Russian Federation

Most eastern European and central Asian countries including Russia use some combinations of ECE and EU regulations, as shown in table 1.21.

Hong Kong

New cars sold in Hong Kong are required to meet either U.S. or Japanese emission standards or the new consolidated European limits (91/441/EEC). Each of these regulations requires the use of three-way catalytic converters with electronic control systems. All new cars were required to meet these standards as of January 1992. Light-duty diesel vehicle emission standards were also tightened effective April 1, 1995. All new passenger cars and taxis must comply with 1990 U.S. standards or equivalent EU and Japanese standards. Similar requirements will apply to medium goods vehicles and light buses. For goods vehicles and buses with a design weight of 3.5 tonnes or more either the 1990 U.S. standards or the Euro 1 standards will apply. Emissions standards have not been established for motorcycles.

Table 1.17 Proposed Exhaust Emission Limits for Gasoline-Powered Heavy-Duty Vehicles, China
(grams per kilowatt hour)

Year	Vehicle	Carbon monoxide (g/kWh)	Hydrocarbons + Nitrogen oxides (g/kWh)
Up to 1997	Certified before 1992	80	32
	Produced after 1992	50	20
	Type approved after 1992	40	16
1997	Certified before 1992	50	20
	Produced after 1992	34	13.6
	Type approved after 1992	28	11

Source: CONCAWE 1994

Table 1.18   List of Revised or New Emission Standards and Testing Procedures, China (Effective 1994)

Number	Title
GB 14761.1-93	Emission standard for exhaust pollutants from light-duty vehicles
GB 14761.2-93	Emission standard for exhaust pollutants from gasoline engine of road vehicles
GB 14761.3-93	Emission standard for fuel evaporative emissions from road vehicle with gasoline engine
GB 14761.4-93	Emission standard for pollutants from crankcase of vehicle engines
GB 14761.5-93	Emission standard for pollutants at idle speed from road vehicle with gasoline engine
GB 14761.6-93	Emission standard for smoke at free acceleration from road vehicles with diesel engine
GB 14761.7-93	Emission standard for smoke at full load from automotive diesel engines
GB/T 14762-93	Measurement method for exhaust pollutants from gasoline engine of road vehicles
GB/T 14763-93	Measurement method of fuel evaporative emissions from road vehicles with gasoline engine
GB/T 3845-93	Measurement method for pollutants at idle speed from road vehicles with diesel engine
GB/T 3846-93	Measurement method for smoke at free acceleration from road vehicles with diesel engine
GB/T 3847-93	Measurement method for smoke at full load from automotive diesel engines
GB 14621-93	Emission standard for exhaust emissions from motorcycles
GB/T 14622-93	Measurement method for exhaust emissions from motorcycles under running mode
GB/T 5466-93	Measurement method for exhaust emissions from motorcycles under idle speed

Source: Walsh 1995

Table 1.19   Emission Limits for Gasoline-Fueled Vehicles for Idle and Low Speed Conditions, Colombia

Model year	Carbon monoxide (%) above msl: 0–1500 m	Carbon monoxide (%) 1501–3000 m	Hydrocarbons (ppm) above msl: 0–1500 m	Hydrocarbons (ppm) 1501–3000 m
2001 or newer	1.0	1.0	200	200
1998–2000	2.5	2.5	300	300
1997–1996	3.0	3.5	400	450
1995–1991	4.5	5.5	750	900
1990–1975	5.5	6.5	900	1000
1974 or older	6.5	7.5	1000	1200

msl = mean sea level
Source: Onursal and Gautam 1996

Table 1.20 Exhaust Emission Standards for Gasoline and Diesel-Fueled Vehicles, Colombia

Vehicle category	Unit	Carbon monoxide	Hydrocarbons	Nitrogen oxides
Light-duty vehicles	g/km	2.3	0.25	0.62
Medium-duty vehicles	g/km	11.2	1.05	1.43
Heavy-duty vehicles	g/bhp–hr	25.0		10[a]

a. Sum of HC and NO_x emissions.
Source: Onursal and Gautam 1996

Table 1.21 Summary of Vehicle Emission Regulations, Eastern Europe

Country	Vehicle type	Implementation date	Regulation	Comments
Czech and Slovak Republics	Passenger cars	Type approval 01.10.92 All vehicles 01.10.93	89/458/EEC	
	Light-duty vehicles	As above	83/351/EEC	
	Heavy-duty vehicles	As above	91/542/EEC	
Hungary	Passenger cars	July 1992	ECE R83	For imported vehicles
	Heavy-duty vehicles	1990	ECE R49	Steady-state CO 14, HC 3.5, and NO_x 18g/kWh
			ECE R24	Full load smoke
			Ordinance 6/1990	Free acceleration smoke
Poland	Passenger cars	July 1995	ECE R83.02; 93/59/EC	
	Heavy-duty vehicles	Oct 1993	ECE R49.02; 91/542/EC	
		1988	ECE R24.03	
	Motorcycles	Nov. 1992	ECE R40.01	
	Mopeds	Nov. 1992	ECE R47	
The Russian Federation	Gasoline passenger cars (without catalytic converters)	1986	OST 37.001 054-86	Similar to ECE R15.04
	Gasoline passenger cars (with catalytic converters)	1986	OST 37.001 054-86	Conforms to ECE R83
	Diesel engines-exhaust emissions	1981	OST 37.001 234-81	CO 9.5, HC 3.4, NO_x 14.35 per bhp-hr (ECE R49 test mode)
	Diesel engines-black smoke emissions	1984	GOST 17.2 01-84	Full load smoke; emission limits as follows: Nominal flow (l/sec) / Smoke limit (opacity %): <42 / 60; 100 / 45; >200 / 34

Source: CONCAWE 1995

India

The union government enacted a revised Motor Vehicle Act in 1990, making emission regulations a federal government responsibility. India has established limits on carbon monoxide emissions (at idle) for gasoline-fueled cars, motorcycles, and three-wheelers; diesel smoke emissions are limited to 75 Hartridge units at full load. New emission standards for gasoline-fueled cars took effect in 1991. Emissions from diesel vehicles came under control in 1992 based on ECE R49 regulations. These limits are similar to the ECE 15-04 limits but with test procedures tailored to Indian driving conditions (table 1.22). Evaporative emissions are not regulated. Conformity of production tests have also been developed. In addition, deterioration factors and endurance tests have been prescribed.

From April 1, 1996, all two-stroke engines in two- and three-wheelers would be required to comply with the tighter emission standards shown below:
- Three-wheelers
  CO: 6.75 g/km; $HC+NO_x$: 5.41 g/km
- Two-wheelers
  CO: 4.50 g/km; $HC+NO_x$: 3.50 g/km

Table 1.22 Exhaust Emission Standards for Gasoline-Fueled Vehicles, India
(grams per kilometer)

Reference mass (kilograms)	Carbon monoxide	Hydrocarbons
Two- and three-wheel vehicles		
Less than 150	12	8
150–350	12+(18*(R-150)/200)	8+(4*(R-150)/200)
More than 350	30	12
Light-duty vehicles		
Less than 1,020	14.3	2.0
1,020–1,250	16.5	2.1
1,250–1,470	18.8	2.1
1,470–1,700	20.7	2.3
1,700–1,930	22.9	2.5
1,930–2,150	24.9	2.7
More than 2,150	27.1	2.9

R = Reference mass.
Source: India 1989

Japan

Japan revised its emissions test procedures for light-duty vehicles in 1991. The new test procedure, resembling the new ECE emissions test cycle, consists of a series of low- and moderate-speed accelerations and decelerations at 20 kilometers per hour to 40 kilometers per hour, as well as a higher-speed component reaching up to 70 kilometers per hour. It will apply to passenger cars and light- and medium-duty trucks (up to 2.5 tons gross weight). The test procedure for heavy-duty engines has also been modified from the previous six-mode test to another steady-state, engine dynamometer test involving 13 operating modes (these modes are different from the ECE 13-mode test). The units of measurement have also been changed, from grams per test and ppm to grams per kilometer and grams per kilowatt hour, making it easier to compare Japanese standards with U.S. and ECE emission standards. In addition to these changes, emission limits on nitrogen oxides (already among the most stringent worldwide) are to be further tightened, and limits on diesel particulate emissions have been introduced. Smoke limits were reduced by 20 percent in 1993 for light- and medium-duty diesel vehicles and more stringent smoke limits were expected for heavy-duty passenger vehicles. Detailed information on Japanese emission standards is available in CONCAWE 1994.

Because of the differences in test procedures, a direct comparison of Japanese emission standards with those applied to the U.S. and Europe cannot be made.

Republic of Korea

Passenger car and light-truck emission standards equal to current U.S. standards have been in effect since 1987. These standards apply to passenger cars using either gasoline or liquified petroleum gas, with engine displacement greater than 0.8 liter and gross vehicle weight less than 2.7 tons. Standards for heavy-duty gasoline and liquified petroleum gas engines, based on the U.S. heavy-duty transient test procedure, are similar to those in effect for heavy-duty gasoline engines in the United States before 1990. Heavy-duty diesel test procedures and emission standards are similar to those of Japan. The Korean government is also moving to discourage the use of diesel engines in medium-duty trucks in favor of gasoline or liquified petroleum gas engines with more effective emissions control. Legislation has also been introduced for two-stroke and four-stroke motorcycles that would require the use of catalytic converters. Emission limits for two- and four-stroke motorcycles are summarized in table 1.23; the test procedure however, is not known (CONCAWE 1994; UNIDO 1990).

Table 1.23 Motorcycle Emission Standards, Republic of Korea

Period	Two-stroke		Four-stroke	
	Carbon monoxide (% v)	Hydrocarbons (ppm)	Carbon monoxide (% v)	Hydrocarbons (ppm)
1/91 to 12/92	5.5	1100	5.5	450
1/93/12/95	4.6	1100	4.5	450
1/96 onward[a]	3.6	450	3.6	400

a. Proposed.
Source: CONCAWE 1994

Malaysia

In Malaysia, vehicle emission regulations based on ECE 15.04 were introduced in September 1992. A further requirement that all new gasoline-fueled vehicles be equipped with catalytic converters has been temporarily postponed.

Mexico

New passenger cars and light-duty trucks sold in Mexico have been subject to exhaust emission standards (generally based on U.S. standards) since 1975 (table 1.24). Until the 1991 model year, however, these standards were loose enough to be met without the use of a catalytic converter or other advanced emission control technology. The 1991 and 1992 model years were a transition during which standards were made sufficiently stringent to require catalytic converters but not the full range of emission control technology required in the United States. Despite the transition period, most Mexican automakers equipped their vehicles with U.S.-model emission controls instead of a less-sophisticated system that would be used for only two years. New cars and light trucks since model year 1993 have been required to meet exhaust emission standards that are equivalent to 1987 U.S. standards. Starting in 1995, all cars, light commercial vehicles and light trucks were required to meet an evaporative emissions standard of 2.0 grams per test as well. Mexico has not yet adopted other elements of the U.S. regulations: emissions durability, emissions warranty requirements, and in-service testing with recall of vehicle models found to be violating emission standards in service. As part of the NAFTA, the United States, Canada, and Mexico are in the process of harmonizing vehicle emission standards.

In the past few years, Mexican authorities have established emission standards for medium- and heavy-duty vehicles (table 1.25). In 1991 authorities and manufac-turers agreed on catalyst-forcing standards for gasoline-fueled microbuses. New microbuses are required to meet these standards. Similar standards will apply to all new medium-duty vehicles in 1994. As of June 1992, new heavy-duty diesel vehicles were required to be equipped with engines meeting 1991 U.S. emission standards at sea level, with an additional test of smoke opacity conducted at Mexico City's altitude (2,000 meters). Work on evaluating the feasibility of meeting 1994 U.S. emission standards at Mexico City's altitude is planned.

Saudi Arabia

Saudi Arabia has adopted emission standards equivalent to ECE R15.03. Annual inspections of vehicle emission control systems is required in Jeddah, Riyadh, and Damman. Evaporative emissions are limited to 6.0 grams per test (SHED).

Singapore

Singapore introduced European (ECE R 15-04) emission standards for passenger cars in 1986. Since July 1, 1992, all new gasoline-fueled vehicles registered in Singapore have been required to comply either with ECE 83 or current Japanese emission regulations. In July 1993, the limits of the Consolidated Emissions Directive replaced ECE 83. New diesel-fueled vehicles have been required to meet the smoke limits of ECE R 24.03 since January 1991, and used diesel vehicles imported to Singapore have been required to meet the same standard since January 1992. New motorcycles have been required since October 1991 to comply with U.S. emission standards stipulated in the U.S. Code of Federal Regulations (40 CFR 86.410-80) before they can be registered for use in Singapore.

Since 1982, all in-use vehicles have been required to undergo a periodic, compulsory mechanical inspection. This is to ensure that vehicles on public roads are maintained properly, are roadworthy, and do not pol-

Table 1.24 Emission Standards for Light-Duty Vehicles, Mexico
(grams per kilometer)

Year	Carbon monoxide	Hydrocarbons	Nitrogen oxides
Pre-1975 (uncontrolled)	54.0	5.5	2.3
1975	29.2	2.5	2.3
1976	24.2	2.1	2.3
1977	24.2	2.6	2.3
1988	22.2	2.0	2.3
1990	18.0	1.8	2.0
1991	7.0	0.7	1.4
1993	2.11	0.25	0.62

Note: Evaporative hydrocarbon emissions are not regulated.
Source: World Bank 1992

Table 1.25   Exhaust Emission Standards for Light-Duty Trucks and Medium-Duty Vehicles by Gross Vehicle Weight, Mexico
(grams per kilometer)

Vehicle type	Carbon monoxide	Hydrocarbons	Nitrogen oxides
Light-duty trucks, less than 2,727 kilograms			
1991–93	22.0	2.0	2.3
1994	8.75	0.63	1.44
Light-duty trucks, 2,728–3,000 kilograms			
1991	35.0	3.0	3.5
1992–93	22.0	2.0	2.3
1994	8.75	0.63	1.44
Medium-duty vehicles, 3,000–3,857 kilograms			
1992	28.0	2.8	2.8
1993	22.0	2.0	2.3
1994	8.75	0.63	1.44
Urban transport minibuses, 3,001–5,500 kilograms[a]			
1991	10.0	0.6	1.5
1992	3.0	0.3	1.0

Note: Evaporative emissions are not regulated.

a. Standards applied only to highly polluted areas through 1992; minibuses outside critical areas were not regulated until 1993.

Source: World Bank 1992

lute the environment. Exhaust emissions are checked during these inspections. Enforcement inspection is also conducted daily by spot checking vehicles on the road. Detailed information on Singapore emission standards is given in CONCAWE 1994.

Taiwan (China)

All new cars sold in Taiwan (China) have been required to meet ECE regulation 15.04 emission standards since July 1987. In July 1990, the regulation was tightened to require compliance with 1983 U.S. emission standards. New models and all imports were required to meet these standards immediately, and existing domestically produced models were allowed waivers of up to three years. Beginning July 1994, all car models sold in Taiwan (China) were required to meet U.S. standards. Emission standards for new motorcycle engines have also been established—1991 standards are some of the most stringent in the world (table 1.26). Compliance with motorcycle standards has required significant engine modifications, including the use of air injection in four-stroke engines and the installation of catalytic converters on two-stroke engines. Electric motorcycles have been available since May 1992 but with modest sales.

Motorcycle durability requirements have been in force since November 1991. All new motorcycles are required to demonstrate that they can meet the relevant standards for a minimum of 6,000 kms. It is expected that the durability requirement will be increased to 20,000 kms from January 1, 1998 and that the share of electric powered motorcycles will be mandated at 5 percent.

Diesel engines have been required to comply with smoke emission limits since 1984. Since July 1993, diesel engines have been required to meet emission limits of 6.0 grams per bhp-hour for nitrogen oxides and 0.7 grams per bhp-hour for particulate matter, based on the U.S. heavy-duty transient cycle. The nitrogen oxides standard is the same as the 1988 U.S. standard, while the standard for particulate matter is slightly more lenient.

Thailand

The rapid growth in the vehicle fleet has compelled the Royal Thai Government to quickly establish emission standards. New gasoline-fueled vehicles have been required to be fitted with catalytic converters since 1993. Thailand has adopted test cycles and emission standards conforming to ECE/EEC regulations for light-duty gasoline and light- and heavy-duty diesel vehicles (table 1.27). Emission standards for motorcycles equivalent to ECE R40 were introduced in August 1993 and soon after revised to comply with ECE R40.01 regulations. Third-phase controls similar to the Taiwanese regulations are being introduced over the period 1994–1997.

In addition to the standards themselves, procedures for verifying compliance and for corrective action to deal with non complying vehicles also need to be developed. The Thai Institute of Standards has established laboratory facilities to measure emissions from light-duty gasoline-and diesel-fueled vehicles and motocycles. Facilities for diesel engines are scheduled to be completed soon after. They will be used only to verify compliance by new vehicles—as in Europe, but unlike

Table 1.26 Exhaust Emission Standards for Motorcycles, Taiwan (China)

		ECE 40 (driving cycle)		ECE 40 (idle)		
Year implemented	Classification	Carbon monoxide (grams per kilometer)	Hydrocarbons + nitrogen oxides (grams per kilometer)	Carbon monoxide (percent by volume)	Hydrocarbons (ppm)	Acceleration smoke (opacity percentage)
1984	New	8.8	6.5	4.5	7,000	—
	In-use	—	—	4.5	9,000	—
1988	Prototype	7.3	4.4	4.5	7,000	15
	Production	8.8	5.5	4.5	7,000	15
	In-use (inspection)	—	—	4.5	9,000	30
1991	Prototype	3.75	2.4	4.5	7,000	15
	Production	4.50	3.0	4.5	7,000	15
	In-use (inspection)	—	—	4.5	9,000	30
1997 (proposed)	Prototype	0.75[a]	0.8[a]	4.5	7,000	15

— Not applicable

a. From January 1994 onwards, the limits are CO (1.5 g/km), HC (1.2 g/km), and NO_x (0.4 g/km). It is proposed to reduce the 1994 limits by half as of January 1997.

Source: Chan and Weaver 1994; CONCAWE 1992

Table 1.27 Exhaust Emission Standards, Thailand

Vehicle	Effective date	Cycle	Equivalent limits
Gasoline	Jan. 1, 1995	ECE 83	ECE 91/411/EEC
Light diesel	Jan. 1, 1995	ECE 83	ECE 91/411/EEC
Heavy diesel	Jan. 1, 1993	ECE 49	ECE 91/542/ (A)/EEC (Euro 1)
	Jan. 1, 2000	ECE 49	ECE 91/542 (B)/EEC (Euro 2)
Motorcycles	Jan. 1, 1994	ECE 40	see table 1.26 footnote (a)
	Jan. 1, 1997	ECE 40	

Source: CONCAWE 1994

in the United States, manufacturers will bear no responsibility for continuing emissions compliance in consumer use.

Compliance with Standards

Stringent emission standards are of little value without a program to ensure that vehicle sellers and buyers comply with them. A comprehensive compliance program should cover both new and in-use vehicles; it should assure that attention is given to emission standards at the vehicle design stage prior to mass production; it should ensure quality on the assembly line; and it should deter the manufacture of non-conforming vehicles through an enforceable warranty and recall system. Furthermore, vehicle owners should be encouraged to carry out maintenance on emission control devices as required by the manufacturer, and the service industry

should be regulated to perform this maintenance properly. A comprehensive compliance program should include the items outlined in the following sections.

Certification or Type Approval

Most countries require some form of certification or type approval by vehicle manufacturers to demonstrate that each new vehicle sold is capable of meeting applicable emission standards. Usually, type approval requires emission testing of prototype vehicles representative of planned production vehicles. Under ECE and Japanese regulations, such compliance is required only for new vehicles (though new regulations require manufacturers to demonstrate in each case the durability of catalytic converters up to 80,000 kilometers). U.S. regulations require that vehicles comply with emission standards throughout their useful lives when maintained according to the manufacturers specifications.

As part of the U.S. certification process, vehicle manufacturers are required to operate a prototype vehicle in an accelerated durability test driving program, for distances ranging from 6,000 kilometers (for small motorcycles) to 160,000 kilometers (for passenger cars) or 192,000 kilometers (for light trucks). Heavy-duty engines are similarly tested, using an accelerated durability procedure on an engine dynamometer. Emission tests on the prototype vehicle or engine undergoing this accelerated testing are used to establish deterioration factors for each pollutant for a given engine family. To determine compliance with the regulations, the emissions from low-mileage prototype vehicles are multiplied by the established deterioration factor; the result is required to be less than the applicable standard. Durability demonstrations, maintenance, and emissions testing are normally conducted by vehicle manufacturers, but national testing authorities will occasionally test certification vehicles on a spot check basis.

The advantage of a certification program is that it can influence vehicle design prior to mass production. Obviously, it is more cost-effective if manufacturers identify and correct problems before production actually begins. As a practical matter, the certification process deals with prototype cars (sometimes almost handmade) in an artificial environment of careful maintenance, perfect driving conditions, and well-trained drivers using ideal roads or dynamometers. As a result, vehicles that fail to meet emission standards during certification will almost certainly fail to meet standards in use. The converse, however, is not true—it cannot be said with confidence that vehicles that pass certification will inevitably perform well in use.

Assembly Line Testing

The objectives of assembly line testing are to enable regulatory authorities to identify certified production vehicles that do not comply with applicable emission standards, to take remedial actions (such as revoking certification and recalling vehicles) to correct the problem, and to discourage the manufacture of noncomplying vehicles.

Assembly line testing provides an additional check on mass-produced vehicles to assure that the designs found adequate in certification are satisfactorily translated into production, and that quality control on the assembly line is sufficient to provide reasonable assurance that cars in use will meet standards. The main advantage of assembly line testing over certification is that it measures emissions from real production vehicles. However, assembly line testing provides no measure of vehicle performance over time or mileage—a substantial and inevitable shortcoming.

In-Use Surveillance and Recall

U.S. emission regulations require vehicles to continue to meet emission standards throughout their useful lives, if maintained according to manufacturers' specifications. U.S. authorities have instituted extensive testing programs to guard against increases in emissions resulting from defective emission controls in customer use. Several hundred vehicles per year are temporarily procured from consumers for testing. A questionnaire and physical examination of the vehicle are used to determine whether the vehicle has been properly maintained. The vehicles are then adjusted to specifications and tested for emissions. Average emissions from all properly maintained vehicles of a given make and model tested must comply with the applicable emissions standard within a range that allows for statistical uncertainty. Otherwise, the manufacturer may be ordered to recall the vehicles for repairs or modifications to bring them within the standard. Hundreds of thousands of vehicles are recalled in this manner each year.

The expense and consumer dissatisfaction generated by emission recalls have induced manufacturers to develop far more durable and effective emission control systems and to establish internal emission targets for new vehicles that are much stricter than legal standards. Most manufacturers design for a margin of at least 30 percent (and preferably 50 percent) between the certified emissions and the standard in order to provide a reasonable allowance for in-use deterioration. Thus, to ensure in-use compliance with a standard of 0.2 grams per mile for nitrogen oxides, most manufacturers would set a development target of 0.1 grams per mile. As a result, the emissions performance of vehicles in use has improved significantly.

Emissions surveillance programs provide extensive data on vehicle emissions in the real world, as opposed to the artificial world of prototype development and certification. They thus provide important information to air quality planning authorities, vehicle regulators, and vehicle manufacturers on the actual effectiveness of emission control programs and technologies. The importance of such real-world data cannot be overestimated. Surveillance programs in the United States have been instrumental in identifying the causes of high vehicle emissions in use, which in turn has led regulators and manufacturers to take actions to correct these problems. In the absence of systematic measurements of actual vehicle emissions in use, government and industry can all too easily assume that no problems exist.

Surveillance and recall programs have had their difficulties. Surveillance programs are expensive and potentially are subject to sampling bias, because citizens must be induced to lend their vehicles to the government for testing. Surveillance testing of heavy-duty engines requires that they be removed from the vehicle and tested on an engine dynamometer—an extremely expensive procedure. As a result, very little of this type of testing has been done to date, and data on in-use emissions from heavy-duty vehicles are correspondingly sparse. Recalls are not fully effective either—in the United States, on average, only 55 percent of the vehicles recalled are actually brought in by their owners for repair. The lag time between identification of a nonconforming class and the manufacturer's recall notice can be well over a year. Mandatory recalls are possible only when a substantial number of vehicles in a class or category exceeds the standards limit; this may preclude recalls of serious but less frequent failures.

Warranty

Warranty programs are intended to provide effective recourse to consumers against manufacturers when individual vehicles fail to meet in-use standards, and to discourage the manufacture of such vehicles. Warranties attempt to assure the remedy of defects in design or workmanship that cause high emissions.

On-Board Diagnostic Systems

Increasingly complicated vehicle engine and emissions control systems have made the diagnosis and repair of malfunctioning systems more difficult. With present inspection and maintenance program designs, many emissions-related malfunctions can go undetected in modern vehicles. This is especially true for malfunctions related to nitrogen oxides, because present inspection and maintenance programs do not test for these emissions. To improve the effectiveness of emissions control diagnosis, the United States has recently adopted second-generation requirements for on-board diagnosis of emissions-related malfunctions.

Alternatives to Emission Standards

Because they are mandatory and universal, traditional vehicle emission standards lack flexibility and may thus impose higher costs in return for lower benefits than more flexible approaches. To avoid disrupting the market, universal standards must be set at a level that nearly all vehicles can meet, which implies that some vehicles are capable of meeting stricter standards but have no incentive to do so. In most cases, standards do not differentiate between vehicles according to use—the taxicab in a highly polluted urban center meets the same standards as a car used in a remote rural area. Emissions control for the latter vehicle has little or no social utility and social resources are wasted.

A number of approaches have been taken to reduce these drawbacks. In the United States, for example, programs for emissions averaging, trading, and banking have been introduced in which manufacturers who are able to do better than the emission standards on one vehicle or engine model can generate credits that can be used to offset higher-than-standard emissions from another model of vehicle or engine. The manufacturer can choose whether to use the credits in the same year (averaging), sell them to another manufacturer (trading), or save them against possible need in a subsequent year (banking).

Another promising strategy for achieving maximum emission reductions at minimum cost is to establish differentiated emission standards for heavily used vehicles in highly-polluted areas. In the United States, for example, fleet vehicles in major urban areas will be covered by the "Clean Fuel Vehicle" program, a special program that requires vehicles certified to lower emission standards. Special strict emission standards have also been established for gasoline-fueled minibuses in highly polluted areas in Mexico. In Chile, buses in Santiago were required to meet emission standards earlier than other vehicles.

An extension of the differentiated emission standard approach would be to offer economic incentives for urban vehicles to adopt more stringent emission controls. By calibrating the size of the incentive to the expected use of the vehicle, the most stringent and expensive emission controls could be applied where they would be most effective.

Some countries in Western Europe, notably Germany, have made effective use of tax incentives to encourage buyers to choose vehicles certified to more stringent emission standards than the minimum requirements. In the United States, consumers are being encouraged to purchase low-emitting vehicles by making these vehicles exempt from transportation control measures, such as mandatory no-drive days. This approach is also being used in Mexico City to encourage owners of commercial vehicles to switch to cleaner fuel systems, such as liquified petroleum gas and natural gas.

From a theoretical standpoint, a vehicle emissions tax would be an ideal economic incentive for controlling emissions. Although such a tax would pose formidable implementation problems, a properly implemented emission tax could encourage vehicle owners to purchase clean cars (leading manufacturers to compete in cleanliness as they now compete in fuel economy), and it would encourage them to maintain their cars properly so that they continue to be clean in use. An alternative to the emissions tax that might be easier to implement would be to impose a high indirect tax (e.g. on fuel), and then to offer a rebate on this tax based on a vehicle's emission performance in a representative test such as the IM240 described in Chapter 4. This would create an incentive for drivers to undergo the test (in order to receive the rebate) rather than to avoid it.

Very similar in theoretical effect to a vehicle emissions tax would be the provision of mobile-source emission reduction credits, which could be traded to stationary sources or other vehicle owners in lieu of meeting emission regulations. If combined with a sufficiently tight limit on overall emissions, such a program would provide an incentive for those who could reduce emissions cost-effectively to do so, in order to sell the resulting reductions (or rights to emit) to others for whom reducing emissions would be more costly. Credit programs of this kind are now being implemented in a number of jurisdictions in the United States.

References

Baines, T.M. 1994. Personal Communication. U.S. Environmental Protection Agency, Washington, D.C.

Boletin Oficial. 1994. *Section No. 27.919*. Buenos Aires, Argentina.

CETESB. 1994. *Relatorio de Qualidade do Ar no Estado de São Paulo - 1993*. Companhia de Tecnologia de Saneamento Ambiental, São Paulo, Brazil.

Chan, L.M. and C.S. Weaver. 1994. "Motorcycle Emission Standards and Emission Control Technology." Departmental Paper Series No. 7, Asia Technical Department, The World Bank, Washington, D.C.

CONCAWE (Conservation of Clean Air and Water in Europe). 1992. *Motor Vehicle Emission Regulations and Fuel Specifications—1992 Update*. Report 2/92, Brussels.

_____. 1994. *Motor Vehicle Emission Regulations and Fuel Specifications—1994 Update*. Report 4/94, Brussels.

_____. 1995. *Motor Vehicle Emission Regulations and Fuel Specifications in Europe and the United States—1995 Update*. Brussels.

CSEPA (China State Environmental Protection Administration). 1989, "Emission Standards for Exhaust Pollution from Light-Duty Vehicles." *National Standard GB11641-89*, Beijing.

Havenith, C., J.R. Needham, A.J. Nicol, and C.H. Such. 1993. "Low Emission Heavy-Duty Diesel Engine for Europe." *SAE Paper* 932959. Warrendale, Pennsylvania.

India, Ministry of Surface Transport. 1989. "Report of The Working Group on Road Transport for the Eighth Plan (1990-95)." Government of India, New Delhi.

Onursal, B. and S. Gautam. 1996. "Vehicular Air Pollution: Experience from Seven LAC Urban Centers." A World Bank Study (forthcoming), Washington, D.C.

Plaskett, L. 1996. "Airing the Differences." *Financial Times*, June 26, 1996, London.

UNIDO (United Nations Industrial Development Organization). 1990. "Control and Regulatory Measures Concerning Motor Vehicle Emissions in the Asia-Pacific Region." Report from a Meeting at the Korea Institute of Science and Technology, Seoul.

van Ruymbeke, C., R. Joumard, R. Vidon, and C. Provost. 1992. "Representativity of Rapid Methods for Measuring Pollutant Emissions from Passenger Cars." INRETS Report LEN9219. Bron, France.

Walsh, M.P. 1995. "Technical Notes." 3105 N. Dinwiddie Street, Arlington, Virginia.

_____. 1996a. "Car Lines." Issue 96-3, 3105 N. Dinwiddie Street, Arlington, Virginia.

World Bank. 1992. "Transport Air Quality Management in the Mexico City Metropolitan Area." Sector Report No. 10045-ME. The World Bank, Washington, D.C.

2

Quantifying Vehicle Emissions

There are several procedures for measuring vehicle emissions for regulatory purposes. The most commonly used are the U.S. federal, the United Nations Economic Commission for Europe (ECE), and the Japanese test procedures. The U.S. and European tests are also used extensively in other countries. These test procedures have many common elements. For light-duty vehicles, including motorcycles, emissions are measured by operating the vehicle on a chassis dynamometer while collecting its exhaust in a constant-volume sampling system. Testing for heavy-duty vehicle engines is done on an engine dynamometer.

The main difference among the procedures is the driving cycle (for vehicles) or operating cycle (for heavy-duty engines). The European and Japanese procedures test in a series of steady-state operating conditions, while the U.S. procedures involve transient variations in speed and load more typical of actual driving. None of the tests fully reflect real-world driving patterns, however. More representative driving cycles have been developed, and are being considered for adoption in the United States (AQIRP 1996).

Vehicle emissions are affected by driving patterns, traffic speed and congestion, altitude, temperature, and other ambient conditions; by the type, size, age, and condition of the vehicle's engine; and, most importantly, by the emissions control equipment and its maintenance. Emission factors are estimates of the pollutant emissions produced per kilometer traveled by vehicles of a given class. Computer models estimate vehicle emission factors as functions of speed, ambient temperature, vehicle technology, and other variables. The U.S. EPA's MOBILE5 is probably the most widely used. It is based on emission tests carried out on in-use vehicles as part of the U.S. EPA's emissions surveillance program. The European Union's COPERT model includes a wide cross-section of European vehicles and is based on an extensive program of emission tests carried out in several European countries.

Vehicle emission factors for a given jurisdiction should ideally be based on emission measurements performed on a representative sample of in-use vehicles from that area. Such data collection is expensive, however, and requires facilities that few countries possess. Without specific data, preliminary planning can be based on MOBILE5 or COPERT estimates for similar technologies, or on data from countries with similar vehicle fleet characteristics.

Emissions Measurement and Testing Procedures

Motor vehicle emissions are highly variable. In addition to differences among vehicles, differences in operating conditions can cause emissions from a given vehicle to change by more than 100 percent. A consistent and replicable test procedure is required if emission regulations or incentive systems are to be enforceable. To ensure improvement in emissions, testing should be representative of in-use conditions or severe enough to ensure effective emission control system performance under all conditions. Current procedures do not always achieve this objective. This chapter includes the key features of the various emission test procedures. Detailed descriptions of these procedures can be found in official documents – Part 86 of the Code of Federal Regulations for the U.S. procedures and the various ECE standards available through the United Nations ECE secretariat in Geneva.

Exhaust Emissions Testing for Light-Duty Vehicles

Three test procedures are presently used to measure the emissions of light-duty vehicles: the U.S. federal test procedure (FTP), the European test procedure established by the United Nations Economic Commission for Europe (ECE) regulation 83, and the Japanese test procedure. The U.S. procedure has now been adopted throughout North America, and is also used in Brazil, Chile, Republic of Korea, Taiwan (China), and some Western European countries. The European test procedure and emissions standards are used in the European Union, most Eastern European countries, China, and India, where it includes an Indian driving cycle. Though primary used in Japan, the Japanese procedure and standards are accepted by several East Asian countries.

All three test procedures measure exhaust emissions produced while the vehicle is driven through a prescribed driving cycle on a chassis dynamometer. Emissions are sampled by means of a constant volume sampling (CVS) system (figure 2.1). The specific driving cycle differs, however. Because emissions in urban areas are the principal concern of control programs, all testing is based on vehicles operating in stop-and-go driving conditions typical of urban areas. The layout of a typical emissions testing laboratory is shown in figure 2.2.

U.S. procedure. In the U.S. federal test procedure (FTP-75), the vehicle is driven on a chassis dynamometer according to a predetermined speed-time trace (driving cycle) while exhaust emission samples are diluted, cooled, and collected in the constant volume sampling system. The driving cycle, lasting 2,475 seconds, reflects the varying nature of urban vehicle operation (figure 2.3). The average driving speed is 31.4 kilometers per hour (excluding the ten-minute "hot soak" between 1,370 and 1,970 seconds when the engine is shut off). The test begins with a cold start (at 20° to 30° Celsius) after a minimum 12-hour soak. The emission results reported are calculated as the weighted average of emissions measured during three phases: cold start, hot stabilized, and hot start.

Although it is intended to represent typical urban driving (based on a transient cycle representative of driving patterns in Los Angeles, California), the driving cycle used for the U.S. test procedure does not cover the full range of speed and acceleration conditions that vehicles experience. When the cycle was adopted in the early 1970s, chassis dynamometers had limited capabilities that made it necessary to use low speeds and acceleration rates. The top speed in the U.S. cycle is 91 kilometers per hour and maximum acceleration is 5.3 kilometers per hour per second (1.47 m/sec^2), both are lower than what most vehicles can achieve on the road.

As the FTP cycle does not cover the full range of possible speed and acceleration conditions, emissions under off-cycle conditions are effectively uncontrolled (AQIRP 1996). Manufacturers can and do take advantage of this to increase the power output and performance of their vehicles under off-cycle conditions. As a result, vehicle emissions may increase dramatically under these conditions. For example, most gasoline passenger car engines use a rich mixture and shut off exhaust gas recirculation at or near full throttle, causing huge increases in CO emissions. Such increases associated with high power and load conditions (such as hard acceleration, high speed operations, or use of accessories), can soar as high as 2,500 times the emission rate noted for stoichiometric operations. Although most vehicles spend less than 2 percent of their total driving time in severe enrichment conditions, this can account for up to 40 percent of total CO emissions (Guensler 1994). NO_x and HC are also increased.

The U.S. EPA and CARB have studied in-use driving patterns and found that high speed and high acceleration driving are not uncommon, and their inclusion in a test procedure can greatly affect overall emission mea-

Figure 2.1 Exhaust Emissions Test Procedure for Light-Duty Vehicles

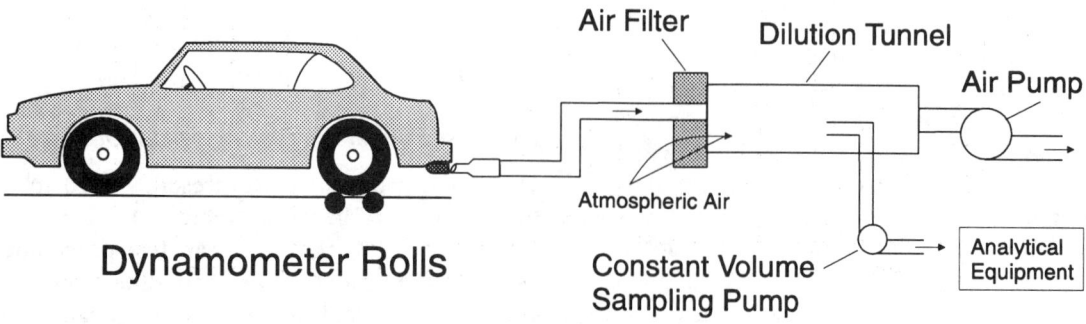

Source: Weaver and Chan 1995

Figure 2.2 Typical Physical Layout of an Emissions Testing Laboratory

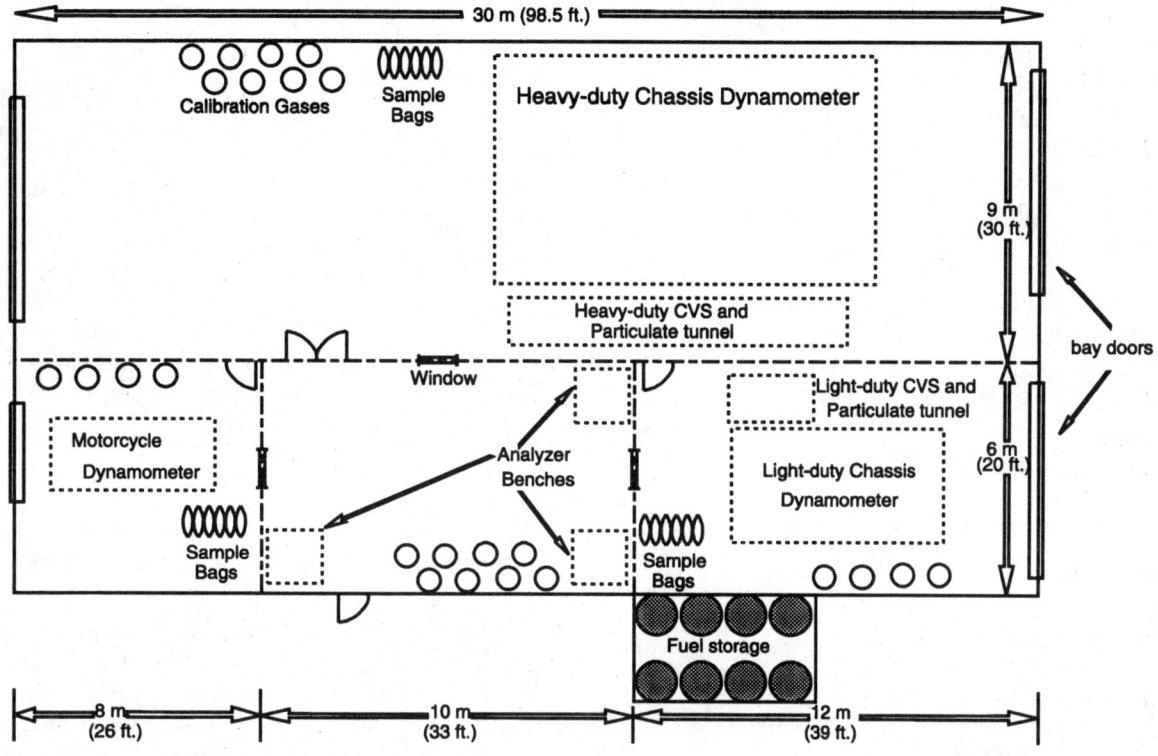

Source: Weaver and Chan 1995

Figure 2.3 U.S. Emissions Test Driving Cycle for Light-Duty Vehicles (FTP-75)

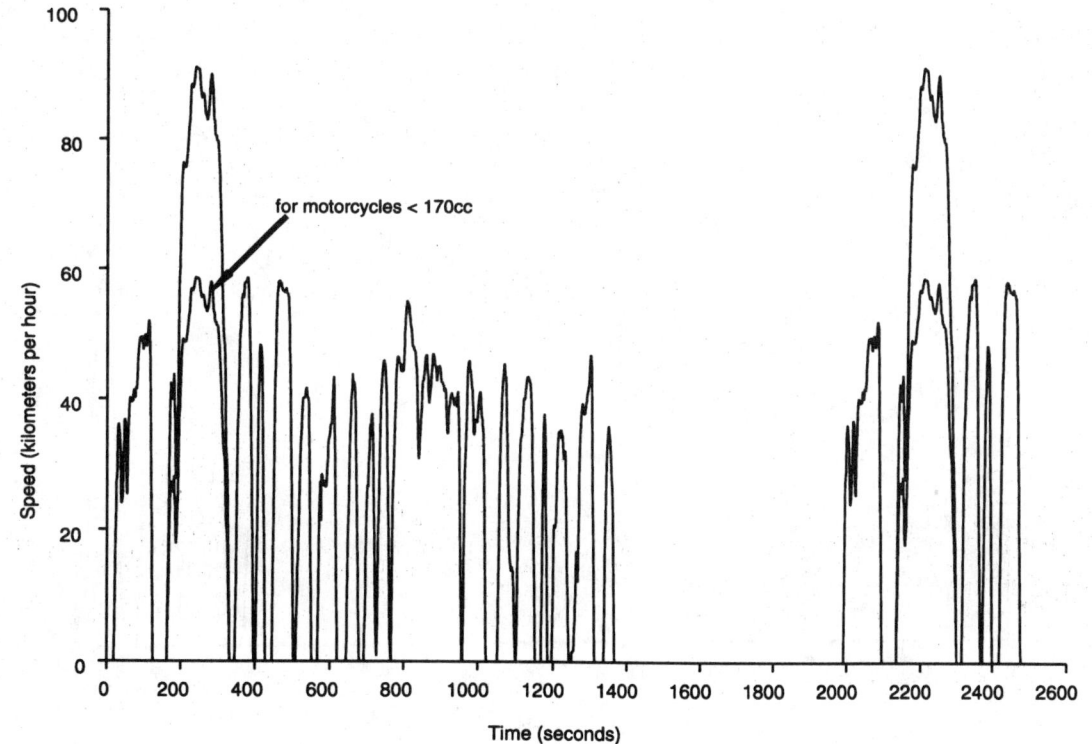

Source: CONCAWE 1994

surements. CARB further concluded that these effects may be responsible for emissions in the South Coast Air Basin of California being as much as 7 percent higher for nitrogen oxides, 20 percent higher for hydrocarbons, and 80 percent higher for carbon monoxide than projected by the current emission models.

Another failing of the FTP-75 is the poor simulation of air conditioner operation. The air conditioner does not run during the FTP; instead, dynamometer load is increased to simulate the additional load on the engine due to the air conditioner compressor. EPA tests of several vehicles with the air conditioner running showed NO_x emissions nearly double those without the air conditioner, although fuel consumption increased only 20 percent. Apparently, manufacturers have not optimized for low emissions with the air conditioner on, since they know that the vehicle will not be tested in that condition.

A revised version of the U.S. test procedure has been developed and proposed for adoption by the U.S. EPA. This procedure includes an additional driving cycle, called the US06, constructed so that when combined with the existing FTP (figure 2.4) it includes the full range of speed and load conditions found in actual driving. The US06 cycle includes higher speeds and more aggressive driving patterns than in the current U.S. pro-

cedure. The proposed revisions also include testing with the air conditioner on, to better reflect actual emissions during warm weather.

Until recently the U.S. emissions test was only carried out at ambient temperatures of about 20 to 30°C. Although it was known that hydrocarbons and especially carbon monoxide emissions increase greatly at low temperatures because of increased cold start enrichment and slower catalyst light-off, these low temperature emissions were unregulated. As a result low temperature carbon monoxide emissions are five to ten times the emissions of carbon monoxide at normal ambient conditions. Recent legislation in the United States, however, requires carbon monoxide emissions testing at both –7°C and normal ambient temperature, bringing this source of excess emissions under control for all new vehicle models by 1996.

European procedure. The emissions test procedure for European passenger cars was defined by ECE regulation 15 and consists of three tests. Like the U.S. procedure, the first test measures the exhaust emissions produced in a driving cycle on a chassis dynamometer. The difference is the driving cycle used. The European test procedure is a simplified representation of the driving cycle in a typical European urban center (such as Rome), with

Figure 2.4 Proposed U.S. Environmental Protection Agency US06 Emissions Test Cycle

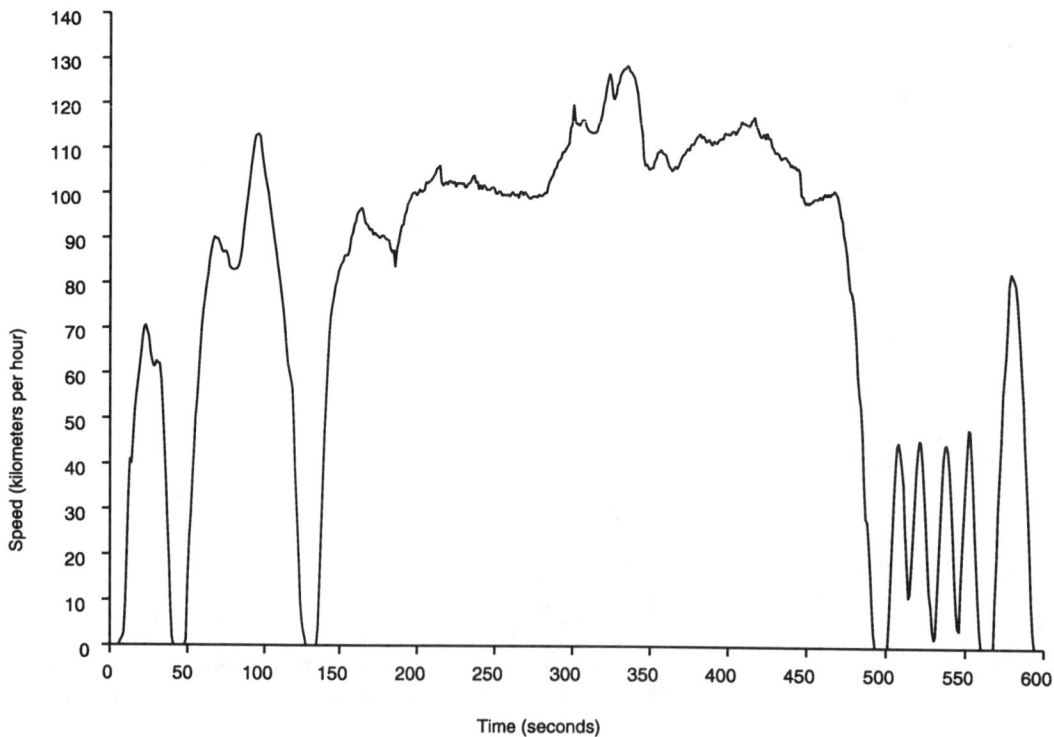

Time (seconds)

Source: Weaver and Chan 1995

15 linked driving modes (figure 2.5). The maximum speed for the European test cycle is 50 kilometers per hour. The vehicle is allowed to soak for at least six hours at a temperature between 20 and 30°C. It is then started and, after idling for 40 seconds, is driven through the test cycle four times without interruption. The cycle lasts 780 seconds and totals 4.052 kilometers, making the average speed 18.7 kilometers per hour. The second test samples tailpipe carbon monoxide concentrations immediately after the last cycle of the first test. The third test measures crankcase emissions.

Compared with the U.S. procedure, the European driving cycle is simple, consisting of stable speeds linked by uniform accelerations and decelerations. The absence of real-world driving effects reduces the severity of the test. The Institut National de Recherche sur les Transports et Leur Securité (INRETS) in France has determined that the European test cycle, characterized by uniform transitions between steady states, underestimates actual emissions by 15-25 percent when compared with more realistic driving at the same average speed (Joumard and others 1990). Also, the maximum acceleration rate in the European procedure is 3.75 kilometers per hour per second (1.04 m/sec^2)—significantly less than the U.S. procedure, and thus even less representative of real driving. This rate is sustained for just four seconds during the first brief peak in the driving cycle; rates for the other two accelerations in the cycle are 2.61 and 1.92 kilometers per hour per second (0.73 and 0.53 m/sec^2, respectively), less than one fifth the rate commonly observed in actual driving.

To account for higher vehicle speeds outside of urban centers, EU emission test procedures now include an extra-urban driving cycle (figure 2.6). The extra-urban driving cycle (EUDC) lasts 400 seconds at an average speed of 62.6 kilometers per hour, with a maximum speed of 120 kilometers per hour. The maximum acceleration rate, however, is only 3 kilometers per hour per second (0.83 m/sec^2). Since the maximum speed and the maximum acceleration rate are still considerably less than in actual driving, emissions at off-cycle conditions remain uncontrolled. As testing is only conducted between 20° and 30°C, low-temperature carbon monoxide emissions are not controlled.

Japanese procedure. The Japanese test procedure is similar to the European one. Until 1991 the main test for light-duty vehicles was a 10-mode driving cycle simulating congested urban driving. This was replaced with the 10.15 mode cycle, which adds a segment reaching 70 kilometers per hour. Both tests are hot-start procedures (i.e. the vehicle is already warmed up). A separate, 11-mode test cycle measures cold-start emissions.

Exhaust Emissions Testing for Motorcycles and Mopeds

The emissions test procedure for motorcycles in the United States is the same as for light-duty passenger cars, except that the maximum speed is reduced for motorcycles with an engine displacement of less than 170 cc engine displacement (figure 2.7). For passenger cars and motorcycles with displacement over 170 cc, the maximum speed is 91 kilometers per hour, while for motorcycles under 170 cc it is 58.7 kilometers per hour. Testing is done on a single-roll dynamometer equipped with a flywheel to simulate the inertia of motorcycle and rider, and with a clamp to hold the motorcycle upright. The European test procedure for motorcycles is the same as for light-duty vehicles, but does not include the extra-urban driving cycle.

As with passenger cars, present emission test cycles for motorcycles are inadequate. The operating characteristics of many motorcycles make acceleration rates particularly significant. Observations in Bangkok indicate that motorcycle acceleration rates in traffic often exceed 12 kilometers per hour per second (3.3 m/sec^2), more than twice the maximum acceleration rate in the U.S. procedure, and three times that in the European procedure (Chan and Weaver 1994).

ECE regulation 47 defines the test procedure for emissions from vehicles with less than 50 cc engine displacement and an unladen weight of less than 400 kilograms (figure 2.7). Such vehicles are almost entirely mopeds. The maximum speed in this test cycle is 50 kilometers per hour, or the maximum speed the vehicle can reach at wide-open throttle if less than 50 kilometers per hour. The United States does not regulate emissions from motorcycles with less than 50 cc engine displacement, but the U.S. EPA has recently proposed doing so. The test cycle would be the same as for motorcycles with an engine displacement between 50 and 170 cc.

Exhaust Emissions Testing for Heavy-Duty Vehicle Engines

Heavy-duty, on-highway vehicles are also tested by U.S., European, and Japanese test procedures. The U.S. procedure (1985) tests under transient conditions while the European (ECE 49) and Japanese (13-mode test) procedures use steady state tests. (These procedures are described in detail in CONCAWE 1994, and in official documents.) All three procedures measure exhaust emissions from the engine alone (removed from the vehicle), operating over a specified cycle on an engine dynamometer. The U.S. procedure involves transient changes in speed and load to mimic actual road operation. The European and Japanese procedures measure emissions at a number of specified steady-state conditions, and combine these according to a weighting scheme. Results of

Figure 2.5 European Emissions Test Driving Cycle (ECE–15)
(repeated four times)

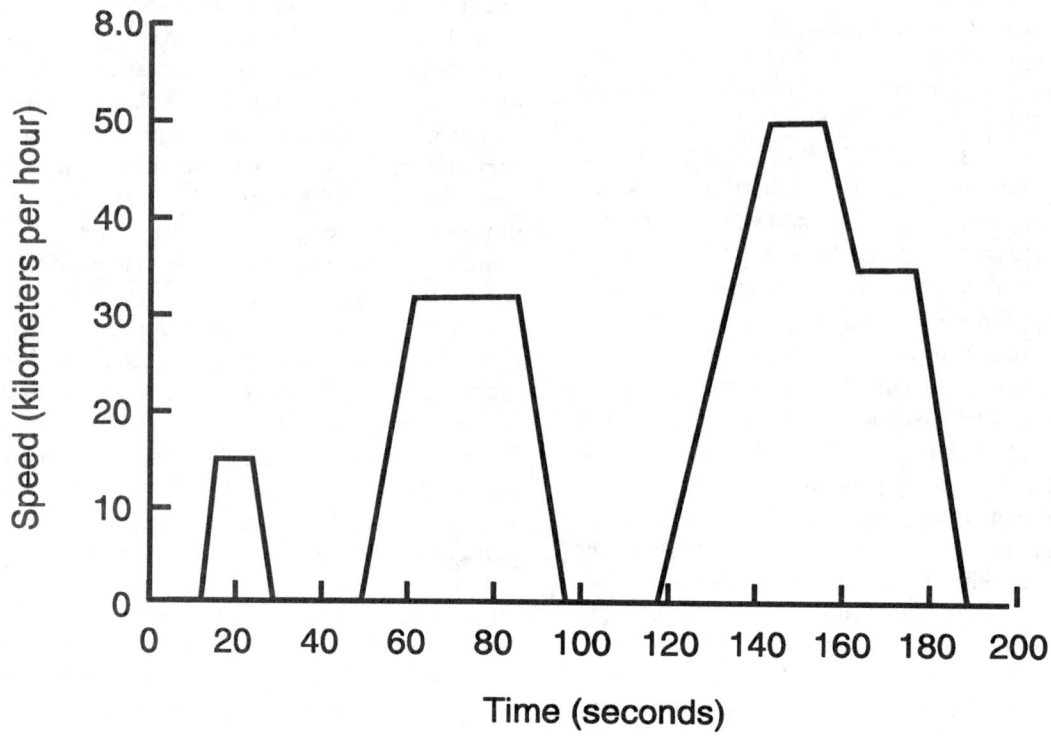

Source: CONCAWE 1994

Figure 2.6 European Extra-Urban Driving Cycle (EUDC)

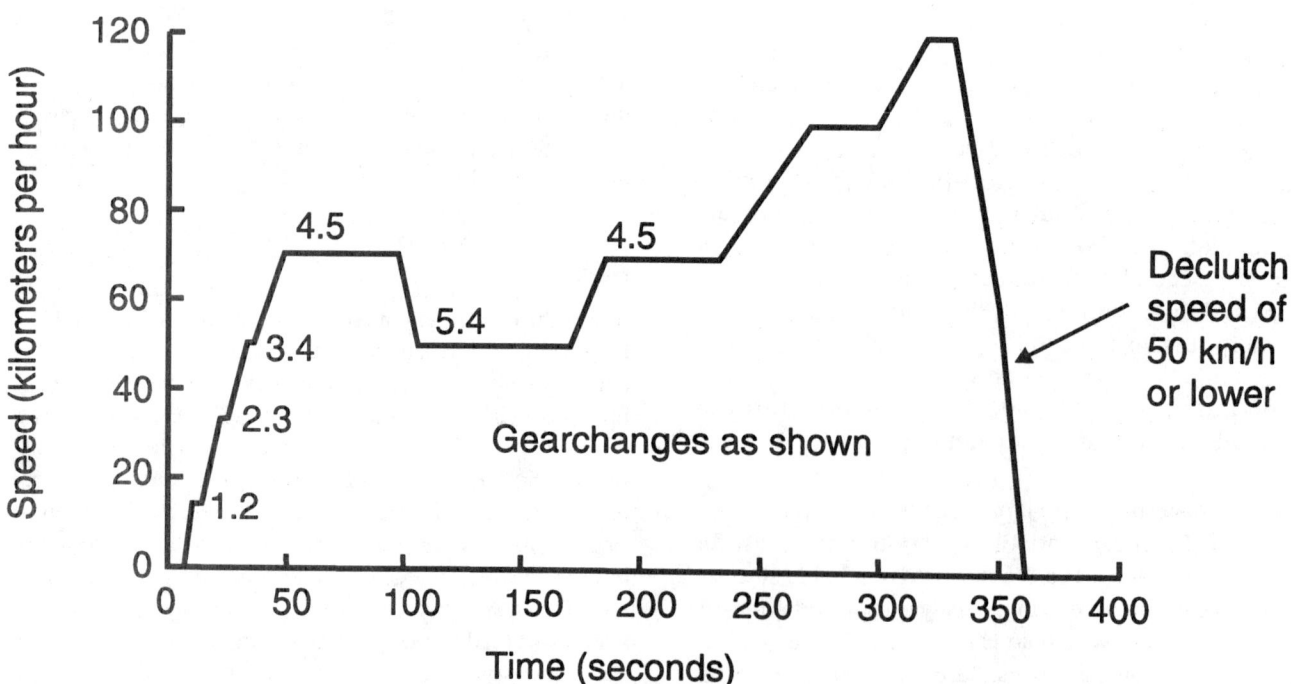

Source: CONCAWE 1994

Figure 2.7 European Emissions Test Driving Cycle for Mopeds

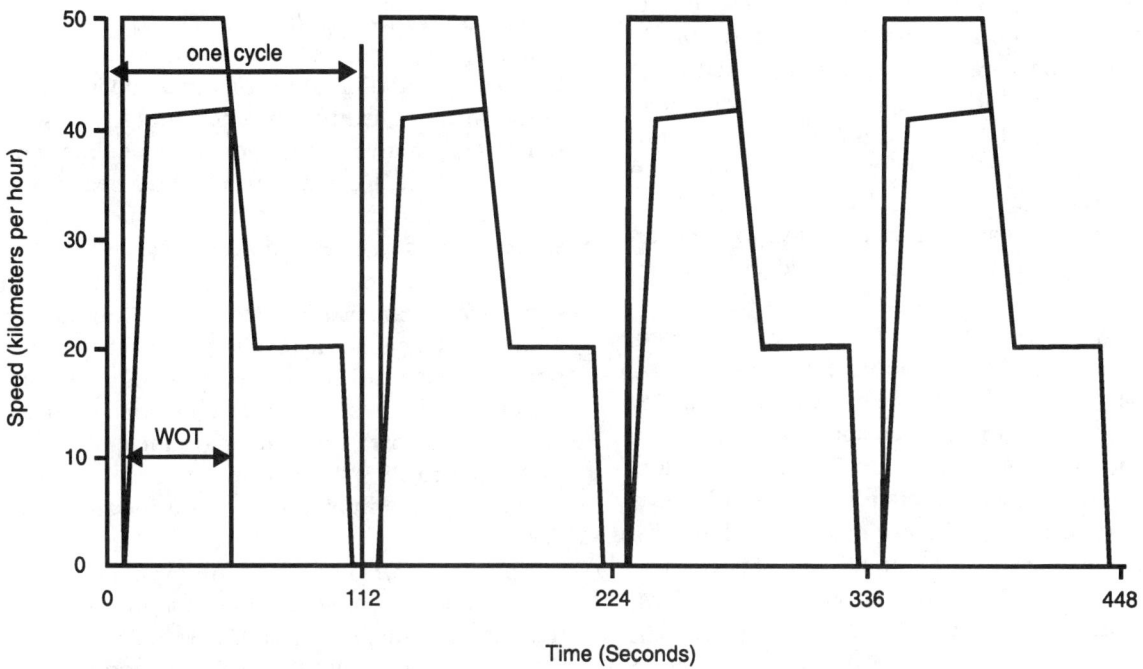

WOT = Wide-open throttle

4 cycles warm up, 4 cycles measurement

Source: CONCAWE 1994

the U.S. and European tests are reported in mass of pollutant emissions per unit of work output rather than emissions per vehicle-kilometer. This is because of the wide range of sizes and applications in heavy-duty vehicles, each of which would otherwise have to be tested individually on a chassis dynamometer. Emissions per kilometer are strongly affected by vehicle size and fuel consumption and would require different standards for each vehicle category. Regulation based on work output allows one set of standards to be applied to engines used in a broad range of vehicles. The Japanese test is reported in terms of pollutant concentration corrected to standard conditions, which has a similar effect.

The U.S. heavy-duty transient test consists of engine speed and load transients that were selected to simulate intracity truck operations. This is because trucks greatly outnumber buses in U.S. urban areas. Engine testing yields pollutant emissions per unit of work output of the engine (in grams per horsepower-hour). The U.S. EPA formula for converting grams per horsepower-hour to grams per mile utilizes a conversion factor of about three horsepower-hours to a mile, based on fuel consumption. However, tests by the U.S. EPA and other organizations suggest that using the U.S. EPA transient test for bus engines may underestimate actual bus chassis emissions by a factor of three to six (Alson, Adler, and Baines 1989).

Testing in-use vehicles with engine dynamometers is difficult since the engine must be removed from the vehicle. Accordingly, emissions from in-service heavy-duty vehicles are usually measured with the whole vehicle operating on a chassis dynamometer. A number of test cycles have been developed for this purpose, including a chassis version of the U.S. heavy-duty transient test cycle and various cycles simulating bus operations. These include the New York City cycle and the Santiago bus cycle developed in Chile (Steiner 1989).

The comparative advantages of transient and steady-state test procedures have often been debated during the past few decades (Cornetti and others 1988). U.S. authorities adopted the transient procedure (replacing an earlier 13-mode steady-state test) in 1985, arguing that steady-state tests do not adequately measure the air-fuel ratio during transient operations such as acceleration. Diesel particulate matter and hydrocarbon emissions are sensitive to the test cycle used, particularly in transient conditions. Particulate matter and hydrocarbon emissions in transient tests are generally found to be higher than in steady state tests. Emissions of nitrogen oxides show better correlation between the two types of tests.

Regulations based on a specific emissions test usually only control emissions in the operating modes experienced during that procedure. Since vehicles operate in a variety of speed-load conditions, it is important that testing procedures reflect these conditions. Emission control strategies based on a procedure that measures emissions at a limited number of operating conditions may be insufficient. Electronic engine control systems can be programmed to undermine a steady state test cycle, by minimizing emissions only at the test points. This is much more difficult in a transient test cycle. In response to these arguments, European authorities are developing an appropriate transient test cycle (Baines 1994).

Adopting the U.S. test cycle in Europe would promote international standardization, but it may not be the optimal solution. Most of the work during the U.S. test cycle is produced near rated speed and load (Cornetti and others 1988). While the U.S. EPA considers this to be typical of truck operations in U.S. cities, it is not typical of European driving patterns or long distance driving patterns in the United States. The development of high torque-rise engines, overdrive transmissions, road speed governors, and the increasing concern for fuel economy are resulting in U.S. trucks spending more time operating near peak torque speed, even in urban areas. Since the transient test contains little operation at this speed, manufacturers are able to calibrate their engines for best performance and fuel economy at peak torque, rather than for least emissions. Trucks that run mostly near peak torque conditions rather than near rated speed will produce more pollutant emissions than estimated from the emission test results. Thus, the argument is strong for a revised test cycle that would give increased emphasis to peak torque operating conditions. Given the significantly higher proportion of trucks and buses in the traffic streams in developing countries, there may be a case to develop test cycles that are more representative of operating conditions (low power to weight ratio and slower speeds) in developing countries.

Crankcase Emissions

There is no common procedure for measuring crankcase emissions across countries. European regulations specify a functional test to confirm the absence of venting from the crankcase, while U.S. regulations simply prohibit crankcase emissions. Uncontrolled crankcase emissions have been estimated from measurements of volatile organic compound concentrations in blowby gases. On vehicles with closed crankcase ventilation systems, crankcase emissions are assumed to be zero.

Adoption of European standards for controlling crankcase emissions may be more appropriate for developing countries, given the similarities of engine technology.

Evaporative Emissions

United States and California. The U.S. procedure for light-duty vehicles measures evaporative emissions from simulated diurnal heating and cooling and evaporation from the carburetor under hot-shutdown conditions. Evaporative emission requirements apply to passenger cars, light-duty trucks, and heavy-duty vehicles using gasoline or other volatile fuels (but not diesel). California also tests motorcycles.

Evaporative emissions are measured by placing the vehicle in an enclosure called Sealed Housing for Evaporative Determination (SHED), which captures all vapors emitted from the vehicle. The diurnal portion of the evaporative test measures emissions from the vehicle as the temperature in the gasoline tank is increased from 15.6 to 28.9°C. This simulates the warming that occurs as the temperature rises during the course of a day.

The *hot soak* portion of the evaporative test measures emissions from the vehicle for one hour following the exhaust emissions test. The vehicle is moved from the dynamometer into the SHED as soon as the driving test is complete. The sum of diurnal and hot-soak emissions total *grams per test*, which is the regulated quantity. Individual diurnal and hot soak test results can also be used to translate grams per test into grams per kilometer. The formula is:

$$\text{grams/kilometer} = \frac{\text{diurnal grams/test} + N * (\text{hot soak grams/test})}{\text{average kilometers driven per day}} \quad (2.1)$$

where N is the average number of trips per day.

Evaporative test procedures have recently been reevaluated in the United States. This is because testing programs indicate that fuel evaporation is a larger source of emissions of volatile organic compounds than had previously been recognized and that under some circumstances significant evaporative emissions were coming even from vehicles equipped with evaporative control systems meeting U.S. EPA and CARB standards. In failing to test the evaporative control system under conditions as severe as in actual use, existing diurnal and hot-soak test procedures may have limited the degree of evaporative control achieved.

These issues have been addressed in the latest California and U.S. EPA regulations, which have established more elaborate evaporative test procedures. These procedures are intended to be more representative of real-world vehicle evaporative emissions. The California procedure includes a 72-hour triple diurnal test cycle in a SHED that ranges between 18 and 41°C. Running losses are measured by operating the vehicle on a chassis dynamometer through three consecutive U.S. driving cycles in the SHED at 41°C. The U.S. EPA test cycle is similar but involves less extreme temperatures. These tests have required manufacturers to design higher-capacity evaporative emissions control systems that achieve better in-use control even under extreme conditions.

Both the CARB and the U.S. EPA limit evaporative hydrocarbon emissions to 2.0 grams per test, which is considered effectively equivalent to zero (a small allowance is needed for other, non-fuel related organic emissions from new cars, such as residual paint solvent). The new test procedures have the same 2.0 grams per test limit, with a separate limit on running-loss emissions of 0.05 gram per mile. Although the standards are nominally the same as before, the more severe testing conditions impose more stringent requirements on manufacturers.

European Union. Until recently the European testing procedures did not provide for evaporative emission measurements as evaporative hydrocarbon emissions were not regulated. This was remedied by the Consolidated Emissions Directive issued by the Council of Ministers of the European Community in June 1991. The directive established evaporative emission limits based on tests similar to the former U.S. SHED test procedure.

Japan. The Japanese evaporative test procedure measures hot-soak emissions only; diurnal emissions, running losses, and resting losses are not measured. The test uses carbon traps connected to the fuel system at points where fuel vapors may escape into the atmosphere. The vehicle is driven at 40 kilometers per hour on a chassis dynamometer for 40 minutes, then the engine is stopped, the exhaust is sealed, and preweighed carbon traps are connected to the fuel tank vent, air cleaner, and any other possible vapor sources. After one hour, the traps are reweighed.

Refueling Emissions

Testing for refueling emissions involves measuring concentrations of volatile organic compounds in the vapors vented from gasoline tanks during the refueling process and observing spillage frequency and volume. Emissions from the underground storage tank vent are monitored by measuring the flow rate and concentrations of volatile organic compounds in gases emitted from the vent.

On-Road Exhaust Emissions

To obtain emissions data that is directly representative of actual traffic conditions and driving patterns, a number of on-board systems have been developed and tested. A system developed by the Flemish Institute for Technological Research includes a miniature constant volume sampler (CVS), gas analyzers, a sensor and measuring system for fuel consumption, an optical sensor to measure vehicle speed and distance travelled, and a data processor on a laptop computer for online collection, with real-time processing and evaluation of data. This system has been used to measure vehicle exhaust emissions in motorway and rural highway traffic. Measured emissions have been within 10 percent of laboratory data. The system can be used with all carbon fuels to directly measure exhaust emissions of carbon monoxide, hydrocarbons, nitrogen oxides, and carbon dioxide in either grams per second or grams per kilometer (Lenaers 1994).

Vehicle Emission Factors

Pollutant emission levels from in-service vehicles vary depending on vehicle characteristics, operating conditions, level of maintenance, fuel characteristics, and ambient conditions such as temperature, humidity, and altitude. The *emission factor* is defined as the estimated average emission rate for a given pollutant for a given class of vehicles. Estimates of vehicle emissions are obtained by multiplying an estimate of the distance traveled by a given class of vehicles by an appropriate emission factor.

Because of the many variables that influence vehicle emissions (see box 2.1), computer models have been developed that estimate emission factors under any combination of conditions. Two of the most advanced models are the U.S. EPA's MOBILE series (the current version is MOBILE5a), and the EMFAC model developed by CARB. Both models use statistical relationships based on thousands of emission tests performed on both new and used vehicles. In addition to standard testing conditions, many of these vehicles have been tested at other temperatures, with different grades of fuel, and under different driving cycles. Relationships have been developed for vehicles at varying emission control levels, ranging from no control to projections of in-use performance of future low-emission vehicle fleets.

Although accurate emission factors and an understanding of the conditions that affect them are obviously important for air quality planning and management, data for in-service vehicles are surprisingly poor. Even in the United States, where systematic emission measurements have been carried out on in-service vehicles for more than a decade, there is considerable uncertainty about the applicability of the results. The most important sources of uncertainty are the sensitivity of vehicle emissions to the driving cycle, the wide variety of driving patterns, and the effects of sampling error, given the highly skewed distribution of emission levels among vehicles equipped with emission controls. The U.S. sampling surveys indicate that a small fraction of "gross" emitters in

---

Box 2.1 Factors Influencing Motor Vehicle Emissions

1. Vehicle/Fuel Characteristics
- Engine type and technology—two-stroke, four-stroke; Diesel, Otto, Wankel, other engines; fuel injection, turbocharging, and other engine design features; type of transmission system
- Exhaust, crankcase, and evaporative emission control systems in place—catalytic converters, exhaust gas recirculation, air injection, Stage II and other vapor recovery systems
- Engine mechanical condition and adequacy of maintenance
- Air conditioning, trailer towing, and other vehicle appurtenances
- Fuel properties and quality—contamination, deposits, sulfur, distillation characteristics, composition (e.g., aromatics, olefin content) additives (e.g., lead), oxygen content, gasoline octane, diesel cetane
- Alternative fuels
- Deterioration characteristics of emission control equipment
- Deployment and effectiveness of inspection/maintenance (I/M) and anti-tampering (ATP) program

2. Fleet Characteristics
- Vehicle mix (number and type of vehicles in use)
- Vehicle utilization (kilometers per vehicle per year) by vehicle type.
- Age profile of the vehicle fleet
- Traffic mix and choice of mode for passenger/goods movements
- Emission standards in effect and incentives/disincentives for purchase of cleaner vehicles
- Adequacy and coverage of fleet maintenance programs
- Clean fuels program

3. Operating Characteristics
- Altitude, temperature, humidity (for NO_x emissions)
- Vehicle use patterns—number and length of trips, number of cold starts, speed, loading, aggressiveness of driving behavior
- Degree of traffic congestion, capacity and quality of road infrastructure, and traffic control systems
- Transport demand management programs

Source: Faiz and others 1995; Faiz and Aloisi de Larderel 1993

---

the vehicle fleet are responsible for a large fraction of total emissions. These are generally vehicles in which emission controls are malfunctioning, tampered with, or damaged. It is difficult to represent this minority accurately in a sample of reasonable size. Another concern is the potential for sampling bias: owners must agree to have their vehicles tested, and owners of the worst vehicles may be less likely to do so. A consensus is developing that the combined effect of these problems has caused existing models to underestimate motor vehicle emission factors by a substantial margin.

A study under the U.S. Air Quality Improvement Research Program (AQIRP 1995) compared real world vehicle emissions to values calculated using the U.S. EPA MOBILE computer models (MOBILE4.1 and MOBILE5). In general, the MOBILE models predicted emissions rates within ± 50 percent although at one site MOBILE5 overpredicted rates to a much greater extent. MOBILE5's predictions were consistently higher than those of MOBILE4.1 at both test sites (Fort McHenry tunnel under Baltimore Harbor and Tuscarora tunnel in the mountains of Pennsylvania). Both MOBILE modes underpredicted light-duty non-exhaust emission rates, which constituted approximately 15-20 percent of the total light-duty non-methane hydrocarbon emissions. For light-duty vehicles,

MOBILE 4.1 predictions of CO/NO_x and $NMHC/NO_x$ ratios were in closer agreement with the observed ratios than were MOBILE5 predictions.

Emission factors calculated by the MOBILE models are based on average speeds, ambient temperature, diurnal temperature range, altitude, and fuel volatility; changes in these input assumptions alter the resulting emission factors. Exhaust pollutant emission factors increase markedly at low temperatures, while evaporative emissions of volatile organic compounds increase with increasing temperature. Evaporative emissions of volatile organic compounds also increase as gasoline volatility and diurnal temperature range increase. Hydrocarbon and carbon monoxide emissions per vehicle-kilometer tend to increase at low average speeds, such as in congested city driving, while emissions of nitrogen oxides tend to increase at high speeds, which corresponds to higher load conditions. The relationship between average speed and emissions estimated by MOBILE5 for uncontrolled motor vehicles is shown in figure 2.8. Low average speeds are due to traffic congestion, and the increase in emissions under these conditions is due to the stop-and-go pattern of traffic flow in congested condition.

Other emission models have been developed, though none incorporates as much data on in-use emissions as

Figure 2.8 Relationship between Vehicle Speed and Emissions for Uncontrolled Vehicles

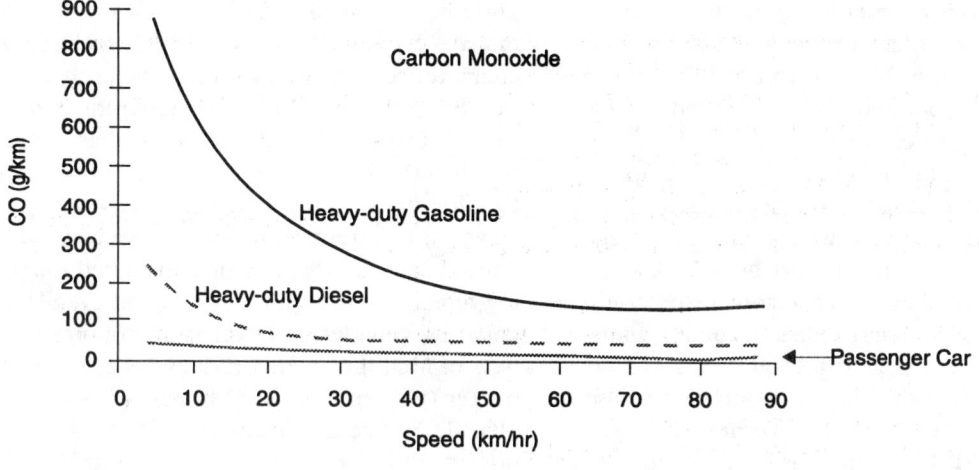

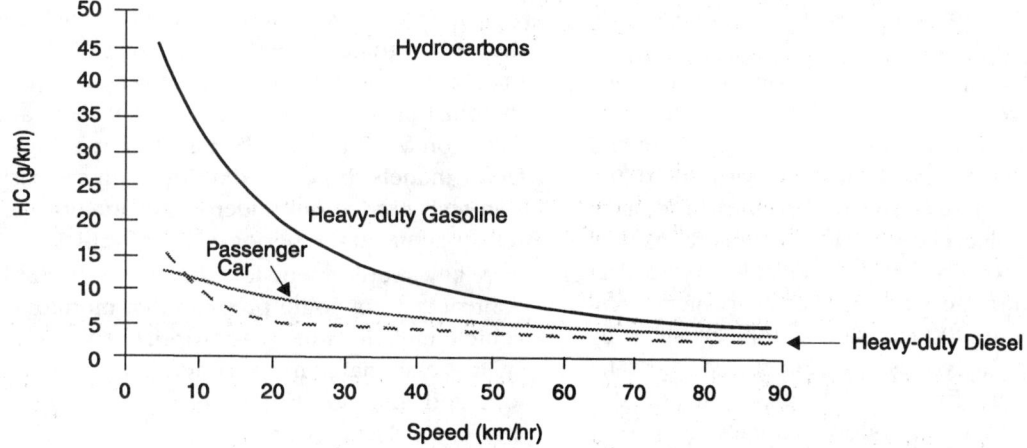

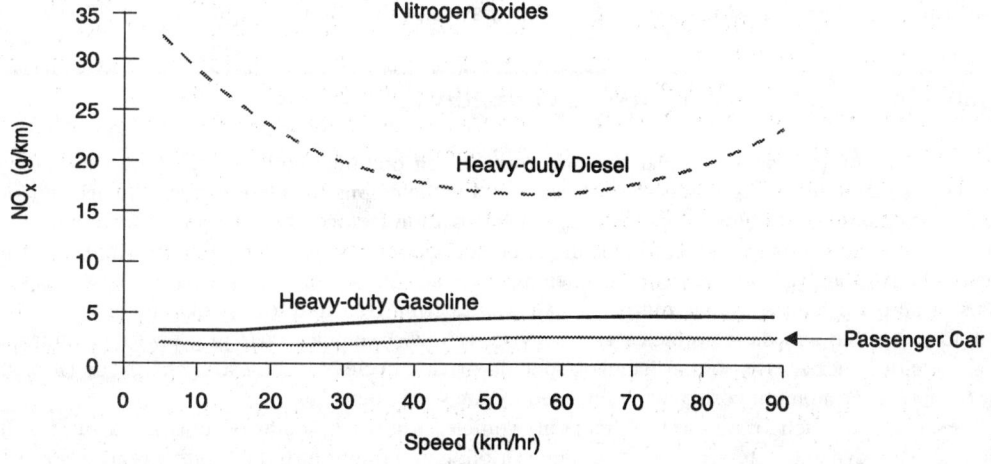

Source: Chan and Weaver 1995

MOBILE5 and EMFAC. The COPERT model (Andrias and others 1992) applies a methodology developed by the CORINAIR working group on emission factors to calculate emissions from road traffic in the EU (Eggleston and others 1991). COPERT can be used to estimate vehicle emission factors for carbon monoxide, non-methane hydrocarbons, methane, oxides of nitrogen, total particulate matter, ammonia, and nitrous oxide. Fuel consumption estimates are also provided. Emission factors are estimated for urban, rural, and highway driving with an average automobile speed of 25 kilometers per hour, 75 kilometers per hour, and 100 kilometers per hour, respectively. COPERT accounts for cold-start emissions and evaporation losses, and uses an average trip length of 12 kilometers. Less extensive emission factor models have also been developed for vehicles in Chile (Turner and others 1993), Indonesia (IGRP 1991), and Thailand (Chongpeerapien and others 1990). The MOBILE4 model has been adapted to estimate emissions from the Mexican vehicle fleet (Radian Corporation 1993).

The Swiss Federal Environment Department has commissioned the development of a data bank of vehicle exhaust and evaporative emissions for both regulated and unregulated pollutants. Information on nearly 300 different compounds, including specific hydrocarbons, aldehydes, phenols, polycyclic aromatic hydrocarbons, and several inorganic compounds, is available in the data bank (Brunner and others 1994). The data bank currently contains about 16,000 emission factors, classified by six main vehicle categories (gasoline-fueled, diesel-fueled, and other light-duty vehicles; gasoline-fueled, diesel-fueled, and other heavy-duty vehicles; and two-wheelers).

Extensive emissions testing of in-use vehicles is required to develop and validate an emission factor model. For lack of better data, the MOBILE5 emissions have frequently been used to estimate vehicle emission factors for uncontrolled vehicles in such countries as Chile, Indonesia, and Mexico. This provides only a rough estimate, however, since the technologies and characteristics of today's vehicles, even without emission controls, are significantly different from the uncontrolled (pre-1970) U.S. vehicles that were used to develop the MOBILE models. In addition, emission factors for motorcycles and heavy-duty vehicles have received little attention in the past and are supported by limited data. They should be considered rough estimates for vehicles in the United States, and even less representative of vehicles in developing countries. One weakness of MOBILE5 is that it does not estimate particulate matter (PM) emissions. Although a PM model based on MOBILE5a became available in 1994, this model does not account for particulate emissions deterioration in use, and thus underestimates real-world emissions.

Since motorcycles and heavy-duty vehicles are among the most significant vehicular sources of air pollution in many developing countries, and PM emissions are among the most pressing concerns, it is clear that continued reliance on MOBILE5 will be insufficient. Vehicle emission factor models should be developed on the basis of emission tests carried out under local conditions and should reflect actual in-use performance of vehicles.

A key requirement for effective long-run emissions control is an ongoing program that monitors in-service vehicle emissions in an appropriate emissions laboratory. It is essential to measure actual vehicle emissions before and after control measures are implemented to know whether they are effective and how they could be improved. As mentioned elsewhere, surveillance testing is an important element of U.S. in-service emissions testing. This testing is performed to verify compliance with emissions durability requirements. Testing programs

---

Box 2.2 Development of Vehicle Emissions Testing Capability in Thailand

The Royal Thai government has adopted an action plan that addresses the air pollution and noise problems caused by road vehicles. Among the measures included in this plan is the provision of a vehicle emissions laboratory. The primary purpose of this laboratory is the measurement and development of in-use vehicle emission factors. Another important function is to develop improved vehicle emission short tests for use in the planned inspection and maintenance program. The laboratory will be capable of measuring (using constant volume sampling under simulated transient driving conditions) exhaust emissions of carbon monoxide, oxides of nitrogen, hydrocarbons, and particulate matter from two- and four-stroke motorcycles, three-wheel taxis (*tuk-tuks*), light-duty gasoline and diesel vehicles, and heavy-duty diesel vehicles weighing up to 21 metric tons. The laboratory will also analyze driving patterns so that Bangkok-specific driving cycles can be established. The laboratory is expected to cost about $2 million. Equipment costs will be financed by the World Bank.

Plans call for three test cells, one each for motorcycle, light-duty vehicle/light truck, and heavy-duty truck/bus testing. Testing equipment will include chassis dynamometers, constant volume sampling and dilution tunnel units, gas analyzer instruments, and data acquisition and control hardware. Each test cell will have its own dynamometer, so the entire weight and power range of vehicles can be tested. Three-wheelers will be tested on the motorcycle dynamometer.

A set of driving cycles representative of Bangkok and other Thai traffic conditions will be developed, to provide representative emission factors for the local vehicle population. The laboratory will provide the flexibility to run standard tests (such as the U.S., European, and Japanese certification cycles) as well as custom-designed cycles.

Source: Chan and Weaver 1994

have also been carried out on in-use vehicles in Chile (Escudero 1991; Sandoval, Prendez, and Ulriksen 1993), Finland (Laurikko 1995), and Greece (Pattas, Kyriakis, and Samaras 1985), among others. Development or expansion of emission laboratories for such testing is proceeding in several developing countries, including Brazil, Iran, Mexico, and Thailand (box 2.2). The results obtained by these laboratories are expected to add significantly to the knowledge of vehicle emission characteristics and compilation of appropriate emission factors for use in developing countries.

Gasoline-Fueled Vehicles

Emission factor estimates for U.S. gasoline-fueled passenger cars and medium-duty trucks equipped with different levels of emission control technology are presented in tables 2.1 and 2.2. These estimates incorporate MOBILE5a results for methane, non-methane

hydrocarbons, nitrogen oxides, and carbon monoxide. Emissions of the greenhouse gases carbon dioxide and nitrous oxide were also estimated: the former based on typical fuel economy and fuel carbon content, the latter based on the data available for different types of emissions control systems.

An inventory of emission factors for gasoline-fueled vehicles derived from the COPERT model and individual studies in Europe and several developing countries is presented in appendix 2.1. These emission factors are likely to be more representative of conditions in developing countries, although there is considerable variation among emission factor measurements, even for similar vehicles and test conditions. This variation indicates the importance of basing emission factor estimates on actual measured emissions for a vehicle fleet rather than relying on data or estimates from other sources. This not only ensures accurate emission factors, it provides an

Table 2.1   Estimated Emission Factors for U.S. Gasoline-Fueled Passenger Cars with Different Emission Control Technologies
(grams per kilometer)

Type of control	Carbon monoxide	Non-methane volatile organic compounds	Methane	Nitrogen oxides	Nitrous oxide	Carbon dioxide	Fuel efficiency (liters per 100 kilometers)
Advanced three-way catalyst control							
Exhaust	6.20	0.38	0.04	0.52	0.019	200	8.4
Evaporative		0.09					
Running loss		0.16					
Resting		0.04					
Total emissions	6.20	0.67	0.04	0.52	0.019	200	
Early three-way catalyst							
Exhaust	6.86	0.43	0.05	0.66	0.046	254	10.6
Evaporative		0.14					
Running loss		0.16					
Resting		0.06					
Total emissions	6.86	0.78	0.05	0.66	0.046	254	
Oxidation catalyst							
Exhaust	22.37	1.87	0.10	1.84	0.027	399	16.7
Evaporative		0.39					
Running loss		0.17					
Resting		0.06					
Total emissions	22.37	2.48	0.10	1.84	0.027	399	
Non-catalyst control							
Exhaust	27.7	2.16	0.15	2.04	0.005	399	16.7
Evaporative		0.70					
Running loss		0.17					
Resting		0.06					
Total emissions	27.7	3.08	0.15	2.04	0.005	399	
Uncontrolled							
Exhaust	42.67	3.38	0.19	2.7	0.005	399	16.7
Evaporative		1.24					
Running loss		0.94					
Resting		0.06					
Total emissions	42.67	5.62	0.19	2.7	0.005	399	

Note: Estimated with the U.S. EPA MOBILE5a model for the following conditions: temperature, 24 °C; speed, 31 kilometers per hour; gasoline Reid Vapor Pressure, 62 kPa (9 PSI); and no inspection and maintenance program in place.
Source: Chan and Reale 1994; Weaver and Turner 1991

Table 2.2 Estimated Emission Factors for U.S. Gasoline-Fueled Medium-Duty Trucks with Different Emission Control Technologies

(grams per kilometer)

Type of control	Carbon monoxide	Non-methane volatile organic compounds	Methane	Nitrogen oxides	Nitrous oxide	Carbon dioxide	Fuel efficiency (liters per 100 kilometers)
Three-way catalyst control							
Exhaust	10.2	0.83	0.12	2.49	0.006	832	34.5
Evaporative		0.38					
Running loss		0.17					
Resting		0.04					
Refueling		0.24					
Total emissions	10.2	1.41	0.04	0.52	0.006	832	
Non-catalyst control							
Exhaust	47.61	2.55	0.21	3.46	0.006	843	35.7
Evaporative		2.16					
Running loss		0.94					
Resting		0.08					
Refueling		0.25					
Total emissions	47.61	5.73	0.21	3.46	0.006	843	
Uncontrolled							
Exhaust	169.13	13.56	0.44	5.71	0.009	1,165	50.0
Evaporative		3.93					
Running loss		0.94					
Resting		0.08					
Refueling		0.32					
Total emissions	169.13	18.50	0.44	5.71	0.009	1,165	

Note: Estimated with the U.S. EPA's MOBILE5a model for the following conditions: temperature, 24 °C; speed, 31 kilometers per hour; gasoline Reid Vapor Pressure, 62 kPa (9 PSI); and no inspection and maintenance program in place.

Source: Chan and Reale 1994; Weaver and Turner 1991

important baseline against which the effectiveness of emission control programs can be measured.

Emission factors are strongly influenced by the way a vehicle is driven—in particular, by the average speed and the degree of acceleration and deceleration in the driving cycle (Joumard and others 1995). The results of emission tests on a number of European vehicles, using a variety of driving cycles, are shown in figure 2.9 (Joumard and others 1990). Average emissions per kilometer increase sharply at the low average speeds typical of highly congested stop-and-go urban driving. Emissions are minimized in free-flowing traffic at moderate speeds, then increase again under the high-speed driving conditions typical of European motorways. Emissions were higher in the transient test cycles; steady-state cycles gave much lower emissions per kilometer. The European test cycle, characterized by uniform transitions between steady states, underestimates actual emissions by about 15 percent.

Pollutant emissions are affected by the vehicle's level of maintenance, and the highest-polluting vehicles are responsible for a disproportionate share of total emissions (chapter 4). Emission-controlled vehicle fleets show the most skewed distribution of emissions, but the uneven distribution is significant even for populations

without emission controls. This effect can be observed in figure 2.10, which shows cumulative distributions for hydrocarbons, carbon monoxide, nitrogen oxides, and particulate matter emissions from a sample of Chilean cars tested in 1989. The lower curve in each plot is the cumulative percentage of cars with emissions greater than a given level, while the upper curve is the percentage of total emissions accounted for by these cars. Only 20 percent of the vehicles, for example, had hydrocarbon emissions above 1.2 grams per kilometer, but these vehicles were responsible for 40 percent of total hydrocarbon emissions. Similar patterns were found for other pollutants: 10 percent of the vehicles accounted for 37 percent of total PM emissions, 20 percent of the vehicles accounted for 43 percent of total carbon monoxide emissions, and 20 percent of the vehicles accounted for 35 percent of total emissions of oxides of nitrogen. Since owners of the worst polluting vehicles may be less likely to volunteer them for testing, the real distribution could be even more skewed. This distribution has important consequences for emissions control strategy. An inspection and maintenance program that identifies the worst 10 to 20 percent of the vehicles and requires that they be repaired or retired could reduce overall emissions significantly.

Figure 2.9 Effect of Average Speed on Emissions and Fuel Consumption for European Passenger Cars without Catalyst (INRETS Driving Cycles; Fully Warmed-up In-use Test Vehicles)

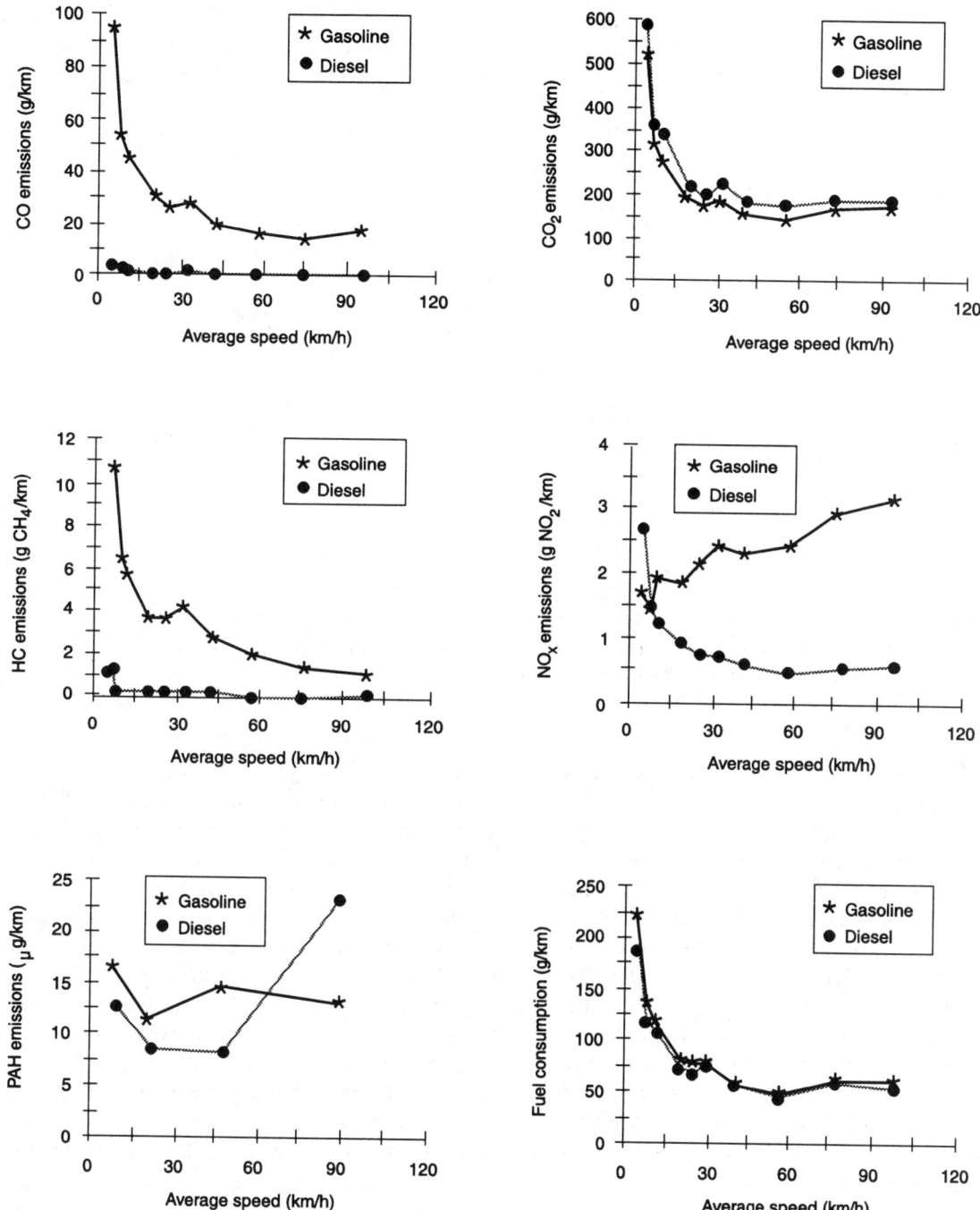

Source: Adapted from Joumard and others 1990

Diesel-Fueled Vehicles

Emission factors for diesel-fueled vehicles are strongly affected by differences in engine technology, vehicle size and weight, driving cycle characteristics, and the state of maintenance of the vehicles. Emission factor estimates for diesel-fueled passenger cars and light-duty trucks in the United States are summarized in table 2.3.

Similar estimates for U.S. heavy-duty diesel-fueled trucks and buses are given in table 2.4. Estimates for diesel-fueled vehicles in Europe and other regions for a variety of emissions control levels are summarized in appendix 2.2. These emission factors may be more representative of diesel-fueled vehicles operating in developing countries. There is, however, considerable variation among

Figure 2.10 Cumulative Distribution of Emissions from Passenger Cars in Santiago, Chile

Distribution of CO emissions

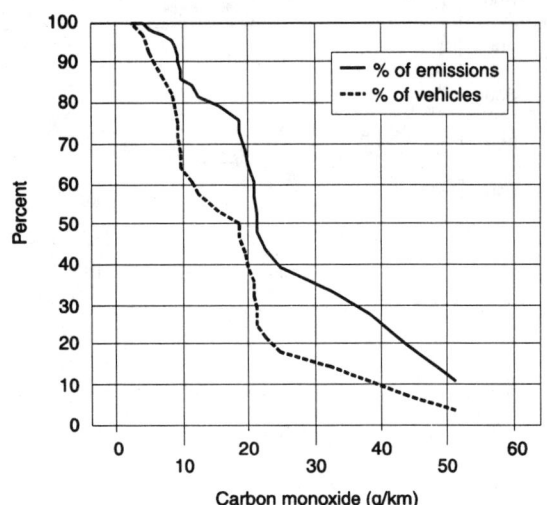

Distribution of PM emissions

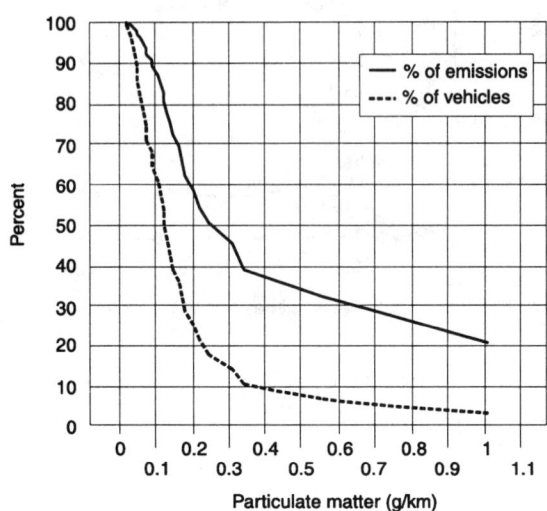

Distribution of HC emissions

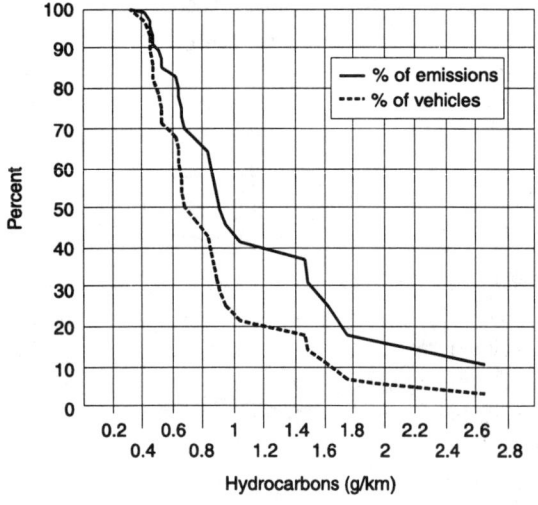

Distribution of NO$_x$ emissions

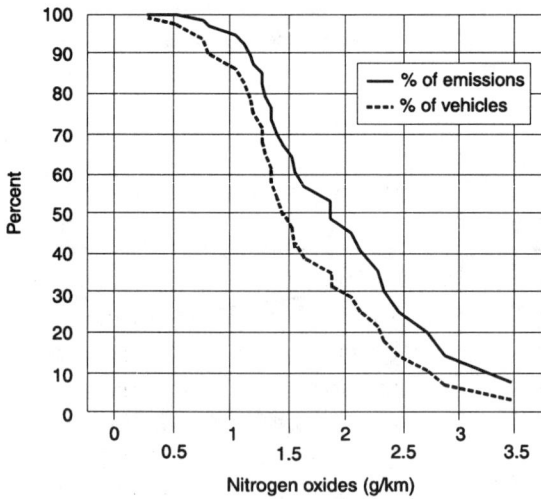

Source: Turner and others 1993

the estimates and measurements in the tables and appendix because of differences in cycle conditions, differences in the sample population, and different estimation techniques. This indicates again the importance of actual emission measurements on the population of interest in order to develop realistic emission factors.

The emission factor estimates for diesel-fueled passenger cars and light trucks in table 2.3 are comparable to the estimates for gasoline-fueled vehicles in table 2.1

and appendix 2.1. This is not true for the estimates for heavy-duty vehicles. Although heavy-duty diesel-fueled and gasoline-fueled vehicles are covered by similar emission standards and test procedures, the average characteristics of the vehicles themselves differ considerably. Unlike heavy-duty gasoline-fueled vehicles, heavy-duty diesel-fueled vehicles are primarily large trucks with gross vehicle weight ratings of 10 to 40 tons. The emission factors in table 2.4 are therefore more representa-

Table 2.3 **Estimated Emission and Fuel Consumption Factors for U.S. Diesel-Fueled Passenger Cars and Light-Duty Trucks**
(grams per kilometer)

Vehicle type	Carbon monoxide	Hydrocarbons	Nitrogen oxides	Particulate matter	Carbon dioxide	Fuel consumption (liters/100 km)
Passenger cars						
Advanced control	0.83	0.27	0.63	—	258	9.4
Moderate control	0.83	0.27	0.90	—	403	14.7
Uncontrolled	0.99	0.47	0.99	—	537	19.6
Light-duty trucks						
Advanced control	0.94	0.39	0.73	—	358	13.0
Moderate control	0.94	0.39	1.01	—	537	19.6
Uncontrolled	1.52	0.77	1.37	—	559	23.3

— Not applicable
Note: MOBILE5 estimates.
Source: Chan and Reale 1994; Weaver and Turner 1991; Weaver and Klausmeier 1988

Table 2.4 **Estimated Emission and Fuel Consumption Factors for U.S. Heavy-Duty Diesel-Fueled Trucks and Buses**
(grams per kilometer)

Vehicle type	Carbon monoxide	Hydrocarbons	Nitrogen oxides	Particulate matter	Carbon dioxide	Fuel consumption (liters/100 km)
U.S. heavy-duty diesel trucks						
Advanced control	6.33	1.32	5.09	—	982	35.7
Moderate control	7.24	1.72	11.56	—	991	35.7
Uncontrolled	7.31	2.52	15.55	—	1,249	45.5
U.S. 1984 measurements						
Single-axle tractors	3.75	1.94	9.37	1.07	1,056	—
Doubleaxle tractors	7.19	1.74	17.0	1.47	1,464	—
Buses	27.40	1.71	12.40	2.46	1,233	—
New York City vehicles						
Medium-heavy trucks	—	2.84	23.28	2.46	—	53.8
Transit buses	—	5.22	34.89	2.66	—	80.7

— Not applicable
Note: MOBILE5 estimates.
Source: Chan and Reale 1994; Weaver and Turner 1991; Weaver and Klausmeier 1988

tive of large trucks (and buses) than smaller medium-duty trucks and vans. The opposite is true for the gasoline-fueled vehicle emission factors in table 2.2.

Average pollutant emissions from heavy-duty diesel-fueled vehicles are especially sensitive to the speed and acceleration characteristics of the driving cycle. Emissions per kilometer vary according to the average cycle speed. For heavy-duty diesel-fueled vehicles, emissions of carbon monoxide, hydrocarbons, and particulate matter increase at low average speeds, due to the stop-and-go driving associated with congested traffic. Emissions of nitrogen oxides and fuel consumption increase between 40 and 50 kilometers per hour (figure 2.11).

The effects of steady-state vehicle speed and road gradient on emissions and fuel consumption of a 40-ton semi-trailer truck are shown in figure 2.12. On negative grades, the emissions of all pollutants are insignificant, but they increase sharply with increasing grades, particularly for nitrogen oxides and hydrocarbons. There is also some correlation of carbon monoxide and particulate emissions with speed and road gradient (Roumegoux 1995).

There are also large variations in particulate and hydrocarbon emission factors for uncontrolled vehicles. This is partly the result of differences in maintenance, which can have a tremendous effect on particulate emissions. In testing buses in Chile, for example, particulate matter emissions from well-maintained buses were more than 80 percent less than the average particulate emissions for the entire bus fleet. Cumulative probability distributions for particulate matter, hydrocarbon, and nitrogen oxides emissions from the Chilean bus fleet illustrate the nature of the problem (figure 2.13). The 10 percent of the buses with highest emissions are responsible for 25 percent of total particulate emissions, while 20 percent of the buses produced 40 percent of the

Figure 2.11 Effect of Average Speed on Emissions and Fuel Consumption for Heavy-Duty Swiss Vehicles

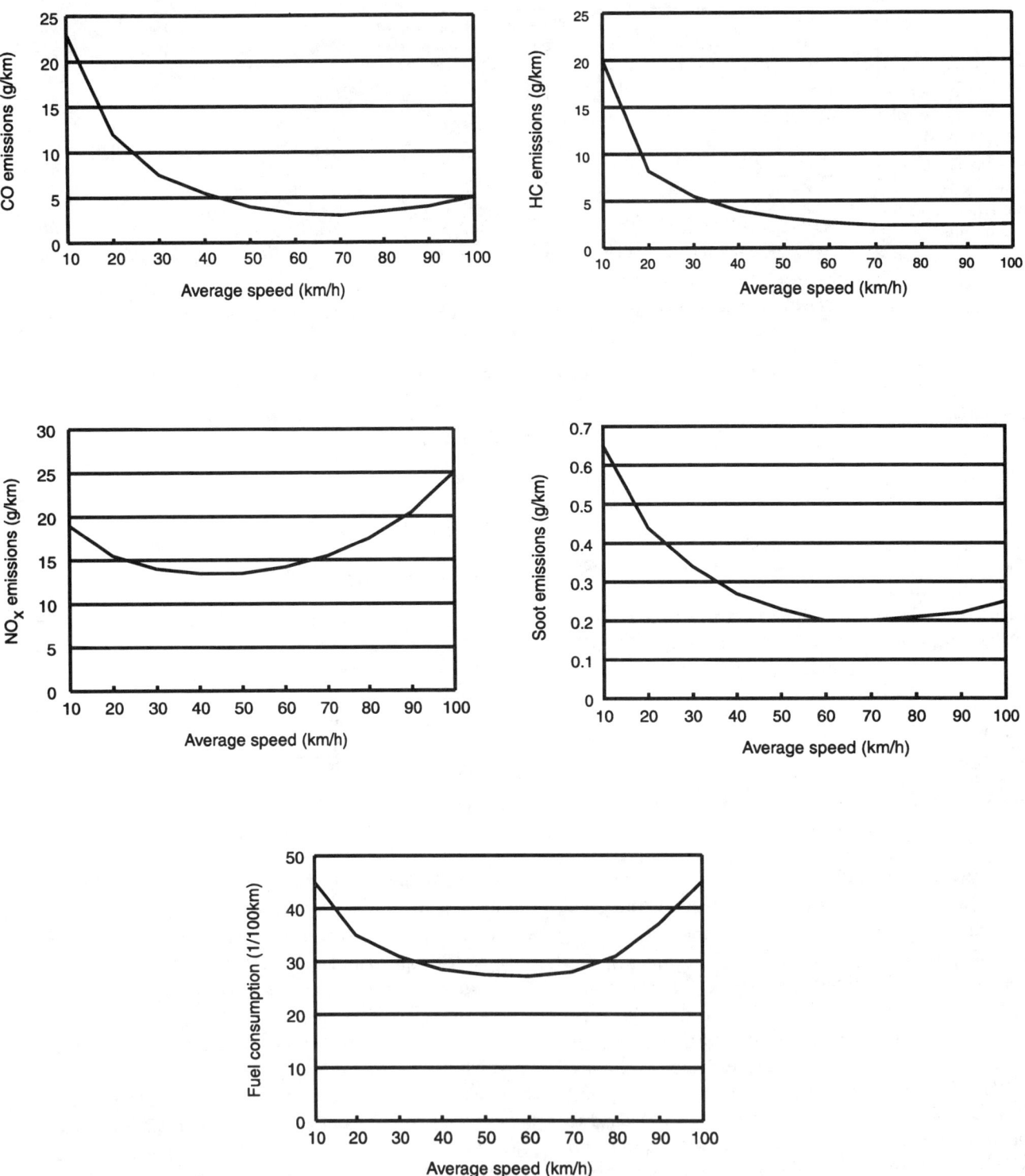

Source: OFPE 1988

Figure 2.12 Effect of Constant Average Speed and Road Gradient on Exhaust Emissions and Fuel Consumption for a 40-ton Semi-Trailer Truck

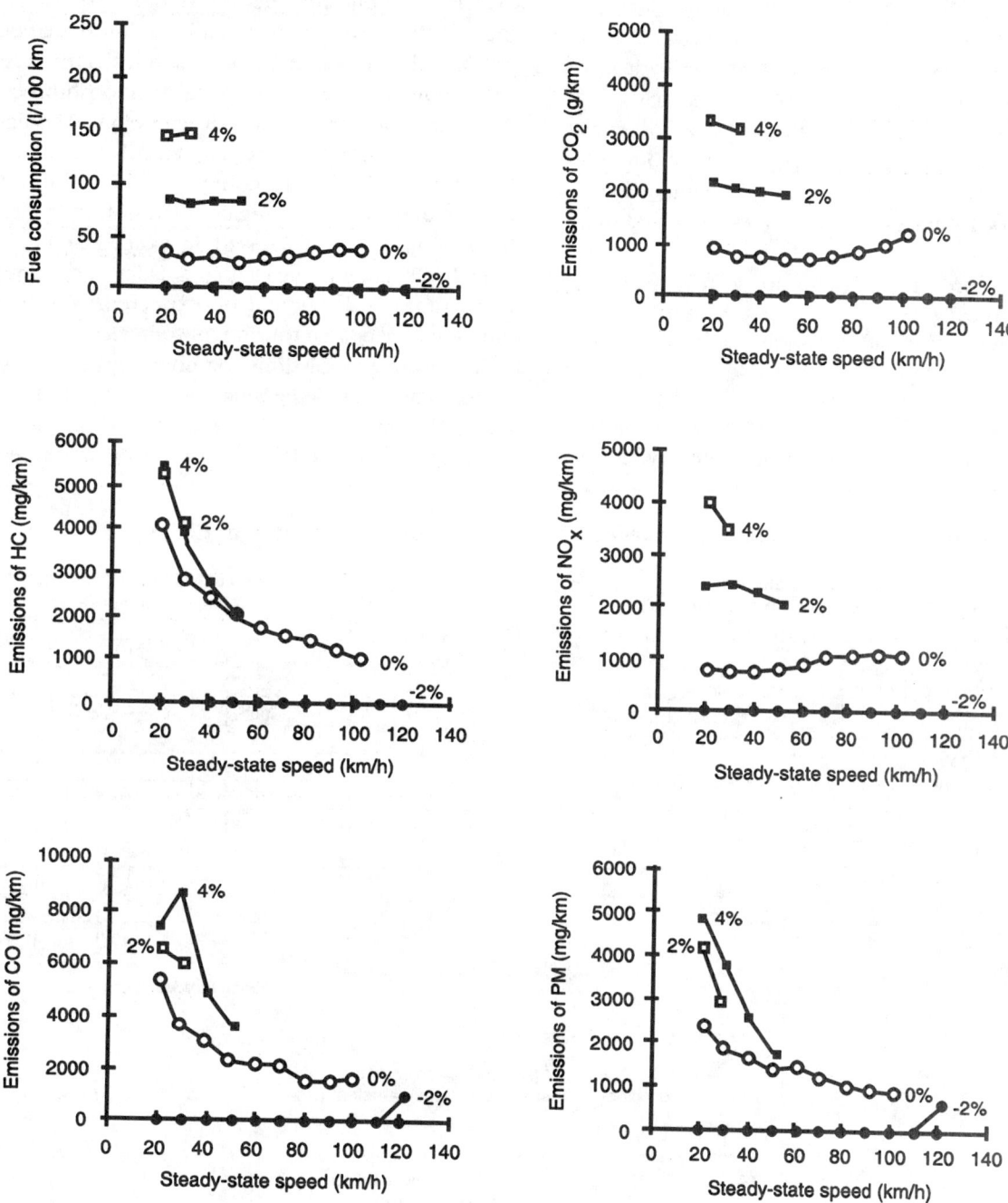

Note: Figures in percent refer to road gradient.
Source: Roumegoux 1995

emissions. The 20 percent of the buses with lowest emissions (generally those with the best maintenance) produced only 7 percent of particulate emissions.

Motorcycles

Two- and three-wheeled vehicles, such as motorcycles and auto-rickshaws, constitute a large portion of motor-

ized vehicles in developing countries, particularly in East and South Asia. While they are responsible for a relatively small fraction of the total vehicle kilometers traveled in most countries, they are major sources of air pollution–particularly two-stroke engines running on a mixture of gasoline and lubricating oil. It has been estimated that uncontrolled motorcycles in industrialized countries emit 22 times the mass of hydrocarbons and 10 times as much carbon monoxide as automobiles controlled to U.S. 1978 levels (OECD 1988). In Taiwan (China), hydrocarbon emissions from two-stroke engine motorcycles were 13 times higher than the hydrocarbon emissions from new four-stroke motorcycles and more than 10 times higher than the hydrocarbon emissions from in-use passenger cars. Carbon monoxide emissions from two-stroke motorcycle engines were similar to those from four-stroke engines (Shen and Huang 1991).

Data on motorcycle emissions are scarce. Available data pertain mostly to uncontrolled U.S. motorcycles, though limited information is also available from Europe and Thailand.

United States. Between 1970 and 1980, considerable study of motorcycle emissions was prompted by motor-

cycle emission regulations developed by U.S. EPA and CARB. Data on uncontrolled emissions from U.S. motorcycles are from this period (table 2.5). These data include the only particulate emission measurements on motorcycles available in the technical literature, and they allow an assessment of the relationship between smoke opacity and particulate emissions. The data are mostly for high- powered, large-displacement touring motorcycles, which are common in the United States but not in most developing countries. Motorcycles used in developing countries seldom exceed 250 cc in engine displacement. Due to their smaller size and weight, these motorcycles would be expected to have lower emissions than large touring motorcycles.

The average emissions for uncontrolled two-stroke motorcycles were 9.9 grams per kilometer of hydrocarbons, 16.1 grams per kilometer of carbon monoxide, 0.022 grams per kilometer of nitrogen oxides, and

Figure 2.13 Cumulative Distribution of Emissions from Diesel Buses in Santiago, Chile

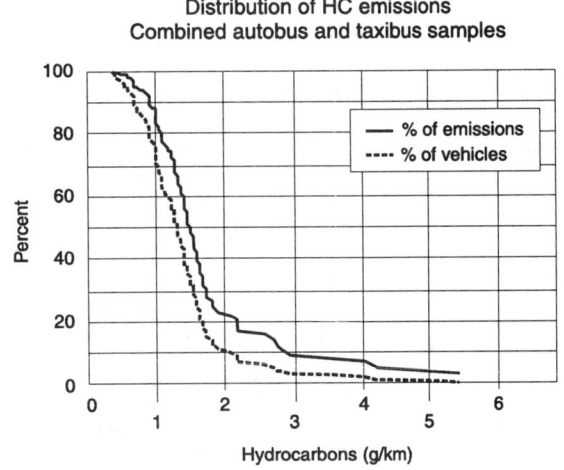

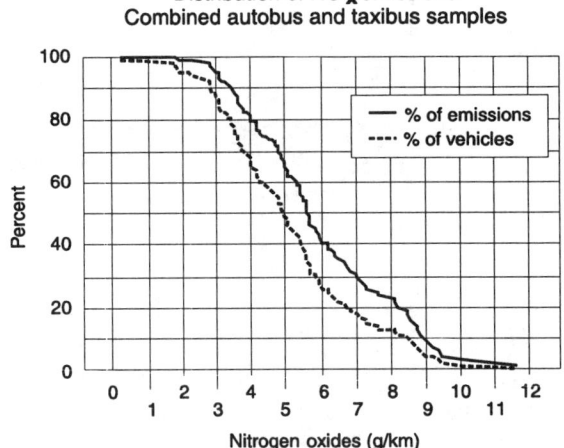

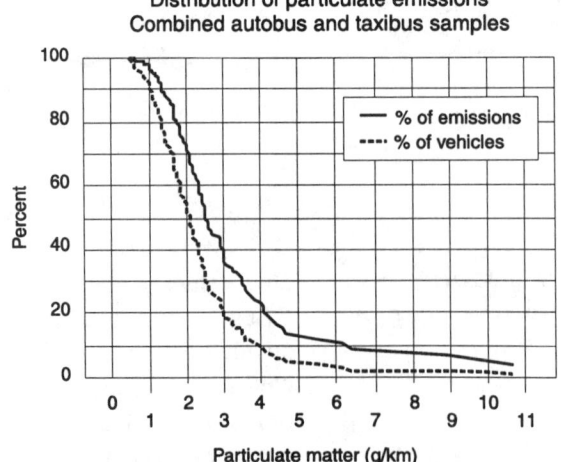

Source: Turner and others 1993

0.281 grams per kilometer of particulate matter. Compared with emissions from four-stroke motorcycles, emissions from uncontrolled two-strokes are about ten times higher for hydrocarbons, carbon monoxide emissions are similar, and emissions of nitrogen oxides are less (though motorcycle emissions of nitrogen oxides are always small compared to those from other vehicles). Particulate matter emissions from the two-stroke motorcycles tested were also about five to ten times those of four-stroke motorcycles. PM emissions data are available for only five motorcycles, however, and cover a wide range, from 0.082 grams per kilometer to 0.564 grams per kilometer. In addition, three of the particulate measurements were not taken under the U.S. procedure but as a weighted combination of measurements under different steady-state operating conditions. Thus, there is some uncertainty about how well these data represent real-world driving conditions. In addition, all measurements were made on new or nearly-new motorcycles that were properly adjusted and maintained. Actual emissions in consumer service would be expected to be significantly higher.

The U.S. emissions data can be used to develop an estimate of the relationship between smoke opacity and particulate emissions for two-stroke motorcycles. A correlation between measured particulate matter emissions and smoke opacity is given by:

$$PM = 0.066 + (0.015 * OP) \qquad (2.2)$$

where PM = particulate matter (g/km),
and OP = smoke opacity for a 3" path length (percent).

Europe. Limited data exist on uncontrolled motorcycle emissions in Europe. A report prepared by the Swiss Office of Environmental Protection in 1986 reviewed several studies on motorcycle emissions and concluded that average exhaust emissions for uncontrolled motorcycles and mopeds were within the range of European standards. The average emissions for 35 uncontrolled four-stroke motorcycles, 40 uncontrolled two-stroke motorcycles, and 141 two-stroke mopeds are shown in table 2.6. The motorcycles tested were in consumer use, so these data might be representative of real motorcycle emissions. The emissions for European two-stroke and four-stroke motorcycles are about twice the levels of new, uncontrolled motorcycles in the United States. Although the data are not strictly comparable because of the differences in the test cycles, they do suggest that

Table 2.5   Emission and Fuel Consumption Factors for Uncontrolled U.S. Two- and Four-Stroke Motorcycles
(grams per kilometer)

Engine type	Carbon monoxide	Hydrocarbons	Nitrogen oxides	Particulate matter	Fuel economy (liters per 100 kilometers)
Two-stroke	16.1	9.9	0.022	0.206	4.7
Four-stroke	23.5	2.0	0.135	0.048	5.2
Four-stroke with displacement less than 250 cc	15.0	1.0	0.206	0.020	2.9

Source: Chan and Weaver 1994

Table 2.6   Emission Factors for Uncontrolled European Motorcycles and Mopeds
(grams per kilometer)

Vehicle type	Engine type	Number tested	Carbon monoxide	Hydrocarbons	Nitrogen oxides
Motorcycle	Four-stroke	35	40.0	5.9	0.2
	Two-stroke	40	24.6	19.0	0.035
Moped	Two-stroke	141	10.0	6.0	0.06

Source: Chan and Weaver 1994

Table 2.7   Emission and Fuel Consumption Factors for Uncontrolled Thai Motorcycles
(grams per kilometer)

Engine type	Carbon monoxide	Hydrocarbons	Fuel economy (liters per 100 kilometers)
Four-stroke	19.0	2.9	1.6
Two-stroke	28.1	14.6	2.5

Source: Chan and Weaver 1994

Figure 2.14 Smoke Opacity Emissions from Motorcycles in Bangkok, Thailand

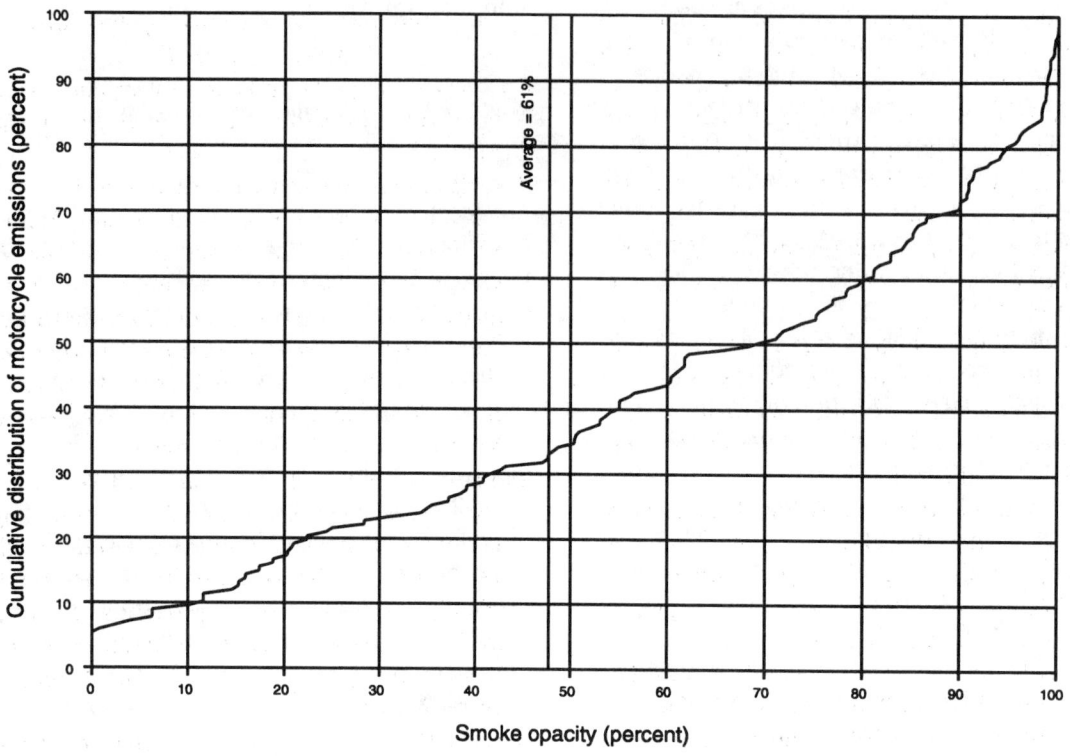

Source: Chan and Weaver 1994

average in-use emissions from uncontrolled motorcycles are higher than emissions from new, properly adjusted motorcycles (Chan and Weaver 1994).

Thailand. A cumulative distribution plot of acceleration smoke opacity for 167 randomly selected motorcycles in Bangkok is shown in figure 2.14. More than 95 percent of the motorcycles tested were equipped with two-stroke engines. The mean smoke opacity, corrected to a three-inch path length, was 61 percent—four times the opacity for the smokiest of the uncontrolled U.S. motorcycles with opacity measurements ranging from 3 to 18 percent. By extrapolating the correlation in equation 2.2 it can be estimated that the 61 percent average smoke opacity in Bangkok is equivalent to average particulate matter emissions of about 1.0 grams per kilometer (figure 2.14).

The Thai Department of Pollution Control has gathered data on hydrocarbon and carbon monoxide emissions for 17 Thai-produced motorcycles based on the European test cycle. These data were obtained from manufacturers and thus probably represent new, properly-adjusted motorcycles (table 2.7). Of the 17 motorcycles, two had four-stroke engines and the rest had two-strokes. Average hydrocarbon and carbon monoxide emissions for the two-stroke motorcycles were 14.6

and 28.1 grams per kilometer, respectively; the averages for the four-stroke motorcycles were 2.9 and 19.0 grams per kilometer, respectively. These values are similar to but lower than the average for European motorcycles reported in table 2.6. Since the Thai data are for new motorcycles and the European data are for motorcycles in use, this is not surprising. In addition to having lower emissions, four-stroke motorcycles also had much better fuel economy than the two-strokes, averaging 1.6 liters per 100 kilometers compared with 2.5 liters per 100 kilometers for the two-stroke motorcycles.

References

Alson, J., J. Adler, and T. Baines. 1989. "Motor Vehicle Emission Characteristics and Air Quality Impacts of Methanol and Compressed Natural Gas." in D. Sperling, ed. *Alternate Transportation Fuels: An Environmental and Energy Solution.* Greenwood Press, Wesport, Connecticut.

Andrias, A., D. Zafiris, Z. Samras, and K-H. Zierock. 1992. "Computer Program to Calculate Emissions from Road Traffic—User's Manual," COPERT, EC Contract No. B4-3045 (91) 10 PH (DG XI/B/3), European Commission, Brussels.

AQIRP. 1995. "Real World Automotive Emissions – Results of Studies in the Fort McHenry and Tuscarora Mountain Tunnels." Air Quality Improvement Research Program, Technical Bulletin No. 14. Auto/Oil Industry Research Council, Atlanta, Georgia.

_____. 1996. "Dynamometer Study of Off-Cycle Exhaust Emissions." Air Quality Improvement Research Program, Technical Bulletin No. 19. Auto/Oil Industry Council, Atlanta, Georgia.

Baines, T. 1994. "Personal Communication." U.S. Environmental Protection Agency, Office of Mobile Sources, Washington, D.C.

Brunner, D., K. Sclapfer, M. Ros, and F. Dinkel. 1994. "Data Bank on Unregulated Vehicle Exhaust and Evaporative Emissions." 3rd International Symposium on Transport and Air Pollution, *Poster Proceedings,* INRETS, Arcueil, France.

Chan, L.M. and M. Reale. 1994. "Emissions Factors Generated from In-House MOBILE5a." Engine, Fuel, and Emissions Engineering. Sacramento, California.

Chan, L.M. and C.S. Weaver. 1994. "Motorcycle Emission Standards and Emission Control Technology." Departmental Paper Series, No. 7, Asia Technical Department, The World Bank, Washington, D.C.

Chan, L.M. and C.S. Weaver. 1995. "Figures Generated with the Use of MOBILE5a," Engine, Fuel, and Emissions Engineering, Sacramento, California.

Chongpeerapien, T., S. Sungsuwan, P. Kritiporn, S. Buranasajja. 1990. "Energy and Environment, Choosing the Right Mix." Resource Management Associates Research Report No. 7, Thailand Development Research Institute, Bangkok.

CONCAWE (Conservation of Clean Air and Water in Europe). 1994. *Motor Vehicle Emission Regulations and Fuel Specifications-1994 Update.* Report 4/94.

Cornetti, G., K. Klein, G. Frankle, and H. Stein. 1988. "U.S. Transient Cycle Versus ECE R.49 13 – Mode Cycle." *SAE Paper* 880715, Society of Automotive Engineers, Warrendale, Pennsylvania.

Eggleston, H., N. Gorissen, R. Joumard, R. Rijkeboeer, Z. Samaras, and K. Zierock. 1991. "CORINAIR Working Group on Emission Factors for Calculating 1990 Emissions From Road Traffic." Volume 1. Methodology and Emission Factors. Final Report, EC Study Contract B4-3045(91)10PH. Commission of the European Communities. Brussels.

Escudero, J. 1991. "Notes on Air Pollution Issues in Santiago Metropolitan Region," Letter #910467, dated May 14, 1991. Comisión Especial de Descontaminación de la Region Metropolitana. Santiago.

Faiz, A., S. Gautam, and E. Burki. 1995. "Air Pollution from Motor Vehicles: Issues and Options for Latin American Countires." *The Science of the Total Environment,* 169:303-310.

Faiz, A., and J. Aloisi de Larderel. 1993. "Automotive Air Pollution in Developing Countries: Outlook and Control Strategies". *The Science of the Total Environment,* 134: 325-344.

Guensler, R. 1994. "Loop Holes for Air Pollution," ITS Review, Volume 18, Number 1. University of California, Berkeley.

IGRP (Indonesian German Research Project). 1991. "Environmental Impacts of Energy Strategies for Indonesia." VWS Report on Assessment of the Emission Coefficients of the Traffic Sector in Jawa. Jakarta.

Joumard, R., L. Paturel, R. Vidon, J. P. Guitton, A. Saber, and E. Combet. 1990. *Emissions Unitaires de Polluants des Véhicules Légers.* Report 116 (2nd ed.) Institut National du Recherche sur les Transports et Leur Securite (INRETS), Bron, France.

Joumard, R., P. Jost, J. Hickman, and D. Hassel. 1995. "Hot Passenger Car Emissions Modelling as a Function of Instantaneous Speed and Acceleration." *The Science of the Total Environment,* 169: 167-174.

Laurikko, J. 1995. "Ambient Temperature Effect on Automotive Exhaust Emissions: FTP and ECE Test Cycle Responses." *The Science of the Total Environment,* 169: 195-204.

Lenaers, G. 1994. "A Dedicated System for On-the-road Exhaust Emissions Measurements on Vehicles." 3rd International Symposium on Transport and Air Pollution, *Poster Proceedings,* INRETS, Arcueil, France.

OECD (Organization for Economic Cooperation & Development). 1988. *Transport and the Environment.* Paris.

OFPE (l'Office federal de la protection de l'environnement). 1988. "Emissions Polluantes du Trafic Routier Prive de 1950 a 2000." *Les cahiers de l'environnement 55,* Berne, Switzerland.

Pattas, K., N. Kyriakis, and Z. Samaras. 1985. "Exhaust Emission Study of the Current Vehicle Fleet in Athens (Phase II)." Final Report. 3 volumes. University of Thessaloniki, Greece.

Radian Corporation. 1993. "Revision of the MOBILE-MEXICO (Mobile Mexico) Model Report." Departamento del Distrito Federal, Mexico City, Mexico.

Roumegoux, J.P. 1995. "Calcul des Emissions Unitaires de Polluants des Véhicules Utilitaires", *The Science of the Total Environment,* 169: 273-82.

Sandoval, H., M. Prendez, and P. Ulriksen eds. 1993. *Contaminacion Atmosferica de Santiago: Estado Actual y Soluciones,* University of Chile, Santiago.

Shen, S., and Huang, K. 1991. "Response to Transport-Induced Air Pollution: The Case of Taiwan." in M.L. Birk and D.L. Bleviss, eds., *Driving New Directions: Transportation Experiences and Options in Developing Countries.* International Institute for Energy Conservation, Washington, D.C.

Steiner, A. 1989. "Operacion y Resultados del Programa de Mediciones de la UMEVE." Paper Presented at the Seminar on Contaminacion Atmosferica Debido a Motores de Combustion Interna, Santa Maria University, Valparaiso, Chile.

Turner, S.H., C.S. Weaver, and M.J. Reale. 1993. "Cost and Emissions Benefits of Selected Air Pollution Control Measures in Santiago, Chile." Engine, Fuel, and Emissions Engineering, Inc. Sacramento, California

Weaver, C.S., and R. Klausmeier. 1988. "Heavy-Duty Diesel Vehicle Inspection and Maintenance Study."

Report to the California Air Resources Board, Engine, Fuel, and Emissions Engineering Inc., Sacramento, California.

Weaver, C.S., and S. Turner. 1991. "Mobile Source Emission Factors for Global Warming Gases." Report to U.S. EPA Contract 68-W8-0113. Sacramento, California.

Weaver, C.S., and L.M. Chan. 1995. "Company Archives" Multiple Sources. Engine, Fuel, and Emissions Engineering Inc., Sacramento, California.

Appendix 2.1

Selected Exhaust Emission and Fuel Consumption Factors for Gasoline-Fueled Vehicles

Table A2.1.1   Exhaust Emissions, European Vehicles, 1970–90 Average
(grams per kilometer)

Traffic type and engine/emission control characteristics	Carbon monoxide	Hydrocarbons	Nitrogen oxides	Particulate matter
Urban				
Two-stroke	32.9	20.2	0.26	0.00
More than 2,000 cc	32.0	3.0	2.00	0.01
1,400–2,000 cc	31.0	2.9	1.80	0.10
Less than 1,400 cc	30.0	2.8	1.70	0.14
Catalyst without O_2 sensor	6.6	0.9	1.00	0.00
Catalyst with O_2 sensor	1.5	0.2	0.27	0.00
Rural highway				
Two-stroke	21.6	13.0	0.33	0.00
More than 2,000 cc	21.0	2.2	2.49	0.01
1,400–2,000 cc	20.0	2.1	2.29	0.01
Less than 1,400 cc	19.0	2.0	2.09	0.01
Catalyst without O_2 sensor	4.5	0.5	1.15	0.00
Catalyst with O_2 sensor	1.2	0.1	0.24	0.00
Motorway				
Two-stroke	22.7	13.0	0.42	0.00
More than 2,000 cc	21.5	2.3	3.03	0.01
1,400–2,000 cc	20.5	2.2	2.83	0.01
Less than 1,400 cc	19.5	2.1	2.63	0.01
Catalyst without O_2 sensor	5.0	0.5	1.35	0.00
Catalyst with O_2 sensor	1.3	0.1	0.32	0.00

Source: Metz 1993

Table A2.1.2   Exhaust Emissions, European Vehicles, 1995 Representative Fleet
(grams per kilometer)

Emission regulations/controls	Carbon monoxide	Hydrocarbons	Nitrogen oxides	Carbon dioxide	Particulate matter
ECE 15-03	31.5	3.57	2.29	188	—
ECE 15-04	24.1	2.97	2.40	192	—
Three-way catalytic converter	5.2	0.32	0.40	247	—
Diesel[a]	0.7	0.13	1.22	188	0.14

— Not applicable
a. Included for comparison.
Source: Joumard and others 1995

Table A2.1.3 Estimated Emissions and Fuel Consumption, European Vehicles, Urban Driving
(grams per kilometer)

Vehicle type	Carbon monoxide (CO)	Non-methane volatile org. compounds (NM-VOC)	Nitrogen oxides (NO_x)	Methane (CH_4)	Nitrous oxide (N_2O)	Ammonia (NH_3)	Evap. emissions	Fuel consumption (liter/100 km)
Passenger Cars								
Two-stroke	20.7	15.2	0.3	0.150	0.005	0.002	1.13	14.93
LPG	6.9	11.8	1.9	0.122	0.000	0.000	—	8.78
Pre-1971 (Pre ECE)								
Less t..han 1,400 cc	56.8	4.2	1.7	0.224	0.005	0.002	1.13	12.99
1,400-2,000 cc	56.8	4.2	1.9	0.224	0.005	0.002	1.13	15.38
More than 2,000 cc	56.8	4.2	2.4	0.224	0.005	0.002	1.13	19.23
1972-76 (ECE15/00-01)								
Less than 1,400 cc	41.6	3.3	1.7	0.224	0.005	0.002	1.13	11.63
1,400-2,000 cc	41.6	3.3	1.9	0.224	0.005	0.002	1.13	13.89
More than 2,000 cc	41.6	3.3	2.4	0.224	0.005	0.002	1.13	15.63
1977-79 (ECE15/02)								
Less than 1,400 cc	35.4	3.3	1.5	0.224	0.005	0.002	1.13	10.64
1,400-2,000 cc	35.4	3.3	1.7	0.224	0.005	0.002	1.13	12.82
More than 2,000 cc	35.4	3.3	2.0	0.224	0.005	0.002	1.13	15.87
1980-84 (ECE15/03)								
Less than 1,400 cc	22.3	3.3	1.6	0.224	0.005	0.002	1.13	10.64
1,400-2,000 cc	22.3	3.3	1.9	0.224	0.005	0.002	1.13	12.82
More than 2,000 cc	22.3	3.3	2.5	0.224	0.005	0.002	1.13	15.87
1985-89 (ECE15/04)								
Less than 1,400 cc	21.4	2.6	1.6	0.224	0.005	0.002	1.13	5.71
1,400-2,000 cc	21.4	2.6	1.9	0.224	0.005	0.002	1.13	10.53
More than 2,000 cc	21.4	2.6	2.3	0.224	0.005	0.002	1.13	18.29
Improved conventional								
Less than 1,400 cc	13.5	1.8	1.4	0.224	0.005	0.002	1.13	7.94
1,400-2,000 cc	7.8	1.5	1.5	0.224	0.005	0.002	1.13	10.53
Oxidizing catalyst without l sensor								
Less than 1,400 cc	15.7	1.6	1.1	0.224	0.005	0.002	1.13	8.70
1,400-2,000 cc	7.6	0.4	0.9	0.224	0.005	0.002	1.13	10.53
3 way catalyst with l sensor								
Less than 1,400 cc	4.1	0.4	0.4	0.062	0.050	0.070	0.23	8.47
1,400-2,000 cc	4.1	0.4	0.4	0.062	0.050	0.070	0.23	10.31
More than 2,000 cc	4.1	0.4	0.4	0.062	0.050	0.070	0.23	12.99
Light-duty vehicles	46.1	4.7	3.1	0.230	0.006	0.002	1.03	17.86
Motorcycles								
Less than 50 cm3	10.0	5.9	0.1	0.100	0.001	0.001	0.28	2.40
More than 50 cm3, two-stroke	22.0	14.9	0.1	0.150	0.002	0.002	0.45	4.00
More than 50 cm3, four-stroke	20.0	2.8	0.3	0.200	0.002	0.002	0.45	5.08

n.a. = Not available
— Not applicable
Notes:
• Average driving speed 25 kilometers per hour.
• Emission factors in g/km are derived from the COPERT model for 1990, utilizing the CORINAIR methodology for road traffic emissions. The pollutants included are: CO, NMVOC, NO_x, CH_4, N_2O, NH_3. Fuel consumption is also estimated. Total VOC (or HC) emission factors may be obtained by adding NM-VOC and CH_4 factors.
• Cold-start emissions are calculated for urban driving conditions only, taking into account the monthly variation of the average minimum and maximum temperatures. For this specific application, average temperatures in Brussels (Belgium) were used, with the following monthly distribution:

Month	01	02	03	04	05	06	07	08	09	10	11	12
Tmin (°C)	-1.2	0.3	2.2	5.1	7.9	10.9	12.1	12.2	10.6	7.3	3.1	0.2
Tmax (°C)	4.3	6.7	10.3	14.2	18.4	22.0	22.7	22.3	20.5	15.4	8.9	5.6

• Evaporative losses are estimated in g/km; REID vapor pressure of gasoline 80 kPa (from October to March) and 65 kPa (from April to September) with monthly temperatures given in note 3.
• For cold-starts and evaporation losses, average trip length equals 12 km.

Source: Samaras 1992

Table A2.1.4 Estimated Emissions and Fuel Consumption, European Vehicles, Rural Driving

(grams per kilometer)

Vehicle type	Carbon monoxide (CO)	Non-methane volatile org. compounds (NM-VOC)	Nitrogen oxides (NO$_x$)	Methane (CH$_4$)	Nitrous oxide (N$_2$O)	Ammonia (NH$_3$)	Evap. emissions	Fuel consumption (liter/100 km)
Passenger Cars								
Two-stroke	7.5	7.24	1.0	0.040	0.005	0.002	0.14	8.77
LPG	3.1	0.6	2.6	0.035	0.000	0.000	–	6.02
Pre-1971 (Pre ECE)								
Less than 1,400 cc	18.5	1.53	2.1	0.025	0.005	0.002	0.14	7.35
1,400-2,000 cc	18.5	1.53	2.8	0.025	0.005	0.002	0.14	8.93
More than 2,000 cc	18.5	1.53	4.3	0.025	0.005	0.002	0.14	10.64
1972-76 (ECE 15/00-01)								
Less than 1,400 cc	14.8	1.23	2.1	0.025	0.005	0.002	0.14	5.88
1,400-2,000 cc	14.8	1.23	2.8	0.025	0.005	0.002	0.14	6.94
More than 2,000 cc	14.8	1.23	4.3	0.025	0.005	0.002	0.14	7.75
1977-79 (ECE 15/02)								
Less than 1,400 cc	7.9	1.03	2.2	0.025	0.005	0.002	0.14	6.02
1,400-2,000 cc	7.9	1.03	2.5	0.025	0.005	0.002	0.14	6.85
More than 2,000 cc	7.9	1.03	2.8	0.025	0.005	0.002	0.14	8.47
1980-84 (ECE 15/03)								
Less than 1,400 cc	8.3	1.03	2.3	0.025	0.005	0.002	0.14	6.02
1,400-2,000 cc	8.3	1.03	2.8	0.025	0.005	0.002	0.14	6.45
More than 2,000 cc	8.3	1.03	3.4	0.025	0.005	0.002	0.14	8.00
1985-89 (ECE 15/04)								
Less than 1,400 cc	4.7	0.83	2.2	0.025	0.005	0.002	0.14	5.78
1,400-2,000 cc	4.7	0.83	2.7	0.025	0.005	0.002	0.14	6.45
More than 2,000 cc	4.7	0.83	2.9	0.025	0.005	0.002	0.14	8.00
Improved conventional								
Less than 1,400 cc	6.5	0.73	2.2	0.025	0.005	0.002	0.14	6.37
1,400-2,000 cc	2.3	0.63	2.7	0.025	0.005	0.002	0.14	7.25
Oxidizing catalyst without l sensor								
Less than 1,400 cc	5.5	0.53	1.7	0.025	0.005	0.002	0.14	6.58
1,400-2,000 cc	3.6	0.13	1.5	0.025	0.005	0.002	0.14	7.52
3-way catalyst with l sensor								
Less than 1,400 cc	1.4	0.12	0.3	0.020	0.050	0.100	0.03	6.25
1,400-2,000 cc	1.4	n.a.	0.3	n.a.	0.050	0.100	0.03	7.30
More than 2,000 cc	1.4	0.12	0.3	0.020	0.050	0.100	0.03	8.85
Light-duty vehicles	15.0	1.7	2.7	0.040	0.006	0.002	0.14	9.03
Motorcycles								
Less than 50 cm^3	10.0	6.00	0.1	0.100	0.001	0.001	0.00	2.40
More than 50 cm^3, two-stroke	22.0	15.05	0.1	0.150	0.002	0.002	0.06	4.00
More than 50 cm^3, four-stroke	20.0	3.00	0.3	0.200	0.002	0.002	0.06	5.08

— Not applicable
n.a. = Not available
Notes:
• Average driving speed, 75 kilometers per hour.
• Emission factors in g/km are derived from the COPERT model for 1990, utilizing the CORINAIR methodology for road traffic emissions. The pollutants included are: CO, NMVOC, NO$_x$, CH$_4$, N$_2$O, NH$_3$. Fuel consumption is also estimated. Total VOC (or HC) emission factors may be obtained by adding NMVOC and CH$_4$ factors.
• Cold-start emissions are calculated for urban driving conditions only, taking into account the monthly variation of the average minimum and maximum temperatures. For this specific application, average temperatures in Brussels (Belgium) were used, with the following monthly distribution:

Month	01	02	03	04	05	06	07	08	09	10	11	12
Tmin (°C)	-1.2	0.3	2.2	5.1	7.9	10.9	12.1	12.2	10.6	7.3	3.1	0.2
Tmax (°C)	4.3	6.7	10.3	14.2	18.4	22.0	22.7	22.3	20.5	15.4	8.9	5.6

• Evaporative losses are estimated in g/km; REID vapor pressure of gasoline 80 kPa (from October to March) and 65 kPa (from April to September) with monthly temperatures given in note 3.
• For cold-starts and evaporation losses, average trip length equals 12 km.
Source: Samaras 1992

Table A2.1.5 Estimated Emissions and Fuel Consumption, European Vehicles, Highway Driving
(grams per kilometer)

Vehicle type	Carbon monoxide (CO)	Non-methane volatile org. compounds (NM-VOC)	Nitrogen oxides (NO$_x$)	Methane (CH$_4$)	Nitrous oxide (N$_2$O)	Ammonia (NH$_3$)	Evap. emissions	Fuel consumption (liter/100 km)
Passenger Cars								
Two-stroke	8.7	0.93	0.7	0.025	0.005	0.002	0.14	7.58
LPG	9.8	0.5	2.9	0.025	0.000	0.000	—	7.23
Pre-1971 (Pre-ECE)								
Less than 1,400 cc	15.5	1.23	2.0	0.026	0.005	0.002	0.14	8.33
1,400–2,000 cc	15.5	1.23	3.1	0.026	0.005	0.002	0.14	10.20
More than 2,000 cc	15.5	1.23	5.5	0.026	0.005	0.002	0.14	11.76
1972-76 (ECE 15/00-01)								
Less than 1,400 cc	18.6	1.13	2.0	0.026	0.005	0.002	0.14	6.49
1,400–2,000 cc	18.6	1.13	3.1	0.026	0.005	0.002	0.14	8.06
More than 2,000 cc	18.6	1.13	5.5	0.026	0.005	0.002	0.14	8.85
1977-79 (ECE 15/02)								
Less than 1,400 cc	8.3	0.93	2.9	0.026	0.005	0.002	0.14	6.85
1,400–2,000 cc	8.3	0.93	3.3	0.026	0.005	0.002	0.14	7.94
More than 2,000 cc	8.3	0.93	3.7	0.026	0.005	0.002	0.14	9.43
1980-84 (ECE 15/03)								
Less than 1,400 cc	7.9	0.93	3.3	0.026	0.005	0.002	0.14	6.85
1,400–2,000 cc	7.9	0.93	3.8	0.026	0.005	0.002	0.14	7.94
More than 2,000 cc	7.9	0.93	4.5	0.026	0.005	0.002	0.14	9.43
1985-89 (ECE 15/04)								
Less than 1,400 cc	4.3	0.73	2.7	0.026	0.005	0.002	0.14	6.37
1,400–2,000 cc	4.3	0.73	3.5	0.026	0.005	0.002	0.14	6.98
More than 2,000 cc	4.3	0.73	3.7	0.026	0.005	0.002	0.14	9.35
Improved conventional								
Less than 1,400 cc	10.5	0.83	2.4	0.026	0.005	0.002	0.14	9.26
1,400–2,000 cc	6.7	0.73	3.7	0.026	0.005	0.002	0.14	10.42
Oxidizing catalyst without λ sensor								
Less than 1,400 cc	8.4	0.53	1.9	0.026	0.005	0.002	0.14	9.01
1,400–2,000 cc	6.7	0.23	1.6	0.026	0.005	0.002	0.14	10.87
3-way catalyst with λ sensor								
Less than 1,400 cc	3.1	0.12	0.5	0.020	0.050	0.001	0.03	8.70
1,400–2,000 cc	3.1	0.12	0.5	0.020	0.050	0.001	0.03	10.20
More than 2,000 cc	3.1	0.12	0.5	0.020	0.050	0.001	0.03	12.99
Light-duty vehicles	12.0	1.0	3.2	0.025	0.002	0.006	0.14	8.54
Motorcycles								
Less than 50 cm^3	10.0	6.00	0.1	0.100	0.001	0.001	0.00	2.40
More than 50 cm^3, two-stroke	22.0	15.05	0.1	0.150	0.002	0.002	0.06	4.00
More than 50 cm^3, four-stroke	20.0	3.00	0.3	0.200	0.002	0.002	0.06	5.08

— Not applicable

Notes:
• Average driving speed, 100 kilometers per hour.
• Emission factors in g/km are derived from the COPERT model for 1990, utilizing the CORINAIR methodology for road traffic emissions. The pollutants included are: CO, NMVOC, NO$_x$, CH$_4$, N$_2$O, NH$_3$. Fuel consumption is also estimated. Total VOC (or HC) emission factors may be obtained by summing up NMVOC and CH4.
• Cold-start emissions are calculated for urban driving conditions only, taking into account the monthly variation of the average minimum and maximum temperatures. For this specific application, average temperatures in Brussels (Belgium) were used, with the following monthly distribution:

Month	01	02	03	04	05	06	07	08	09	10	11	12
Tmin (°C)	-1.2	0.3	2.2	5.1	7.9	10.9	12.1	12.2	10.6	7.3	3.1	0.2
Tmax (°C)	4.3	6.7	10.3	14.2	18.4	22.0	22.7	22.3	20.5	15.4	8.9	5.6

• Evaporative losses are estimated in g/km; REID vapor pressure of gasoline from 80 kPa (from October to March) and 65 kPa (from April to September) with monthly temperatures given in note 3.
• For cold-starts and evaporation losses, average trip length equals 12 km.
Source: Samaras 1992

Table A2.1.6 Automobile Exhaust Emissions, Chile

(grams per kilometer)

Vehicle type	Carbon monoxide	Hydrocarbons	Nitrogen oxides	Particulate matter
Private car	26.0	1.00	1.2	0.07
Taxi	28.0	1.50	1.4	0.06

Note: Measured using the U.S. federal testing procedure.
Source: Escudero 1991

Table A2.1.7 Automobile Exhaust Emissions as a Function of Test Procedure and Ambient Temperature, Finland

(grams per kilometer)

53	5	Hydrocarbons	Nitrogen oxides
ECE-15			
22°C	2.60	0.27	0.27
-7°	11.22	1.09	0.65
-20°C	17.81	2.79	0.62
FTP-75			
22°C	1.40	0.13	0.16
-7°C	5.34	0.50	0.29
-20°C	8.58	1.25	0.31

Note: Based on emission tests on cars with a three-way catalytic converter.
Source: Laurikko 1995

Table A2.1.8 Automobile Exhaust Emissions as a Function of Driving Conditions, France

(grams per kilometer)

Traffic type	Carbon monoxide	Hydrocarbons	Nitrogen oxides	Carbon dioxide
Congested urban	94.1	10.70	1.6	520.73
Free-flowing urban	29.3	3.52	1.8	189.54
Highway	19.4	2.45	2.2	149.05
Motorway	16.0	1.09	3.0	153.48

Source: Joumard and others 1990

Table A2.1.9 Automobile Exhaust Emissions and Fuel Consumption as a Function of Driving Conditions and Emission Controls, Germany

(grams per kilometer)

Vehicle type	Carbon monoxide	Hydrocarbons	Nitrogen oxides	Carbon dioxide	Fuel consumption (liters/100km)
European test procedure (low-speed urban driving with cold start)					
Catalyst with O_2 sensor	6.27	0.81	0.59	274	11.90
Catalyst without O_2 sensor	15.34	1.93	0.94	234	9.80
No catalyst	17.67	2.62	1.29	243	10.50
U.S. federal test procedure (mixed urban driving)					
Catalyst with O_2 sensor	3.02	0.27	0.39	204	8.90
Catalyst without O_2 sensor	11.76	1.35	0.88	180	7.80
No catalyst	12.14	1.84	1.63	182	7.90
Rural highway					
Catalyst with O_2 sensor	0.98	0.06	0.28	136	5.90
Catalyst without O_2 sensor	3.98	0.42	1.15	130	5.60
No catalyst	4.89	0.76	1.94	126	5.50
Motorway					
Catalyst with O_2 sensor	5.13	0.14	0.75	193	8.30
Catalyst without O_2 sensor	12.72	0.60	1.74	189	8.20
No catalyst	12.78	0.94	3.18	178	7.70

Source: Hassel and Weber 1993

Table A2.1.10   Exhaust Emissions, Light-Duty Vehicles and Mopeds, Greece
(grams per kilometer)

Vehicle type	Carbon monoxide	Hydrocarbons	Nitrogen oxides
Light-duty	45.8	1.6	1.60
Motorcycle	21.4	3.4	0.11
Moped	14.0	10.4	0.05

Note: Measured using the ECE-15 testing procedure.
Source: Pattas, Kyriakis, and Nakos 1993

Table A2.1.11   Hot-Start Exhaust Emissions, Light-Duty Vehicles, Greece
(grams per kilometer)

Engine capacity/Emission standard	Carbon monoxide	Hydrocarbons	Nitrogen oxides
Less than 1,400 cc			
Pre-control (1971)	60.02	5.15	1.23
ECE 15-00 (1971–75)	58.17	4.80	1.42
ECE 15-02 (1975–79)	44.37	3.77	1.58
ECE 15-03 (1980–84)	31.33	3.60	1.81
ECE 15-04 (1985–present)	25.50	2.08	2.08
1,400–2,000 cc			
Pre-control	73.61	4.87	1.06
ECE 15-00	54.51	4.67	1.82
ECE 15-02	51.20	3.65	1.75
ECE 15-03	35.94	3.27	1.95
ECE 15-04	27.99	2.11	2.06
More than 2,000 cc			
Pre-control	77.12	5.60	1.40
ECE 15-00	91.97	5.85	0.77
ECE 15-02	21.28	1.62	2.28
ECE 15-03	81.50	3.32	1.02
ECE 15-04	30.18	2.14	2.04

Note: Measured using the ECE-15 testing procedure, average speed 18.7 kilometers per hour.
Source: Pattas, Kyriakis, and Nakos 1993

Table A2.1.12   Exhaust Emissions, Light-Duty Vehicles and 2-3 Wheelers, India
(grams per kilometer)

Vehicle type	Carbon monoxide	Hydro-carbons	Nitrogen oxides	Sulfur dioxide
Car/Jeep	23.8	3.5	1.6	0.1
Taxi	29.1	4.3	1.9	0.1
Two-wheeler	8.2	5.1	—	—
Auto-rickshaw (3-wheeler)	12.5	7.8	—	0.0
Light-duty vehicles	40.0	6.0	3.2	0.08
Motorcycles	17.0	10.0	0.07	0.02

— Not applicable
n.a. = Not available
Sources: Biswas and Dutta 1994; Bose 1994; Gargava and Aggarwal 1994

References

Biswas, D. and S.A. Dutta. 1994. "Strategies for Control of Vehicular Pollution in Urban Areas of India." Proceedings of a Workshop on The Energy Nexus—Indian Issues & Global Impacts (April 22–23, 1994), University of Pennsylvania, Philadelphia.

Escudero, J. 1991. "Notes on Air Pollution Issues in Santiago Metropolitan Region", Letter # 910467, dated May 14, 1991. Comisión Especial de Descontaminación de la Region Metropolitana. Santiago, Chile.

Gargava, P., and A. Aggarwal. 1994. "Prediction of Impact of Air Environment on Planning the Control Strategies for Automobile Pollution in an Indian Coastal City." *Poster Proceedings*, 3rd International International Symposium on Transport and Air Pollution, INRETS, Arcueil, France.

Hassel, D. and F-J. Weber. 1993. "Mean Emissions and Fuel Consumption of Vehicles in Use with Different Emission Reduction Concepts." *The Science of the Total Environment* 134: 189-95.

Joumard, R., L. Paturel, R. Vidon, J. P. Guitton, A. Saber, and E. Combet. 1990. *Emissions Unitaires de Polluants des Véhicules Légers.* Report 116 (2nd ed.) Institut National du Recherche sur les Transports et leur Securite (INRETS), Bron, France.

Joumard, R., P. Jost, J. Hickman, and D. Hassel. 1995. "Hot Passenger Car Emissions Modelling as a Function of Instantaneous Speed and Acceleration." *The Science of the Total Environment*, 169: 167-74.

Laurikko, J. 1995. "Ambient Temperature Effect on Automotive Exhaust Emissions: FTP and ECE Test Cycle Responses." *The Science of the Total Environment*, 169: 195-204.

Metz, N. 1993. "Emission Characteristics of Different Combustion Engines in City, on Rural Roads, and on Highways." *The Science of the Total Environment*, 134: 225-35.

Pattas, K., N. Kyriakis, and C. Nakos. 1993. "Time Dependence of Traffic Emissions in the Urban Area of Thessaloniki." *The Science of the Total Environment* 134: 273-84.

Samaras, Z. 1992. "COPERT Emission Factors." Informal Communication. Directorate-General for Environment, Nuclear Safety, and Civil Protection, Commission of the European Communities, Brussels.

Appendix 2.2

Selected Exhaust Emission and Fuel Consumption Factors for Diesel-Fueled Vehicles

Table A2.2.1 Exhaust Emissions, European Cars
(grams per kilometer)

Traffic type/emission control	Carbon monoxide	Hydrocarbons	Nitrogen oxides	Particulate matter
Urban				
with catalyst	0.05	0.08	0.70	0.20
without catalyst	1.30	0.10	0.90	0.30
Rural				
with catalyst	0.02	0.10	0.61	0.18
without catalyst	0.60	0.10	0.79	0.29
Highway				
with catalyst	0.20	0.10	0.77	0.27
without catalyst	0.70	0.10	0.97	0.37

Source: Metz 1993

Table A2.2.2 Estimated Emissions and Fuel Consumption, European Cars and Light-Duty Vehicles
(grams per kilometer)

Traffic and vehicle type	Carbon monoxide	Hydrocarbons	Nitrogen oxides	Particulate matter	Fuel consumption (liters/100km)
Urban					
Passenger cars					
Less than 2,000 cc	1.0	0.306	0.7	0.362	10.0
More than 2,000 cc	1.0	0.306	1.0	0.362	10.0
Light-duty vehicles	2.4	0.506	1.7	0.333	14.08
Rural					
Passenger cars					
Less than 2,000 cc	0.5	0.105	0.4	0.131	5.05
More than 2,000 cc	0.5	0.105	0.7	0.131	5.05
Light-duty vehicles	0.8	0.205	1.2	0.131	8.40
Motorway					
Passenger cars					
Less than 2,000 cc	0.4	0.105	0.5	0.170	6.17
More than 2,000 cc	0.4	0.105	0.9	0.170	6.17
Light-duty vehicles	0.6	0.105	1.3	0.160	7.87

Notes:
- Average driving speeds for urban, rural and motorway are 25 km/hour, 75 km/hour and 100 km/hour, respectively.
- Emission factors in g/km are derived from the COPERT model for 1990, utilizing the CORINAIR methodology for road traffic emissions. The pollutants included are: CO, HC, NO_x, TPM. Fuel consumption is also estimated.

Source: Samaras 1992

Table A2.2.3 Estimated Emissions, European Medium- to Heavy-Duty Vehicles
(grams per kilometer)

Vehicle type	Carbon monoxide	Hydrocarbons	Nitrogen oxides	Particulate matter	CH_4	N_2O	NH_3	Fuel consumption (liters/100km)
Urban								
3.5-16.0 tons	18.8	2.79	8.7	0.95	0.085	0.030	0.003	27.03
More than 16.0 tons	18.8	5.78	16.2	1.60	0.175	0.030	0.003	43.48
Rural								
3.5-16.0 tons	7.3	0.76	7.4	0.82	0.010	0.030	0.003	22.22
More than 16.0 tons	7.3	2.58	14.8	1.40	0.080	0.030	0.003	38.46
Motorway								
3.5-16.0 tons	4.2	0.62	6.0	1.67	0.020	0.030	0.003	18.18
More than 16.0 tons	4.2	2.27	13.5	1.25	0.070	0.030	0.003	34.48

Notes:
• Average driving speed for urban: 25 km/h; rural: 75 km/h; and highway: 100 km/h.
• Emission factors in g/km are derived from the COPERT model for 1990, utilizing the CORINAIR methodology for road traffic emissions. The pollutants included are: CO, NO_x, TPM. Fuel consumption is also estimated.
Source: Samaras 1992

Table A2.2.4 Exhaust Emissions, European Heavy-Duty Vehicles
(grams per kilometer)

Engine type and vehicle loading	Carbon monoxide	Hydrocarbons	Nitrogen oxides	Particulate matter
Natural aspiration, 3.5–16.0 tons	3.41	0.61	6.58	0.55
Turbo-charged				
3.5–16.0 tons	2.00	0.57	13.07	0.37
16.0–38.0 tons	4.21	1.06	26.90	0.71
Turbo-charged with inter-cooling, 16.0–38.0 tons	5.37	1.00	16.90	0.61

Source: Sawer 1986

Table A2.2.5 Exhaust Emissions and Fuel Consumption, Utility and Heavy-Duty Trucks, France
(grams per kilometer)

Vehicle type	Avg. Speed km/h	Fuel Consmp. km/l	Carbon monoxide	Hydrocarbons	Nitrogen oxides	Particulate matter
Empty						
Utility truck (3.5t), IDI	76.5	11.0	1.0	0.6	1.6	0.5
	123.7	6.1	1.7	1.6	1.9	2.0
Heavy-duty truck (19t), DI	68.9	4.3	2.6	0.8	15.5	0.5
	88.4	3.9	2.8	0.7	12.0	0.4
Semi-trailer (40t), DI	69.2	4.0	1.9	1.1	6.7	0.9
	88.0	3.7	1.7	1.0	7.4	0.9
Loaded						
Utility truck (3.5t), IDI	74.2	9.3	1.2	0.9	1.6	0.9
	117.7	5.8	1.7	1.8	1.9	2.3
Heavy-duty truck (19t), DI	66.8	3.5	3.1	0.8	16.4	0.5
	84.7	3.4	3.8	0.7	13.4	0.5
Semi-trailer (40t), DI	62.2	2.3	3.2	1.1	10.7	1.4
	75.6	2.4	3.0	1.0	10.1	1.3

Source: Roumegoux 1995

Table A2.2.6 Exhaust Emissions, Santiago Buses, Chile
(grams per kilometer)

Testing procedure	Carbon monoxide	Hydrocarbons	Nitrogen oxides	Particulate matter
Santiago cycle (CADEBUS)	5.70	1.40	5.40	2.50

Source: Escudero 1991

Table A2.2.7 Exhaust Emissions, London Buses, United Kingdom
(grams per kilometer)

Testing procedure	Carbon monoxide	Hydrocarbons	Nitrogen oxides	Particulate matter
London Bus Limited				
In-service simulation				
Laden	7.09	1.19	28.89	1.69
Unladen	6.61	0.92	32.37	1.47
Test cycle				
Laden	5.64	0.67	22.50	1.36
Unladen	5.64	0.62	14.07	0.58

Source: Gore 1991

Table A2.2.8 Exhaust Emissions, Utility and Heavy-Duty Vehicles, Netherlands
(grams per kilometer)

Testing procedure and loading	Carbon monoxide	Hydrocarbons	Nitrogen oxides	Particulate matter
Urban				
Less than 3.5 tons	3.0	1.3	1.3	1.2
3.5–5.5 tons	4.0	2.0	6.0	1.5
5.5–12.0 tons	10.0	7.0	10.0	3.5
12.0–15.0 tons	13.0	9.0	13.0	5.0
More than 15.0 tons	16.0	12.0	20.0	7.0
Rural				
Less than 3.5 tons	1.5	0.7	1.3	0.6
3.5–5.5 tons	2.0	1.0	6.0	1.0
5.5–12.0 tons	4.0	2.5	10.0	2.0
12.0–15.0 tons	4.5	3.0	13.0	2.5
More than 15.0 tons	5.0	3.5	20.0	3.0
Highway				
Less than 3.5 tons	0.9	0.5	1.4	0.5
3.5–5.5 tons	1.0	0.8	7.0	0.9
5.5–12.0 tons	1.5	2.0	13.0	1.8
12.0–15.0 tons	2.0	2.3	15.0	2.0
More than 15.0 tons	2.0	2.5	25.0	2.5

Source: Veldt 1986

Table A2.2.9 Automobile Exhaust Emissions as a Function of Driving Conditions, France
(grams per kilometer)

Traffic type	Carbon monoxide	Hydrocarbons	Nitrogen oxides	Particulate matter	Carbon dioxide
Congested urban	3.29	1.04	2.70	0.68	588
Free-flowing urban	1.05	0.29	0.76	0.29	225
Highway	0.61	0.16	0.57	0.19	179
Motorway	0.61	0.09	0.56	0.25	166

Source: Joumard and others 1990

Table A2.2.10   Automobile Exhaust Emissions and Fuel Consumption as a Function of Testing Procedures, Germany
(grams per kilometer)

Testing procedure	Carbon monoxide	Hydrocarbons	Nitrogen oxides	Particulate matter	Carbon dioxide	Fuel consumption (liters/100km)
European (ECE-15)	1.00	0.17	0.91	0.115	215	38.30
With cold start (ETK)	1.00	0.17	0.19	0.115	215	38.30
Extra urban driving cycle (EUDC)	0.27	0.05	0.55	0.081	128	4.90
ETK + EUDC	0.54	0.10	0.68	0.093	160	6.70
U.S. federal (FTP-75)	0.61	0.10	0.70	0.092	166	6.40
Highway	0.25	0.04	0.48	0.052	115	4.40
Motorway	0.33	0.05	0.83	0.119	179	6.90

Source: Hassel and Weber 1993

Table A2.2.11   Exhaust Emissions, Cars, Buses, and Trucks, Greece
(grams per kilometer)

Vehicle type	Carbon monoxide	Hydrocarbons	Nitrogen oxides
Passenger car	1.34	1.81	0.69
Urban bus	21.16	5.57	10.40
Other buses	4.95	2.15	5.94
Trucks	6.19	2.68	7.43

Source: Pattas, Kyriakis, and Nakos 1993

Table A2.2.12   Exhaust Emissions, Light-Duty Vehicles and Trucks, India
(grams per kilometer)

Vehicle type	Carbon monoxide	Hydrocarbons	Nitrogen oxides	Sulfur dioxide	Particulate matter
Light-duty vehicles	1.1	0.28	0.99	0.39	2.0
Heavy-duty truck	12.70	2.10	21.0	1.50	3.0

Source: Biswas and Dutta 1994; Gargava and Aggarwal 1994

References

Biswas, D. and S.A. Dutta. 1994. "Strategies for Control of Vehicular Pollution in Urban Areas of India," Proceedings of a the Workshop on The Energy Nexus—Indian Issues & Global Impacts (April 22–23, 1994), University of Pennsylvania, Philadelphia.

Escudero, J. 1991. "Notes on Air Pollution Issues in Santiago Metropolitan Region", Letter# 910467, dated May 14, 1991, Comisión Especial de Decontaminación de la Region Metropolitana, Santiago.

Gargava, P., and A. Aggarwal. 1994. "Prediction of Impact of Air Environment on Planning the Control Strategies for Automobile Pollution in an Indian Coastal City." 3rd International Symposium on Transport and Air Pollution, *Poster Proceedings*, INRETS, Arcueil, France.

Gore, B.M. 1991 (draft). "Vehicle Exhaust Emission Evaluations." Office of the Group of Engineers, London Buses Limited, London.

Hassel, D. and F-J. Weber. 1993. "Mean Emissions and Fuel Consumption of Vehicles in Use with Different Emission Reduction Concepts." *The Science of the Total Environment,* 134: 189–95.

Joumard, R., L. Paturel, R. Vidon, J. P. Guitton, A. Saber, and E. Combet. 1990. *Emissions Unitaires de Polluants des Véhicules Légers.* Report 116 (2nd ed.) Institut National du Recherche sur les Transports et leur Securite (INRETS), Bron, France.

Metz, N. 1993. "Emission Characteristics of Different Combustion Engines in City, on Rural Roads, and on Highways." *The Science of the Total Environment* 134: 225–35.

Pattas, K., N. Kyriakis, and C. Nakos. 1993. "Time Dependence of Traffic Emissions in the Urban Area of Thessaloniki." *The Science of the Total Environment* 134: 273-84.

Roumegoux, J.P. 1995. "Calcul des Emissions Unitaires de Polluants des Véhicules Utilitaires," *The Science of the Total Environment,* 169: 273-82.

Samaras, Z. 1992. "COPERT Emission Factors." Informal Communication. Directorate-General for Environment, Nuclear Safety, and Civil Protection, Commission of the European Communities, Brussels.

Sawer, J.M. 1986. "A Review of Diesel Engine Emissions in Europe," Report DP 86/1946, Ricardo Consulting Engineers, Shoreham-by-Sea, England.

Veldt, C. 1986. "Emissions from Road Transport", Discussion Paper for the OECD Workshop on Comparison of Emission Inventory Data, MT-TNO, Apeldoorn, Netherlands (October 22–24).

3

Vehicle Technology for Controlling Emissions

The principal pollutant emissions from vehicles equipped with spark-ignition gasoline engines include unburned hydrocarbons, carbon monoxide, and nitrogen oxides in the exhaust. Emissions of respirable particulate matter (PM) can also be considerable, particularly from two-stroke engines. Lead aerosol emissions from combustion of leaded gasoline are also significant and have important impacts on public health. Evaporation of gasoline in the fuel system, the escape of gasoline vapors during refueling, and the escape of blowby losses from the crankcase contribute additional hydrocarbon emissions.

In new automobiles, carbon monoxide, hydrocarbon, and nitrogen oxide emissions can be reduced by 50 percent or more from uncontrolled levels through engine modifications, at a cost of about U.S.$130 per car. Fuel consumption may increase slightly. Hydrocarbon and carbon monoxide reductions of 90 to 95 percent and nitrogen oxide reductions of 80 to 90 percent are possible with three-way catalysts and electronic engine control systems that cost about U.S.$600 to U.S.$800 per car. Such devices have little impact on fuel economy. Lean-burn techniques combined with an oxidation catalytic converter can achieve comparable hydrocarbon and carbon monoxide reductions, a 60 to 75 percent reduction in nitrogen oxides, and a 10 to 15 percent improvement in fuel economy.

Two-stroke gasoline engines—used in motorcycles and three-wheelers, predominantly in Asia and Europe, and formerly in some automobiles in Eastern Europe— are a special case. Hydrocarbon emissions from two-stroke engines are high because a significant part of the air-fuel mixture escapes unburned into the exhaust. Particulate emissions from two-strokes are also excessive because oil is mixed with the fuel, and recondenses into oil particles in the exhaust. Hydrocarbon emissions from a single two-stroke motorcycle can exceed those from three uncontrolled passenger cars and particulate matter emissions can exceed those from a heavy-duty diesel

truck. These emissions can be controlled by substituting a four-stroke engine or an advanced two-stroke design that uses fuel injection, at a cost of about U.S.$60 to U.S.$80 per vehicle. This change also reduces fuel consumption by 30 to 40 percent. Further control of motorcycle and three-wheeler emissions can be achieved with catalytic converters.

The most significant emissions from diesel-fueled vehicles are particulate matter, nitrogen oxides, and hydrocarbons. Particulate matter emissions from uncontrolled diesel engines are six to ten times those from gasoline engines. Diesel smoke is also a visible public nuisance. Emissions of other pollutants from diesel engines are generally lower than those for comparable gasoline engines. Compared with similar vehicles with uncontrolled gasoline engines, light-duty diesel vehicles without emission controls emit about 90 percent less hydrocarbons and carbon monoxide and about 50 to 70 percent less nitrogen oxides. Heavy-duty diesel vehicles emit 50 to 100 percent more nitrogen oxides than their gasoline counterparts, but 90 to 95 percent less hydrocarbons and 98 percent less carbon monoxide. Both light- and heavy-duty diesel- fueled vehicles are considerably more fuel efficient than their gasoline counterparts (15 to 40 percent for light-duty diesels, as much as 100 percent for heavy-duty ones) and therefore emit less carbon dioxide.

Diesel engine emissions of nitrogen oxides and hydrocarbons can be reduced by more than 50 percent and emissions of particulate matter by more than 75 percent from uncontrolled levels through engine design changes, improved fuel injection systems, turbocharging, and charge air cooling. These changes improve fuel economy (diesel fuel-efficiency has improved by 30 percent since the 1970s) but increase engine costs. Particulate matter and hydrocarbon emissions can be further reduced through the use of low-sulfur fuel and an oxidation catalytic converter. The lowest particulate matter emissions (a 95 percent reduction from uncontrolled levels) are possible with the use of trap-oxidizers, but the reliability of these systems has not been proven conclusively.

Automotive Engine Types

Pollutant emissions from motor vehicles are determined by the vehicle's engine type and the fuel it uses. Spark-ignition and diesel engines are the two most common engines. Engine technology, emission characteristics, and emission control technologies for these two basic engine types are discussed in detail in appendices 3.1 and 3.2. Measures to improve fuel economy and directly reduce emissions of carbon dioxide are presented in appendix 3.3. Other engine technologies and advanced vehicle propulsion systems have been reviewed by Watkins (1991), Brogan and Venkateswaran (1993), Kimbom (1993), Mason (1993), MacKenzie (1994), and OTA (1995).

Spark-Ignition (Otto) Engines

Most passenger cars and light-duty trucks use spark-ignition (Otto cycle) gasoline engines. These engines are also used in heavy-duty trucks and buses in some countries, including China, Mexico, the Russian Federation and other republics of the former Soviet Union, and the United States. Other fuels used in spark-ignition engines include natural gas, liquified petroleum gas, alcohols, and hydrogen (see chapter 5).

Spark-ignition gasoline engines have either a two-stroke or four-stroke design. Two-stroke engines are cheaper, lighter, and can produce greater power output per unit of displacement, so they are widely used in small motorcycles, outboard motors, and small power equipment. Two-stroke engines emit 20 to 50 percent of their fuel unburned in the exhaust, resulting in high emissions and poor fuel economy. Because the crankcase pumps the air-fuel mixture through the engine, two-stroke engines require that oil be mixed with the air-fuel mixture to lubricate bearings and pistons. Some of this oil appears as white smoke in the exhaust, resulting in high emissions of particulate matter.

All gasoline engines currently used in automobiles and larger vehicles use the four-stroke design, although advanced two-stroke engines are being developed. These advances pertain to fuel injection, combustion, and the lubrication system. Advanced two-stroke engines under development would achieve lower emissions and fuel consumption than four-stroke engines and retain the two-stroke's advantages of lower weight and cost per unit of power output.

The main pollutant emissions from spark-ignition gasoline engines are hydrocarbons, carbon monoxide, and nitrogen oxides. Carbon monoxide and nitrogen oxides are only emitted in the vehicle exhaust, while hydrocarbon emissions occur in the vehicle exhaust, the engine crankcase, the fuel system, and from atmospheric venting of vapors during fuel distribution and dispensing.

Particulate matter emissions from gasoline engines are caused by the condensation of oil vapor in the exhaust. These particulate matter emissions are usually small for four-stroke engines. Two-stroke engines and four-strokes with excess oil consumption can exhibit high particulate matter emissions.

The use of lead as an antiknock additive in gasoline is being discontinued in many countries for environmental reasons. Where these compounds are still in use, lead aerosol emissions from gasoline engines are the major source of airborne lead in the environment. A review of lead additives in gasoline is presented in an appendix to chapter 5.

Diesel Engines

Most heavy-duty trucks and buses have diesel engines, as do some light-duty vehicles and passenger cars. Diesel engines in light-duty vehicles are common in Europe, (about 20 percent of the light-duty fleet) and in parts of southeast Asia. Diesel engines, unlike spark-ignition engines, do not premix fuel with air before it enters the cylinder. Instead, the fuel is injected at high pressure near the top of the compression stroke. Once injected, the fuel is heated to ignition by the compressed air in the cylinder, eliminating the need for a separate spark-ignition system.

There are two types of diesel engine: indirect and direct injection. In an *indirect injection* (IDI) diesel engine the fuel is injected into a pre-chamber where ignition occurs and combustion then spreads to the main combustion chamber. Indirect injection technology is mainly used for small, high-speed applications such as passenger cars, where low noise and high performance are important.

In a *direct injection* (DI) engine the fuel is sprayed directly into and ignited in the combustion chamber. These engines are generally used in medium and large trucks and give higher power output and better fuel economy but they are considerably noisier. Developments in reducing noise and improving performance have led to the use of these engines in passenger cars, although there is considerably less experience with these engines in small applications (Holman 1990).

Compared with gasoline spark-ignition engines, heavy-duty diesel engines have lower carbon monoxide and hydrocarbon emissions but higher nitrogen oxide emissions. They are up to 100 percent more fuel-efficient, resulting in lower emissions of carbon dioxide. Light-duty diesels exhibit better fuel efficiency and lower carbon monoxide, hydrocarbon, and nitrogen oxide emissions than their gasoline counterparts.

Particulate matter emissions from diesel engines are considerably higher than from gasoline engines. Diesel emissions—in the form of black smoke—are a major source of high ambient concentrations of particulate matter in most large cities of the developing world. Other pollutant emissions from diesel vehicles include sulfur dioxide and noise. Particulate matter and noise emissions result from the combustion process. These emissions can be reduced by modifying the engine and combustion system (appendix 3.1). Diesel engines meeting current U.S. and future European emission standards are smokeless when properly maintained, have better fuel efficiency, are less noisy, and emit less nitrogen oxides and hydrocarbons than the uncontrolled diesel engines sold in developing countries.

Rotary (Wankel) Engines

A rotary engine utilizes a triangular rotor which turns within an elliptical combustion chamber. The motion of the triangular rotor varies the volume of the space between the rotor and the chamber wall and performs the compression and expansion functions of a piston in a conventional engine. Rotary engines are smaller, lighter and simpler than reciprocating piston engines. Carbon monoxide and hydrocarbon emissions are significantly higher from rotary engines as compared to conventional engines; emissions of nitrogen oxides are about the same. Production models of passenger cars and motorcycles have been built with rotary engines.

Gas-Turbine (Brayton) Engines

Gas-turbine engines are used in aircraft, stationary applications, high-speed trains, and marine vessels. These engines have high output in relation to engine size and low emissions because of a low-pressure combustion process. Gas-turbine engines have been tested in road vehicles since the 1960s, but no commercially viable vehicle system has been developed. Drawbacks of gas turbines for road vehicles include high costs, poor transient response, and inefficiency, particularly at light loads. The problem of poor efficiency at light loads is especially severe in passenger vehicles, which commonly use less than 10 percent of the maximum power output in highway cruise conditions.

Steam (Rankine) Engines

Steam engines were used in early automobiles. These engines lost favor to spark-ignition engines because of the efficiency of gasoline engines, the time required to raise steam pressure, the need to refill with both high-purity water and fuel, and safety concerns over the high-pressure boiler. Like other engines using external combustion, steam engines exhibit low pollutant emissions compared with uncontrolled internal combustion engines. Some research has been done on closed-circuit, low-temperature Rankine engines as a bottoming cycle for internal combustion engines in heavy-duty trucks. These engines would use the wasted heat in the truck's exhaust to produce additional power, thereby increasing overall efficiency.

Stirling Engines

Stirling engines have been of interest for many years. They are theoretically capable of achieving high fuel efficiency, and have demonstrated low emission levels. Currently available Stirling engines are not practical for automotive use, however, because of their high cost, poor transient response, and poor power-to-weight and power-to-volume ratios.

Electric and Hybrid Vehicles

Electric vehicles have been pursued because of their mechanical simplicity and the absence of direct pollutant emissions, although emissions from the power source should be taken into account. The potential to recover kinetic energy during braking can contribute to increased fuel efficiency. Electric vehicles are used in specialized applications, but current battery technology is inadequate for electric vehicles to compete with internal combustion vehicles in most applications. Although improved batteries are being researched, a breakthrough that would make electric vehicles competitive appears unlikely in the near future (OTA 1995).

Hybrid vehicle designs, in which an internal combustion Brayton or Stirling engine would supplement the batteries, are being developed. In this design, the engine supplies average power while batteries supply surge power for acceleration and absorb power during braking. Running under steady state conditions at its most efficient point, the engine in a hybrid vehicle could have very low emissions, and such vehicles could more than double the fuel efficiency of present vehicle designs. Electric and hybrid vehicles are discussed in appendix 5.2.

Control Technology for Gasoline-Fueled Vehicles (Spark-Ignition Engines)

Emissions from spark-ignition engines can be reduced through changes in engine design, combustion conditions, and catalytic aftertreatment. Some of the engine and combustion variables that affect emissions are the air-fuel ratio, ignition timing, turbulence in the combustion chamber, and exhaust gas recirculation. Of these, the most important is the air-fuel ratio. These topics are discussed briefly below, and in detail in appendix 3.1.

Engine-out pollutant emissions can be reduced substantially from uncontrolled levels through appropriate

Figure 3.1 Effect of Air-Fuel Ratio on Spark-Ignition Engine Emissions

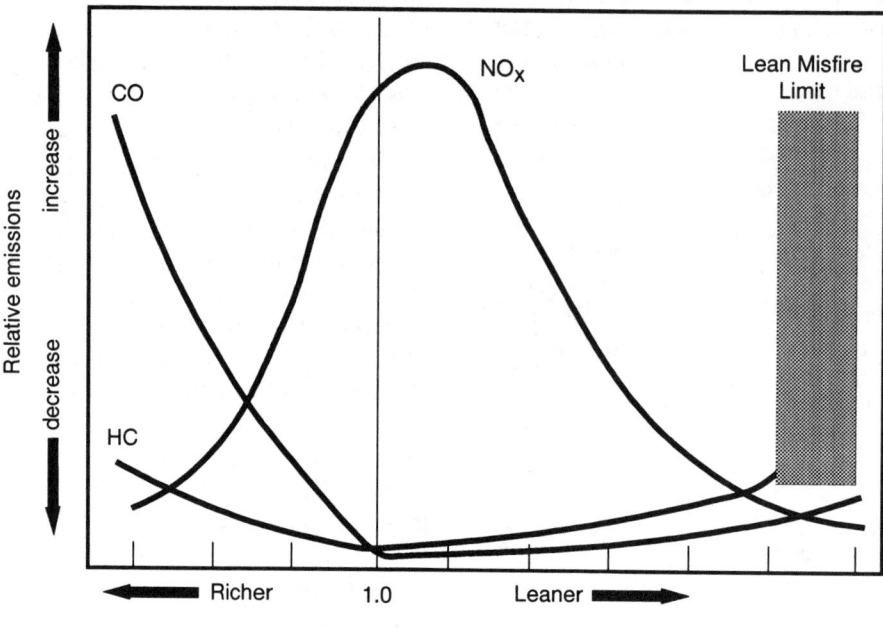

Source: Weaver and Chan 1995

engine design and control strategies. This involves tradeoffs among engine complexity, fuel economy, power, and emissions. The use of catalytic aftertreatment allows a further order-of-magnitude reduction in pollutant emissions and, by reducing the need for engine-out control, an improvement in power and fuel economy at a given emissions level.

Air-Fuel Ratio

The air-fuel ratio has an important effect on engine power, efficiency, and emissions. The ratio of air to fuel in the combustible mixture is a key design parameter for spark-ignition engines. An air-fuel mixture that has exactly enough air to burn the fuel, with neither air nor fuel left over, is *stoichiometric*, and has a normalized air-fuel ratio (λ) of 1.0[1]. Mixtures with more air than fuel are lean, with λs higher than 1.0; those with more fuel are rich, with λs less than 1.0. A mixture with a λ of 1.5 has 50 percent more air than needed to burn all the fuel. Engines using lean mixtures are more efficient than those using stoichiometric mixtures. There are a number of reasons for this, including less heat loss, higher compression ratios (lean mixtures knock less readily), lower throttling losses at part load, and favorable thermodynamic properties in burned gases. Engines designed to burn very lean mixtures—λ more than 1.2 (numeric air-fuel

ratio of 17.60:1)—are *lean burn*. The generalized variation of emissions with air–fuel ratio for a spark-ignition engine is shown in figure 3.1.

Electronic Control Systems

Electronic control technology for stoichiometric engines using three-way catalysts has been extensively developed. Nearly all engine emission control systems used in the United States since 1981 incorporate computer control of the air-fuel ratio. Similar systems have been used in Japan since 1978 and in Europe since the late 1980s. These systems measure the air-fuel ratio in the exhaust and adjust the air-fuel mixture going into the engine to maintain stoichiometry. In addition to the air-fuel ratio, computer systems control features that were controlled by vacuum switches or other devices in earlier emission control systems. These include spark timing, exhaust gas recirculation, idle speed, air injection systems, and evaporative canister purging.

The stringent air-fuel ratio requirements of three-way catalysts made advanced control systems necessary. But the precision and flexibility of the electronic control system can reduce emissions even in the absence of a catalytic converter. Many control systems can self-diagnose engine and control system problems. Such diagnostics are mandatory in the United States. The ability to warn the driver of a malfunction and assist the mechanic in its diagnosis can improve maintenance quality. Self-diagnostic capabilities are becoming increasingly sophisticated and important as engine control systems become more

---

1.　The numerical value of the stoichiometric air-fuel ratio for gasoline is 14.7:1, corresponding to a λ of 1.00.

complex. Computer-controlled engine systems are also more resistant to tampering and maladjustment than mechanical controls. The tendency for emissions to increase over time is thus reduced in computer-controlled vehicles.

Catalytic Converters

The catalytic converter is one of the most effective emission control devices available. The catalytic converter processes exhaust to remove pollutants, achieving considerably lower emissions than is possible with in-cylinder techniques. Vehicles with catalytic converters require unleaded fuel, since lead forms deposits that "poison" the catalytic converter by blocking the access of exhaust gases to the catalyst. A single tank of leaded gasoline can significantly degrade catalyst efficiency. Sulfur and phosphorous in fuel can also poison the catalytic converter. Converters can also be damaged by excessive temperature, which can arise from excess oxygen and unburned fuel in the exhaust.

The catalytic converter comprises a ceramic support, a washcoat (usually aluminum oxide) to provide a very large surface area and a surface layer of precious metals (platinum, rhodium, and palladium are most commonly used) to perform the catalyst function. Catalysts containing palladium are more sensitive to the sulfur content of gasoline than platinum/rhodium catalysts (ACEA/EUROPIA 1995).

Two types of catalytic converters are commonly used in automotive engines: oxidation (*two-way*) catalysts control hydrocarbon and carbon monoxide emissions and oxidation–reduction (*three-way*) catalysts control hydrocarbons, carbon monoxide, and nitrogen oxides (figure 3.2). A new type of catalytic converter is the *lean nitrogen-oxide* catalyst, which reduces nitrogen oxide emissions in lean conditions where a three-way catalyst is ineffective.

Two-way catalysts. Oxidation catalysts use platinum, palladium, or both to increase the rate of reaction between oxygen, unburned hydrocarbons, and carbon monoxide in the exhaust. This reaction would normally proceed slowly. Catalyst effectiveness depends on its temperature, the air-fuel ratio of the mixture, and the mix of hydrocarbons present. Highly reactive hydrocarbons such as formaldehyde and olefins are oxidized more effectively than less-reactive ones. Short-chain paraffins like methane, ethane, and propane are among the least reactive hydrocarbons and are difficult to oxidize.

Three-way catalysts. Three-way catalysts generally use a combination of platinum, palladium, and rhodium. In addition to promoting the oxidation of hydrocarbons and carbon monoxide, palladium and rhodium promote the reduction of nitric oxide (NO) to nitrogen and oxygen. For efficient NO reduction, a rich or stoichiometric air-fuel ratio is required. At optimal conditions a three-way catalyst can oxidize hydrocarbons and carbon monoxide and reduce nitrogen oxides. The window of air-fuel ratios in which this occurs is narrow, and there is a tradeoff between nitrogen oxide and hydrocarbon/carbon monoxide control even within this window. The variation of three-way catalyst efficiency with the normalized air–fuel ratio (λ) is shown in figure 3.3. To maintain the precise air-fuel ratio required, gasoline cars use exhaust (λ) sensors (also known as oxygen sensors) with electronic control systems for feedback control of the air-fuel ratio.

Lean nitrogen-oxide catalysts. Conventional three-way catalysts are ineffective at reducing nitrogen oxides under lean conditions. This has restricted the use of advanced lean-burn engines in passenger vehicles. Because of their superior fuel efficiency and low carbon monoxide emissions, lean-burn engines are otherwise an attractive technology. Researchers have developed zeolite catalytic materials that reduce nitrogen oxide emissions, using unburned hydrocarbons in the exhaust as the reductant. Although the lean nitrogen-oxide catalyst is typically about 50 percent effective—considerably less than a three-way catalyst under stoichiometric conditions—the benefit is still significant. A few automobile models using lean nitrogen-oxide catalysts have been introduced in Japan.

Crankcase Emissions and Control

The blowby of compressed gases past the piston rings consists mostly of unburned or partly-burned hydrocarbons. In uncontrolled vehicles, the blowby gases were vented to the atmosphere. Crankcase emission controls involve closing the crankcase vent port and venting the crankcase to the air intake system via a check valve. Control of these emissions is no longer considered a significant technical issue.

Evaporative Emissions and Control

Gasoline is a relatively volatile fuel. Even at normal temperatures, significant gasoline evaporation occurs if gasoline is stored in a vented tank. The most common measure of gasoline volatility is the Reid vapor pressure (RVP), which is the vapor pressure measured under standard conditions at an air-to-liquid ratio of 4:1 and a temperature of 37.8°C. Gasoline volatility is normally adjusted to compensate for variations in ambient temperature. When temperatures are below freezing 0°C, gasoline is usually adjusted to an RVP of about 90 kPa (13 psi) to increase fuel vaporization. This level of volatility would cause vapor lock in vehicles at tempera-

Figure 3.2 Types of Catalytic Converters

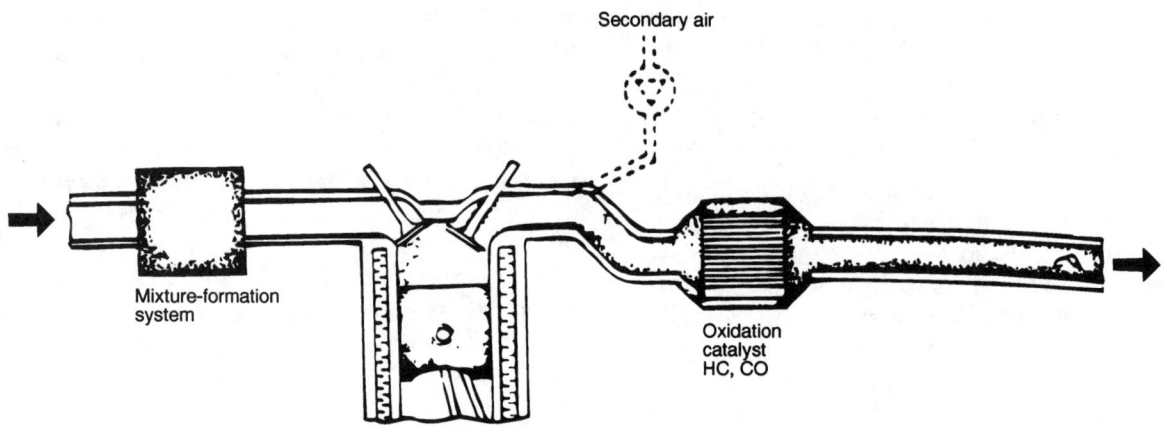

SINGLE BED OXIDATION CATALYTIC CONVERTER

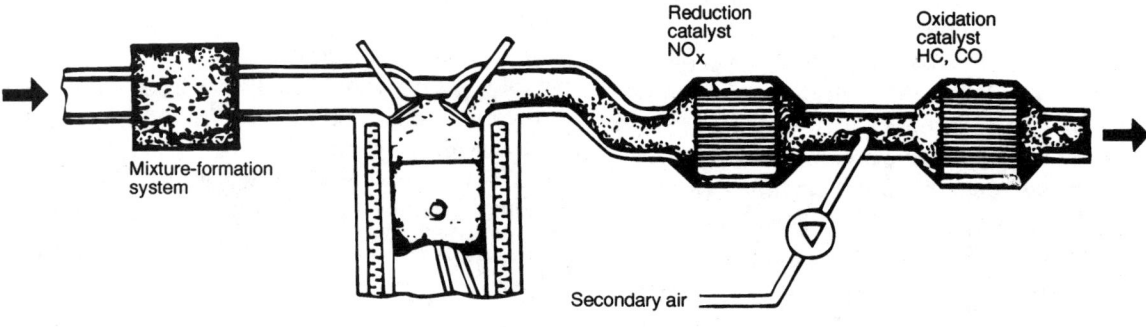

DUAL BED OXIDATION CATALYTIC CONVERTER

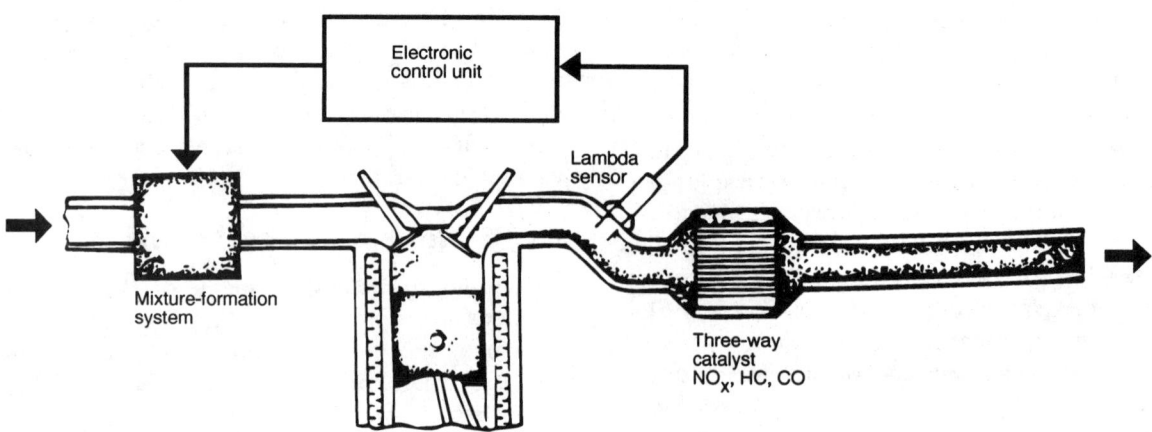

SINGLE BED THREE-WAY CATALYTIC CONVERTER

Source: Wijetilleke and Karunaratne 1992

Figure 3.3 Effect of Air-Fuel Ratio on Three-Way Catalyst Efficiency

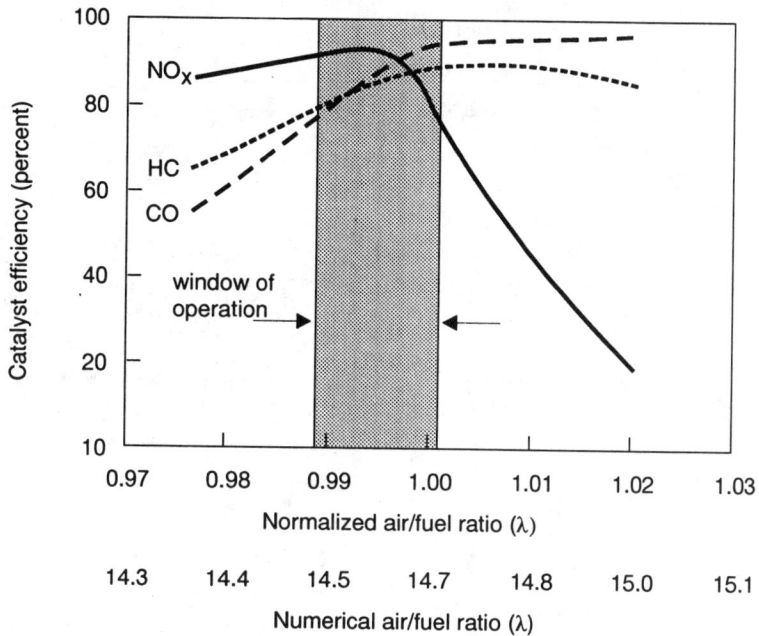

Source: Weaver and Chan 1995

tures exceeding 30°C. At these temperatures, gasoline RVP is ideally kept below 70 kPa (10 psi). Gasoline with an RVP of 75 kPa (11 psi) will produce about twice the evaporative emissions of gasoline with an RVP of 60 kPa (8.7 psi).

The four primary sources of evaporative emissions from vehicles are diurnal (daily) emissions, hot-soak emissions, resting losses, and running losses.

Diurnal and hot-soak emissions have been controlled for some time in the United States, and such controls were included in the Consolidated Emissions Directive adopted by the European Community in 1991. Evaporative emissions are controlled by venting the fuel tank (and, in carbureted vehicles, the carburetor bowl) to the atmosphere through a canister of activated charcoal. Hydrocarbon vapors are adsorbed by the charcoal, so little vapor escapes to the air. The charcoal canister is regenerated or "purged" by drawing air through it into the intake manifold when the engine is running. Adsorbed hydrocarbons are stripped from the charcoal and burned in the engine.

Fuel Dispensing/Distribution Emissions and Control

As with evaporative emissions, emissions from fuel distribution are significant only for vehicles using volatile fuels, such as gasoline. These emissions result from fuel vapor contained in the headspace of the vehicle fuel tank. This vapor is displaced as fuel is added during refueling. Vapor emissions also occur when the service station tank is refilled from a tank truck, and when the

tank truck is refilled at the bulk terminal. Sources and magnitude of hydrocarbon vapor emissions from gasoline distribution and dispensing are shown in figure 3.4.

Technology to reduce gasoline distribution emissions involves two types of controls. One method controls vapors displaced from the receiving tank by venting them to the delivery truck tank. This is known as *Stage I* and is about 95 percent effective, reducing vapor emissions from 1.14 grams per liter dispensed to 0.06 grams per liter. International experience shows that the cost of retrofitting fuel storage tanks, delivery trucks, and service stations with vapor recovery devices is small and the payback period based on the benefit of fuel savings alone, is two to three years. Two alternatives are available to control fuel vapors displaced from the vehicle tank during refueling (*Stage II* control). One alternative modifies the gasoline dispensing system to capture vapors. The other alternative captures vapor on board the vehicle in a charcoal canister similar to that used for controlling evaporative emissions.

Control Technology for Diesel-Fueled Vehicles (Compression–Ignition Engines)

The principal pollutants emitted by diesel engines are nitrogen oxides, sulfur dioxide, particulate matter, and hydrocarbons. Diesels also produce carbon monoxide, smoke, odors, and noise. Fuel quality affects diesel emissions, the main factors being fuel density, sulfur content, aromatic content, and certain distillation char-

Figure 3.4 Hydrocarbon Vapor Emissions from Gasoline Distribution

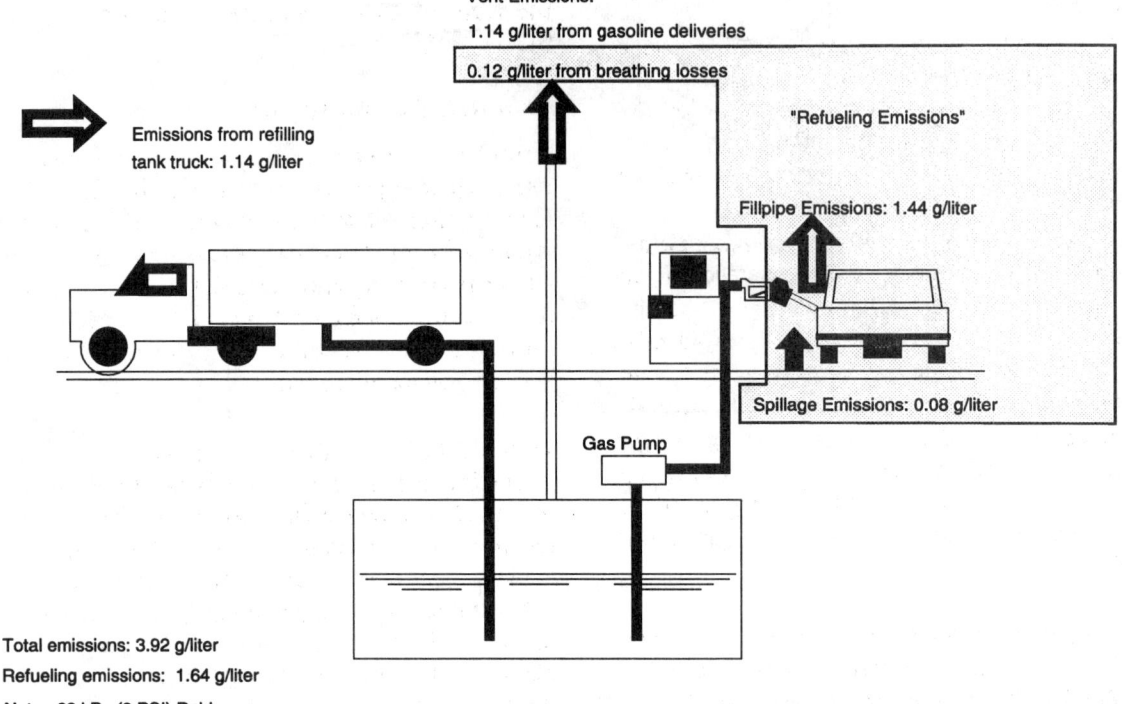

Vent Emissions:

1.14 g/liter from gasoline deliveries

0.12 g/liter from breathing losses

Emissions from refilling
tank truck: 1.14 g/liter

"Refueling Emissions"

Fillpipe Emissions: 1.44 g/liter

Spillage Emissions: 0.08 g/liter

Gas Pump

Total emissions: 3.92 g/liter

Refueling emissions: 1.64 g/liter

Note: 62 kPa (9 PSI) Reid vapor pressure

Source: Weaver and Chan 1995

acteristics. Engine variables with the greatest effect on diesel emission rates are the combustion chamber design, air-fuel ratio, rate of air-fuel mixing, fuel injection timing, compression ratio, and the temperature and composition of the charge in the cylinder. These factors are discussed in detail in appendix 3.2.

Engine Design

There is a tradeoff between nitrogen oxide and particulate control measures in diesel vehicles. This tradeoff is shown in figure 3.5 for three different levels of diesel technology. The tradeoff is not absolute—both nitrogen oxides and particulate matter emissions can be reduced simultaneously. There are limits on the extent to which either can be reduced, however, without increasing the other. To minimize all pollutants simultaneously requires optimization of fuel injection, fuel-air mixing, and combustion processes over the range of operating conditions.

Reduced nitrogen oxide and particulate emissions have resulted from an improved understanding of the diesel combustion process and the factors affecting pollutant formation and destruction in the cylinder. Modifying the diesel combustion process is complex; it has a direct impact on cost, fuel economy, power and torque output, cold starting, and visible smoke, and it

involves complex tradeoffs among nitrogen oxide, hydrocarbon, and particulate matter (PM) emissions.

Most engine manufacturers have followed a broadly similar approach to reducing diesel emissions, although the specific techniques used differ considerably from one manufacturer to the next. This typical approach includes the following major elements:

*   Reducing parasitic hydrocarbon and PM emissions (those not directly related to the combustion process) by minimizing injection nozzle sac volume and oil consumption
*   Reducing PM emissions and improving fuel efficiency and power output through turbocharging and by refining the match between the turbocharger and the engine
*   Reducing emissions of PM and nitrogen oxides by cooling the compressed-charge air with aftercoolers
*   Further reducing nitrogen oxides to meet regulatory targets by retarding fuel injection timing over most of the speed–load range. A flexible timing system minimizes the adverse effects of retarded timing on smoke, starting, and light-load hydrocarbon emissions
*   Further reducing nitrogen oxides in light-duty vehicles by recirculating exhaust gas under light-load conditions

- Reducing the PM increase resulting from retarded timing by increasing the fuel injection pressure and injection rate
- Improving air utilization (and reducing hydrocarbon and PM emissions) by minimizing parasitic volumes in the combustion chamber–such as the clearance between piston and cylinder head and the clearance between the piston and the walls of the cylinder
- Optimizing in-cylinder air motion through changes in combustion chamber geometry and intake air swirl to provide adequate mixing at low speeds (to minimize smoke and PM) without over-rapid mixing at high speeds (which would increase hydrocarbons, nitrogen oxides, and fuel consumption); and
- Controlling smoke and PM emissions in full-power operation and transient accelerations by improving the governor curve shape and limiting transient smoke (frequently through electronic governor controls).

These changes have reduced PM emissions from diesel engines by more than 80 percent and emissions of nitrogen oxides by 50 to 70 percent compared with uncontrolled levels. Fuel efficiency has increased markedly compared with older engines. These emission reductions and fuel economy improvements have required complete redesign of large parts of the engine and combustion system, and costs to manufacturers and engine purchasers have been sizable. But the benefits in the form of cleaner and more efficient engines have been significant.

Exhaust Aftertreatment

Another approach to reducing pollutant emissions is to use a separate process to eliminate pollutants from the exhaust after it leaves the engine, but before it is emitted into the air. Aftertreatment systems include particulate trap-oxidizers and diesel catalytic converters, both of which have been used in vehicles. Work is under way on lean nitrogen oxide catalysts for diesel engines, but success has been limited.

Trap-oxidizers. A trap-oxidizer system has a particulate filter (the *trap*) in the engine exhaust stream and some means of burning (*oxidizing*) collected particulate matter from the filter. Manufacturing a filter capable of collecting soot and other particulate matter from the exhaust stream is straightforward, and effective trapping media have been developed and demonstrated. The main problem of trap-oxidizer system development is how to remove the soot effectively and regenerate the filter. Diesel particulate matter consists of solid carbon coated with heavy hydrocarbons. This mixture ignites at 500 to 600°C, well above the normal range of diesel engine exhaust temperatures (150–400°C). Special means

Figure 3.5 Nitrogen Oxide and Particulate Matter Emissions from Diesel-Fueled Engines

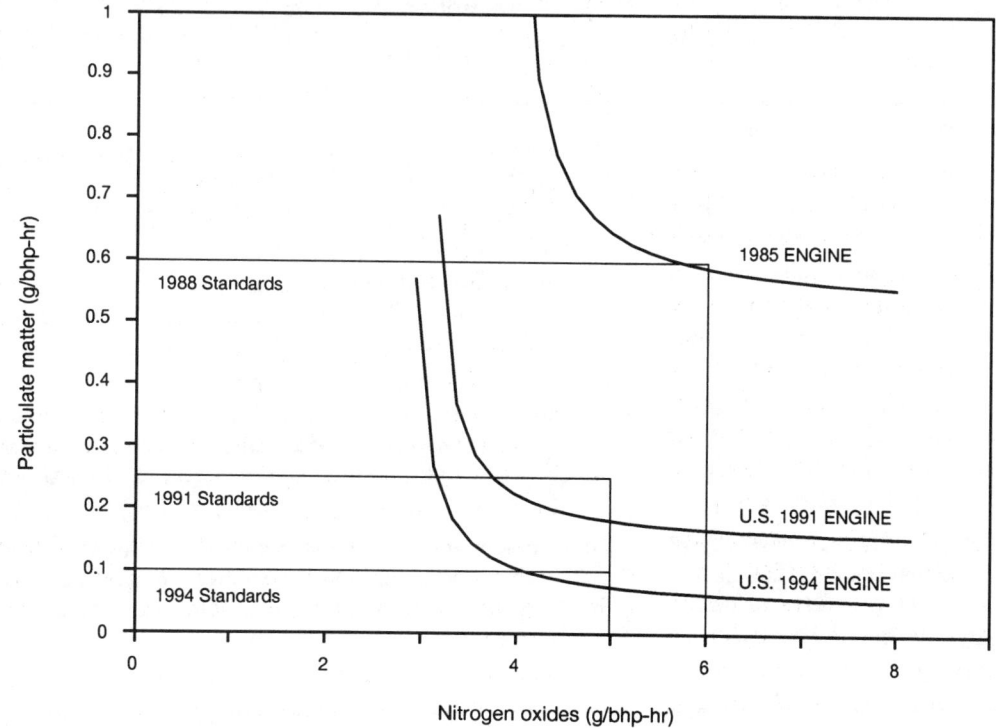

Source: Weaver and Chan 1995

are therefore needed to ensure ignition. Once ignited, however, this material burns at temperatures that can melt or crack the particulate filter. Initiating and controlling regeneration without damaging the trap is the central problem of trap-oxidizer development.

A number of trapping media have been tested or proposed, including cellular ceramic monoliths, woven ceramic-fiber coils, ceramic foams, corrugated multi-fiber felts, and catalyst-coated, stainless-steel wire mesh. The most successful trap-oxidizer systems use either the ceramic monolith or the ceramic-fiber coil traps (Feutlinske 1989; Holman 1990; Knecht 1991).

Many techniques for regenerating particulate trap-oxidizers have been proposed, and much development effort has been invested. Regeneration techniques can be divided into passive and active approaches. *Passive* systems attain the conditions required for regeneration as a result of normal vehicle operation. This requires a catalyst (as either a coating on the trap or a fuel additive) to reduce the ignition temperature of the collected particulate matter. Regeneration temperatures of 420 °C have been reported with catalytic coatings, and lower temperatures can be achieved with fuel additives. *Active* systems monitor particulate matter in the trap and trigger specific actions to regenerate it when needed. A variety of approaches to trigger regeneration have been proposed, including diesel-fuel burners, electric heaters, and catalyst injection systems.

Passive regeneration is difficult on heavy-duty vehicles. Regeneration temperatures must be attained in normal operation, even under light-load conditions. Currently, no purely passive regeneration system is under consideration for heavy-duty applications. Some

manufacturers are working on quasi-passive systems, in which the system usually regenerates passively without intervention, but the active system remains as a backup.

No catalytic coating has sufficiently reduced trap regeneration temperature to permit reliable passive regeneration in heavy-duty diesel service. But catalyst coatings have a number of advantages in active systems. The reduced ignition temperature and increased combustion rate resulting from the catalyst imply that less energy is needed from the regeneration system. Regeneration will also occur spontaneously under most duty cycles, greatly reducing the number of times the regeneration system must operate. Spontaneous regeneration also provides insurance against regeneration system failure. Finally, a trap catalyst may simplify a regeneration system.

To date, trap-oxidizer systems have been used in only a few engine and vehicle models. Traps were installed on one diesel passenger car model sold in California in the early 1990s, but the systems were not durable and were withdrawn after two years. Trap-oxidizer systems were standard devices on new, heavy-duty, U.S. bus engines certified to meet 1993 PM standards, and they have been retrofitted to buses in a number of U.S. urban areas, including New York City and Philadelphia. They have also been deployed in a number of demonstration projects in Europe, including transit buses in Germany and Athens, Greece (Pattas and others 1990). The Athens program was so successful that proposals were made to fit trap-oxidizers to all buses in the city (box 3.1).

Addition of a trap-oxidizer would add substantially to the initial cost of a diesel engine, and would increase fuel consumption and maintenance costs. Engine manufacturers anticipate strong market resistance to this

---

Box 3.1 Trap-Oxidizer Development in Greece

Urban buses are responsible for more than half the traffic-produced smoke in downtown areas of major Greek cities. Since the service life of these vehicles often approaches 15 years, the possibility of retrofitting urban buses with trap-oxidizer systems was considered by the Athens Bus Corporation, utilizing the following features:

- Wall-flow ceramic monoliths for filtration, with more than 90 percent filtration efficiency,
- A regeneration system using cerium-based fuel additives and exhaust gas throttling, and
- Bypass control of the regeneration system for protection against filter melting or cracking.

A pilot phase was initiated in 1989 to determine the service life of the filters and the feasibility of the trap-oxidizer system. A Greek manufacturer produced the systems, and 110 buses were retrofitted and put in normal service. Two years of pilot operation indicated that the normal service life of the filter exceeded 100,000 kilometers—more than a year of bus operation. The trap-oxidizer system represented 3 to 5 percent of the market price of a new bus. And the operational cost (fuel penalty plus fuel additive) was 2 percent of the cost of fuel.

On the basis of these findings, the Greek Ministry for the Environment recommended retrofitting the entire Athens urban bus fleet with traps, sponsoring similar actions in other major Greek cities, and adopting U.S. regulations for particulate emissions for all new urban buses sold in Athens. Unfortunately, the Greek Ministry of Transport was unable to raise the 1 billion drachmas needed to fit the remaining 1,700 city buses.

Source: Pattas and others 1990; Hope 1992

technology. Their success in reducing engine-out PM emissions from new diesel engines has greatly diminished the interest in trap-oxidizers.

In-cylinder diesel PM control has greatly reduced PM emission levels. Progress has been most effective in reducing the soot portion of PM emissions, so the soluble organic portion of particulate matter now accounts for a much larger share. Depending on engine and operating conditions, the soluble organic portion may account for 30 to 70 percent of PM emissions.

Oxidation catalysts. Like a catalytic trap, a diesel catalytic converter oxidizes a large portion of the hydrocarbons present in the soluble organic portion of PM emissions, as well as gaseous hydrocarbons, carbon monoxide, odor (from organic compounds such as aldehydes), and mutagenic emissions. Unlike a catalytic trap, the oxidizing catalytic converter does not collect solid particulate matter, which passes through in the exhaust. This eliminates the need for a regeneration system.

Oxidation catalytic converters have been used in light-duty vehicles and demonstrated to be effective for heavy-duty applications. They have little effect on nitrogen oxide emissions, but can reduce volatile organic compound and carbon monoxide emissions by up to 80 percent. The durability of oxidation catalytic converters on heavy-duty engines has yet to be determined, but it is likely to be acceptable. These catalysts have a negligible effect on fuel consumption.

The main difficulty with using oxidation catalytic converters on heavy-duty diesel engines is that they can cause the formation of sulfuric acid and sulfates from sulfur dioxide in the exhaust. If fuel sulfur levels are significant, these compounds can add considerably to particulate mass. Fuel with less than 0.05 percent sulfur by weight is required for diesel catalysts to perform well. However, the linkage between fuel sulfur content and the higher conversion rates of SO_2 (gaseous) to SO_3 (particulate matter) in vehicles equipped with oxidation catalysts may be due to the high operating temperature of the catalyst (above 350°C) obtained during static dynometer tests. Actual on-the-road operating conditions tend to result in lower catalyst temperature and hence a lower rate of sulphate particle emissions (CONCAWE Review 1994).

Lean nitrogen-oxide catalysts. Since diesel engines operate with lean air-fuel ratios, three-way catalytic converters do not reduce the emissions of nitrogen oxides. Research is underway on zeolite-based, lean nitrogen-oxide catalysts that reduce nitrogen oxide emissions using unburned hydrocarbons in the exhaust. A 20 percent reduction in nitrogen oxide emissions has been

achieved, but results have been discouraging, partly because water in the exhaust inhibits the catalyst and partly because the sulfur in diesel poisons the catalyst. Current catalyst formulations also require that the exhaust be hydrocarbon-enriched to achieve reasonable efficiency, thus increasing fuel consumption and possibly emissions. Despite these problems, some analysts expect viable lean nitrogen-oxide catalysts for diesel engines to be developed by the late 1990s.

Vertical exhausts. The exhaust pipes on heavy-duty vehicles are either vertical (so that the exhaust is emitted above the vehicles) or horizontal. Although the choice of exhaust location does not affect overall pollutant emissions, it can have a significant effect on local concentrations of pollutants. A vertical exhaust pipe reduces the concentration of exhaust pollutants at breathing level, reducing human exposure to high local concentrations. Vertical exhausts can reduce exposure to high local concentrations of pollutants by 65 to 87 percent (Weaver and others 1986). Vertical exhausts also make it easier to enforce on-road smoke limitations.

Many heavy-duty trucks and some buses are designed with vertical exhausts from the beginning. There is no technical reason why all trucks and buses could not be so equipped. Retrofitting vertical exhausts to trucks originally equipped with horizontal exhausts is feasible in many cases, but can be impractical because of limits imposed by truck design or use. Retrofitting buses is more complex, but also feasible. In Santiago, Chile, regulations adopted in 1987 required vertical exhausts on all buses, and led to a large number of retrofits.

Emission Control Options and Costs

This section discusses motor vehicle emission controls achievable with current and foreseeable technology, and estimates the costs of achieving these controls. The focus is on technology rather than regulations (see Chapter 1 for a discussion of vehicle emission standards and regulations).

Gasoline-Fueled Passenger Cars and Light-Duty Trucks

Many technologies that improve automotive fuel efficiency such as fuel injection, electronic control of spark timing, advanced choke systems, and improved transmissions, also reduce exhaust emissions. And some emission control requirements have improved fuel efficiency. In the absence of tight emission standards and controls, it is unlikely that these advanced engine technologies would have been applied to automobiles.

In industrialized countries, passenger cars and light-duty trucks are responsible for a larger share of total mobile source pollutant emissions than any other vehicle category. Many jurisdictions have adopted strict limits on emissions from new light-duty vehicles (chapter 1). As a result of these emission regulations, several levels of control technology have been developed that can be classified according to effectiveness, complexity, and cost. Emission controls range from those achievable through simple air-fuel ratio and timing adjustments to standards requiring feedback-controlled fuel injection systems with exhaust gas recirculation and three-way catalysts.

The costs of emission control systems are controversial. Industry estimates of cost and fuel consumption changes for gasoline vehicles in various engine and treatment configurations are given in table 3.1. Estimates by other parties give somewhat different results (compare table 3.2, based on estimates developed by Christopher Weaver for the U.S. EPA).

As vehicle technology is pushed to achieve low pollution levels, common international elements are emerging. In every case, the least polluting vehicles use catalytic converters. Since these systems are poisoned by lead and the phosphorous in most engine oils, they foster the introduction of unleaded gasoline and cleaner oils, reducing overall lead pollution. To optimize these systems, better air–fuel and spark management systems have evolved, leading to increased use of both electronics and fuel injection. These advances also increase fuel efficiency and lower carbon dioxide emissions.

Improved emissions in the United States have been accompanied by improved fuel economy. The weighted fleet average in 1967 was 14.9 miles per gallon, compared with 27.3 miles per gallon in 1987, an increase of 83 percent. Correcting for vehicle weight reductions, the improvement was about 47 percent. The introduction of unleaded fuel and catalytic converters in 1975 coincided with substantial fuel economy gains.

Table 3.2 summarizes six possible levels of emission control for light-duty vehicles which range from the simple controls used in the United States and Japan in the 1970s (and, until the early 1990s, in Europe) to the most sophisticated systems now available.

Non-catalyst controls. Emissions standards at this control level can be met by four-stroke gasoline engines without emissions aftertreatment and correspond to U.S. standards of the early 1970s. Exhaust emission controls for gasoline vehicles involve modification in carburetor design and setting the air-fuel ratio to minimize carbon monoxide and hydrocarbon emissions, while nitrogen oxide control is achieved by retarding ignition timing, exhaust gas recirculation, or both. Diesel vehicles require no modifications to meet these standards. Two-stroke gasoline engines would either be eliminated or forced to adopt fuel injection. Crankcase and evaporative emission controls for gasoline vehicles are also needed.

The non-catalyst approach avoids the complexities and costs of catalysts and unleaded fuel. Total system cost would be about U.S.$130 for a passenger car—U.S.$60 for air injection, U.S.$20 for evaporative controls, and U.S.$5 for crankcase controls. The remaining U.S.$45 covers engine modifications and cold-start emission controls.

Oxidation catalyst. Unleaded gasoline increases the technological feasibility of more stringent exhaust emissions control. With two-way catalysts, air injection, and mechanical air-fuel ratio controls (such as a standard carburetor), total system cost is about U.S.$380–$205 for the catalytic converter, U.S.$60 for the air injection system, and U.S.$25 for crankcase and evaporative controls. The

Table 3.1 Automaker Estimates of Emission Control Technology Costs for Gasoline-Fueled Vehicles
(percent)

Technology	Engine cost increase	Fuel consumption change
Lean-burn engine with carburetor and conventional ignition	1.0	–2
Pulse air and exhaust gas recirculation	4.5	3
Lean-burn engine with carburetor and programmed ignition	2.0	1
Recalibrated conventional engine with electronic fuel injection	8.0	2
Lean-burn engine with electronic fuel injection	9.0	–7
Lean-burn engine with oxidation catalyst	4.5	–3
Open loop, three-way catalyst carburetor	4.1	2
Lean-burn engine, closed loop, electronic fuel injection, variable intake oxidation catalyst	15.0	–7
Closed loop, electronic fuel injection, three-way catalyst	13.0	3

Note: Baseline is a small vehicle with 1.4-liter conventional carburetor engine meeting ECE 15/04 standard.
Source: ECMT 1990

Table 3.2 Exhaust Emission Control Levels for Light-Duty Gasoline-Fueled Vehicles

| Control level | Emission standard | | Controls required | Fuel economy (percent) | Estimated cost per vehicle (U.S. dollars) |
	Grams per kilometer[a]	Percent controlled[b]			
Non-catalyst controls	Hydrocarbons—1.5 Carbon monoxide—15 Nitrogen oxides—1.9	66 63 11	Ignition timing Air-fuel ratio Air injection Exhaust gas recirculation	–5	130
Oxidation catalyst	Hydrocarbons—0.5 Carbon monoxide—7.0 Nitrogen oxides—1.3	89 83 39	Oxidation catalyst Ignition timing Exhaust gas recirculation	–5	380
Three-way catalyst	Hydrocarbons—0.25 Carbon monoxide—2.1 Nitrogen oxides—0.63	94 95 71	Three-way catalyst Closed-loop carburetor or electronic fuel injection	–5 (carburetor) 5 (electronic fuel injection)	630
Lean-burn engine	Hydrocarbons—0.25 Carbon monoxide—1.0 Nitrogen oxides—0.63	94 98 71	Oxidation catalyst Electronic fuel injection Fast-burn combustion chamber	15	630
U.S. tier 1	Hydrocarbons—0.16 Carbon monoxide—1.3 Nitrogen oxides—0.25	96 97 88	Three-way catalyst Electronic fuel injection Exhaust gas recirculation	5	800
California low-emission vehicle standard	Hydrocarbons—0.047 Carbon monoxide—0.6 Nitrogen oxides—0.13	99 99 94	Electric three-way catalyst Electronic fuel injection Exhaust gas recirculation	Unknown	More than 1,000

a. At 80,000 kilometers.
b. Compared with uncontrolled levels.
Source: U.S. EPA 1990

remaining U.S.$90 covers electronic ignition timing, high-energy ignition, and cold-start emission controls.

Three-way catalyst/lean-burn engine. This level is equivalent to 1981 U.S. emissions standards. It is essentially the world standard of the early 1990s—many other countries have adopted them, and current Japanese and ECE regulations require a similar level of control. In gasoline-fueled vehicles, a three-way catalyst (in conjunction with exhaust gas recirculation, control of spark timing, and other measures) reduces emissions of carbon monoxide, hydrocarbons, and nitrogen oxides. This is achievable using a stoichiometric carburetor system with closed-loop electronic trim, though the trend is to use fully electronic systems with fuel injection[2]. These levels have also been met with lean-burn technology (with an oxidation catalyst) at a similar cost, but with better fuel economy and lower carbon monoxide.

Total system cost for a gasoline-fueled passenger car or light-duty truck is estimated at U.S.$630—U.S.$265

for the catalytic converter, U.S.$60 for the air injection system, and U.S.$25 for crankcase and evaporative controls. The remaining U.S.$280 is allocated for the fuel injection system and electronic controls, which also help to improve performance and fuel economy.

U.S. tier 1. These standards, adopted in amendments to the U.S. Clean Air Act, reflect the state-of-the-art in emission control for light-duty vehicles. These standards were implemented in California in the early 1990s and in the rest of the United States in the mid-1990s. These standards require a three-way catalyst, with the air-fuel ratio controlled through electronic fuel injection and air-fuel ratio feedback. This system costs about U.S.$800. Compared with vehicles meeting the 1981 U.S. standards, vehicles meeting this emission level have more precise air-fuel ratio control, more precious metal in the catalytic converter, better evaporative emission controls, and better durability and reliability to meet in-use requirements. Contemporary light-duty diesel vehicles in the United States are capable of meeting the particulate matter, hydrocarbon, and carbon monoxide standards, but require a higher nitrogen oxide standard—about 0.5 grams per kilometer.

---

2.   Diesel vehicles only require engine modifications, control optimization, and possibly exhaust gas recirculation to achieve these emission levels.

California low-emission vehicle (LEV). These standards, to be implemented in California beginning in 1997, and in other parts of the United States in 2001, exceed the current state-of-the-art in emission control for gasoline vehicles. Compliance with these standards will require even more precise control of the air-fuel ratio, heavier catalyst loadings, and better engine-out emission controls than the current state-of-the-art, and possibly new technologies or alternative fuels. For large vehicles, it appears that preheated catalytic converters may be required to reduce emissions when the vehicle is started cold. This will pose a significant challenge to automakers. The total cost is unknown, but will most likely exceed U.S.$1,000.

Heavy-Duty Gasoline-Fueled Vehicles

Heavy-duty vehicles with gasoline engines can use the emission control technologies outlined above. Because of differences in vehicle size and testing procedures, however, emissions standards cannot be directly compared. Because of their smaller numbers, heavy-duty gasoline vehicles receive less regulatory attention, and emissions standards are less strict. Emission standards for these vehicles are also lax because they typically operate at high loads, making it difficult to ensure catalyst durability. Design solutions for the catalyst durability problem exist, and recent California regulations will require vehicles under 14,000 pounds gross vehicle weight to meet low-emission vehicle standards. The cost of meeting emissions standards in larger vehicles is expected to be 50 to 100 percent more than for passenger cars because of the larger size of the equipment required.

Motorcycles

In the past, motorcycles were subject to lenient emission controls or none at all. This reflects their minor contribution to emissions in most industrial countries and the difficulty and expense of installing emission controls on small, heavily-loaded engines. In many Asian cities, however, motorcycles and three- wheelers are responsible for a large fraction of hydrocarbon and PM emissions. Recommended emission control levels based on an assessment of motorcycle emissions in Thailand are summarized in table 3.3 (Weaver and Chan 1994).

The first step in controlling emissions from motorcycles vehicles is eliminating the excessive emissions from two-stroke engines. This can be done by switching to a four-stroke design or to a two- stroke design incorporating timed fuel injection and crankcase lubrication. This would reduce hydrocarbon and particulate matter emissions by about 90 percent, at a cost of about U.S.$60 per vehicle. Additional emission reductions are possible with improved four-stroke engine design and calibration and through the use of catalytic converters. Catalytic converters are used on two-stroke motorcycles in Taiwan (China) and on mopeds in Austria and Switzerland. Catalyst-forcing standards for four-stroke engines have not been adopted in any jurisdiction.

Diesel-Fueled Vehicles

As with gasoline-fueled vehicles, diesel engine emission reductions have accompanied improvements in fuel efficiency. Although measures such as retarding injection timing increase fuel consumption, these have been offset by gains from turbocharging, charge-air

Table 3.3   Recommended Emission Control Levels for Motorcycles in Thailand

Control level	Emission standard Grams per kilometer[a]	Percent controlled[b]	Controls required	Fuel economy (percent)	Estimated cost per vehicle (U.S. dollars)
Eliminate two-stroke	Hydrocarbons—5.0	66	Four-stroke engine or advanced two-stroke	30–40	60–80
	Carbon monoxide—12.0	50			
	Nitrogen oxides—NR	—			
	Particulate matter—0.15	50–90			
Non-catalyst controls	Hydrocarbons—1.0	90	Four-stroke or two stroke with catalyst, ignition timing, air-fuel ratio control	0	80–100
	Carbon monoxide—12.0	50			
	Nitrogen oxides—0.5	200			
	Particulate matter—0.15	50–90			
Oxidation catalyst or advanced technology	Hydrocarbons—0.5	98	Four-stroke or advanced two- stroke, ignition timing, air-fuel ratio control, catalytic converter or electronic fuel injection	–5	80–100
	Carbon monoxide—2.0	80			
	Nitrogen oxides—0.5	200			
	Particulate matter—0.05	85			

— Not applicable
NR = Not regulated.
a. At 80,000 kilometers.
b. Compared with uncontrolled two-stroke.
Source: Weaver and Chan 1994

Table 3.4 Industry Estimates of Emission Control Technology Costs for Diesel-Fueled Vehicles
(percent)

Technology	Engine cost increase
Baseline engine, no emission control equipment	0
Injection timing retard	0
Low sac volume and valve covering nozzle	Minimal
Turbocharging	3–5
Charge cooling	5–7
Improved fuel injection	13–15
High-pressure fuel injection with electronic control	14–16
Variable geometry turbocharging	1–3
Particulate trap	4–25

Source: ECMT 1990 (based on Tonkin and Etheridge 1987)

cooling, and improved fuel injection equipment. The costs of these features are offset by lower fuel consumption. Industry estimates of cost increases for engine modifications to meet emission standards are shown in table 3.4.

Uncontrolled emissions of nitrogen oxides from heavy-duty diesel engines range from 12 to 21 grams per kilowatt-hour (9 to 16 grams per brake horsepower-hour) when measured using the U.S. transient or European 13-mode cycle. Particulate matter emissions on the transient cycle are typically 1–5 g/kWh (0.75–3.7 g/bhp-hr), but are significantly lower on the European steady-state cycle. Engines that have been tampered with or poorly maintained may emit higher PM emissions in the form of smoke.

By moderately retarding fuel injection timing from the optimal point, nitrogen oxides can be reduced to less than 11.0 g/kWh (8 g/bhp-hr). This may require upgraded fuel-injection equipment. Particulate matter emissions can be limited through smoke opacity standards under acceleration and full-load conditions. Achievable peak acceleration smoke opacity levels range from 25 to 35 percent, while steady- state smoke opacity of less than 5 percent (the limit of visibility) is readily achievable. This minimal control level is comparable to that required of California engines in the 1970s and European engines until the early 1990s.

Moderate control of 8 g/kWh (6 g/bhp-hr) of nitrogen oxides and 0.7 g/kWh (0.5 g/bhp-hr) of particulate matter requires further optimization of the injection timing and the overall combustion system. This corresponds to 1990 U.S. federal standards. The next level of control corresponds to the 1991 U.S. standards, which have also been adopted in Canada and Mexico. Achieving this level of nitrogen oxides control while meeting PM emission standards requires major engine design modifications. These include variable fuel-injection timing, high-pressure fuel injection, combustion optimiza-

tion, and charge- air cooling. A tighter U.S. particulate matter standard (0.13 g/kWh; 0.10 g/bhp-hr) took effect in 1994. This standard applied to diesel engines in urban buses in 1993, and a limit of 0.07 g/bhp-hr was adopted for 1994 and 1995. After 1995, U.S. urban buses will be required to meet an emission standard of 0.07 g/kWh (0.05 g/bhp-hr).

Further reductions are being contemplated. The 1990 U.S. Clean Air Act amendments require reduced emissions of nitrogen oxides for all heavy-duty truck and bus engines (5.4 g/kWh; 4.0 g/bhp-hr) in 1998.

The stringent nitrogen oxides and PM standards adopted by the United States would not have been possible for diesel engines until very recently. Developments in fuel-injection rate-shaping, and the potential use of exhaust gas recirculation mean that nitrogen oxide emission levels of 2.6 g/kWh (2.0 g/bhp-hr) may now be achievable, in combination with low PM emissions. Using selective exhaust gas recirculation and extensive engine optimization, nitrogen oxide levels as low as 2.6 g/kWh (2.0 g/bhp-hr) have been achieved in the laboratory (Needham, Doyle, and Nicol 1991). Translating research results into marketable engines takes time, but this may be feasible by the end of the 1990s.

Further reductions in nitrogen oxide and particulate matter emissions are possible with alternative fuels. Pre-production methanol direct-injection engines using glow-plug assisted compression ignition have produced nitrogen oxide emissions below 2.9 g/kWh (2.1 g/bhp-hr), with efficiency comparable to that of a regulated diesel engine. Heavy-duty, lean-burn, natural-gas engines have achieved nitrogen oxide levels below 2.5 g/kWh (1.9 g/bhp-hr), with energy efficiency about 10 percent worse than the diesel. Spark-ignition engines using natural gas, liquefied petroleum gas, and gasoline with three-way catalysts, stoichiometric air-fuel ratios, and closed-loop control have achieved nitrogen oxide emissions below 1.5 g/kWh (1.1 g/bhp-hr) at low mile-

Table 3.5   Emissions Control Levels for Heavy-Duty Diesel Vehicles

| Control level | Emissions limit at full useful life | | Controls required | Fuel economy[a] (percent) | Estimated cost per engine (U.S. dollars) |
	Grams per kilowatt-hour	Grams per brake horsepower-hour			
Uncontrolled	Nitrogen oxides—12.0 to 21.0 Particulate matter—1.0 to 5.0	9.0 to 16.0 0.75 to 3.70	None (PM level depends on smoke controls & maintenance level)	0	0
Minimal control	Nitrogen oxides—11.0 Particulate matter—0.7 to 1.0 Peak smoke—20 to 30 percent opacity	8.0 0.5 to 0.75	Injection timing Smoke limiter	-3 to 0	0–200
Moderate control	Nitrogen oxides—8.0 Particulate matter—0.7	6.0 0.5	Injection timing Combustion optimization	-5 to 0	0–1,500
1991 U.S. standard (Euro 2)	Nitrogen oxides—6.7 (7.0) Particulate matter—0.34 (0.15)[b]	5.0 0.25	Variable injection timing High-pressure fuel injection Combustion optimization Charge-air cooling	-5 to 5	1,000–3,000
Lowest diesel standards under consideration	Nitrogen oxides—2.7 to 5.5[c] Particulate matter—0.07 to 0.13	2.0 to 4.0 0.05 to 0.10	Electronic fuel injection Charge-air cooling Combustion optimization Exhaust gas recirculation Catalytic converter or particulate trap	-10 to 0	2,000–6,000
Alternative-fuel forcing	Nitrogen oxides—less than 2.7 Particulate matter—less than 0.07	2.0 0.04	Gasoline/three-way catalyst Natural gas lean-burn Natural gas/three-way catalyst Methanol-diesel	-30 to 0	0–5,000

Note: Kilowatt-hours are converted to brake horsepower-hours by multiplying by 0.7452.

a. Potential fuel economy improvements result from addition of turbocharging and intercooling to naturally aspirated engines.

b. Euro-2 emissions are measured on a steady-state cycle that underestimates PM emissions in actual driving. Actual stringency of control requirements is similar to that of U.S. 1991.

c. Not yet demonstrated in production vehicles.

Source: Weaver 1990

Table 3.6   Emission Control Levels for Light-Duty Diesel Vehicles

Control level	Emissions limit at full useful life (grams per kilometer)	Reduction[a] (percent)	Controls required	Fuel economy (percent)	Estimated cost per engine (U.S. dollars)
Uncontrolled	Nitrogen oxides—1.0 to 1.5 Particulate matter—0.6 to 1.0	0 0	None (PM level depends on smoke controls & maintenance level)	0	0
Moderate control	Nitrogen oxides—0.6 Particulate matter—0.4	40 33	Injection timing Combustion optimization	-5 to 0	0–500
1988 U.S. standard (EU Directive 91/441/EEC)	Nitrogen oxides—0.6 (HC+NO$_x$:0.97) Particulate matter—0.13 (0.14)	40 78	Variable injection timing Combustion optimization Exhaust gas recirculation	-5 to 0	100–200
Advanced diesel technology	Nitrogen oxides—0.5 Particulate matter—0.05–0.08	40 92	Electronic fuel injection Combustion optimization Exhaust gas recirculation Catalytic converter or particulate trap	-10 to 0	200–500

a. Compared with uncontrolled levels.

Source: Weaver 1990

age, and acceptable catalyst durability has been demonstrated in some cases. Particulate matter emissions with these fuels are also low. Since they do not form soot or condensable organic compounds, particulate emissions derive only from the lubricating oil. With a catalytic converter, these are less than 0.05 g/kWh (0.04 g/bhp-hr).

The emission control levels achievable for heavy-duty diesel engines, the costs of achieving these levels, and the corresponding effects on fuel economy are shown in table 3.5. These estimates prepared for an OECD study differ from those based on industry sources.

Potential emission controls for light-duty diesel vehicles range from none to moderate control (applied in Europe until recently) to the stringent controls typical of California and the rest of the United States. Given the collapse in demand for diesel vehicles in the U.S., no manufacturer considered it worthwhile to develop an emissions control system to meet the California limits. Similar emission standards for diesel-fueled passenger cars have recently been adopted in Europe, where diesel cars are a large part of the market. Vehicles meeting these control levels are thus likely to be developed in the near future. Estimates of cost and emission control effectiveness for light-duty diesel-fueled automobiles are provided in table 3.6.

References

ACEA/EUROPIA (European Automobile Manufacturer's Association/European Petroleum Industry Association). 1995. *European Program on Emissions, Fuels, and Engine Technologies.* EPEFE Report, Brussels.

Brogan, J.J. and S.R. Venkateswaran. 1993. "Fuel-Efficient Low-Emission Propulsion Technology" in *Towards Clean & Fuel Efficient Automobiles.* Proceedings of an International Conference held in Berlin (March 25-27, 1991), OECD, Paris.

CONCAWE Review. 1994. "Oxidation Catalysts for Diesel Cars," Volume 3, Number 2, October, Brussels.

ECMT (European Conference of Ministers of Transport). 1990. *Transport Policy and the Environment.* Organisation for Economic Cooperation and Development (OECD). Paris.

Feutlinske, H. 1989. "Possibilities for Reducing Pollutant Emissions from Diesel Engines–Comments on the Development Status of Particle Filters." Report 4, 48th International Congress, International Union of Public Transport. Brussels.

Holman, C. 1990. "Pollution from Diesel Vehicles." Briefing Document, Friends of the Earth, London.

Hope, K. 1992. "Urban Air Pollution: Greeks Battle to Defeat the Nefos." *Financial Times*, March 11, 1992. London.

Kimbom, G. 1993. "Some Examples of Alternative Engines" in *Towards Clean & Fuel Efficient Automobiles.* Proceedings of an International Conference held in Berlin (March 25-27, 1991), OECD, Paris.

Knecht, W. 1991. "Reduction of Heavy-duty Diesel Emissions and Application of Particulate Traps and Oxidation Catalysts." in *The Diesel Engine–Energy Stake and Environment Constraints,* TÜV Brussels, Rheinland/OPET.

MacKenzie, J.J. 1994. *The Keys to the Car—Electric and Hydrogen Vehicles for the 21st Century.* World Resources Institute, Washington, D.C.

Mason, J.L. 1993. IC Engines & Fuels for Cars & Light Trucks: 2015." in *Transportation & Global Climate Change,* eds. L. Greene & D.J. Santini, American Council for an Energy Efficient Economy, Washington, D.C.

Needham, J.R., D.M. Doyle, and A.J. Nicol. 1991. "The Low Nitrogen Oxide Truck Engine." *SAE Paper* 910731, Society of Automotive Engineers, Warrendale, Pennsylvania.

OTA (Office of Technology Assessment, U.S. Congress). 1995. *Advanced Automotive Technology: Visions of a Super-Efficient Family Car.* OTA-ETI-638. U.S. Government Printing Office, Washington, D.C.

Pattas, K., Z. Samaras, N. Patsatzis, C. Michalopoulous, O. Zogou, A. Stamatellos, and M. Barkis. 1990. "On-Road Experience with Trap Oxidiser Systems Installed on Urban Buses." *SAE Paper* 900109, Presented at the Society of Automotive Engineers (SAE) International Congress and Exposition, Detroit, Michigan.

Tonkin, P.R. and P. Etheridge. 1987. "A New Study of the Feasibility and Possible Impact of Reduced Emission Levels from Diesel Engined Vehicles," Report No. DP87/0927, Shoreham by Sea, Ricardo Consulting Engineers, U.K.

U.S. EPA (United States Environmental Protection Agency). 1990. "Volatile Organic Compounds from On-Road Vehicles: Sources and Control Options." Report prepared for the Long-range Transboundary Air Pollution Convention. United Nations Economic Commission for Europe. Washington, D.C. (Draft)

Watkins, L.H. 1991. *Air Pollution from Motor Vehicles.* State-of-the-Art Review/1. Transport and Road Research Laboratory. HMSO. London.

Weaver, C.S., R.J. Klausmeier, L.M. Erickson, J. Gallagher, and T. Hollman. 1986. "Feasibility of Retrofit Technology for Diesel Emissions Control." *SAE Paper* 860296, Society of Automotive Engineers, Warrendale, Pennsylvania.

Weaver, C.S. 1990. "Emissions Control Strategies for Heavy-Duty Diesel Engines." EPA report 460/3- 90-001. U.S. Environmental Protection Agency. (Draft) Ann Arbor, Michigan.

Weaver, C.S. and L.M. Chan. 1994. "Motorcycle Emission Standards and Emission Control Technology." Report to the Royal Thai Ministry of Science, Technology, and the Environment, and The World Bank.

Sacramento, California, Engine, Fuel, and Emissions Engineering, Inc.

Weaver, C.S. and L.M. Chan. 1995. "Company Archives" Multiple Sources. Engine, Fuel and Emissions Engineering, Inc. Sacramento, California.

Wijetilleke, L and S. Karunaratne. 1992. "Control and Management of Petroleum Related Air Pollution." The World Bank, Washington, D.C. (Draft)

Appendix 3.1

Emission Control Technology for Spark-Ignition (Otto) Engines

This appendix provides a more detailed technical discussion to supplement the general information on spark-ignition engine emissions and control technologies in chapter 3.

Combustion and Pollutant Formation in Spark-Ignition Engines

Gasoline engines emit carbon monoxide (CO), nitrogen oxides (NO_x), hydrocarbons (HC), particulate matter (PM), and lead (where leaded gasoline is used), as well as other toxics such as benzene, 1,3 butadiene, and formaldehyde.

Exhaust Emissions

Exhaust emissions are caused by the combustion process. Figure A3.1.1 shows the combustion process in an Otto-cycle engine. After the initial spark, there is an ignition delay while the flame kernel created by the spark grows. The flame then spreads through the combustion

Figure A3.1.1 Combustion in a Spark-Ignition Engine

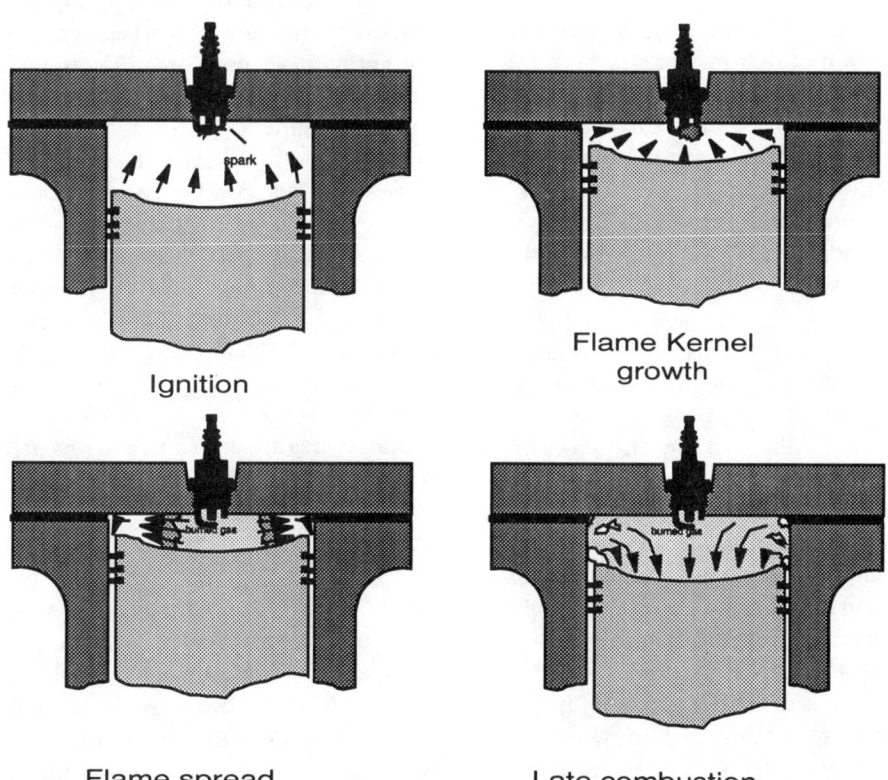

Ignition

Flame Kernel growth

Flame spread

Late combustion

Source: Weaver and Chan 1995

81

chamber. The rate of spread is determined by the flame speed, which is a function of air-fuel ratio, temperature, and turbulence level. The increase in volume of the hot burned gases behind the flame front presses the unburned charge outward. Overall cylinder pressure increases, increasing the temperature of the unburned charge. Finally, the remaining elements of the unburned mixture burn out as the piston descends. The main exhaust pollutants are discussed below.

Nitrogen oxides. The two main nitrogen oxides emitted from combustion engines are nitric oxide (NO) and nitrogen dioxide (NO_2). Most nitrogen oxides from combustion engines—90 percent—are nitric oxide. This gas is formed from nitrogen and free oxygen at high temperatures. The rate of formation is a function of oxygen availability, and is exponentially dependent on the temperature. Most nitrogen oxide emissions form early in the combustion process, when the piston is near the top of its stroke (*top-dead-center*) and temperatures are highest. Nitrogen oxide emissions are controlled by reducing the flame temperature (by retarding combustion, diluting the reacting mixture, or both) and by minimizing the time that burned gases stay at high temperatures.

Carbon monoxide. Carbon monoxide emissions are caused by the combustion of rich mixtures, where the air-fuel ratio, λ is less than 1.0. In such mixtures, there is insufficient oxygen to convert all the carbon to carbon dioxide. A small amount of carbon monoxide is also emitted under lean conditions because of chemical kinetic effects. Carbon monoxide emissions are controlled in the engine by adjusting the air-fuel ratio of the charge entering the cylinder.

Hydrocarbons. Hydrocarbon emissions result from elements of the air-fuel mixture that have not finished burning at the time the exhaust valve opens. Hydrocarbon emissions are composed of unburned fuel and products of partial combustion, such as ethylene and formaldehyde. Hydrocarbon sources include crevice volumes, such as the space between the piston and cylinder wall above the piston ring, and the quenched layer immediately next to the combustion chamber walls. Unburned mixture is forced into these crevices during compression and combustion, and emerges late in the expansion and during the exhaust stroke. This is the major source of hydrocarbon emissions from four-stroke engines. In two-stroke engines, fuel mixing with the exhaust during scavenging and misfire at light loads are sources of hydrocarbons. Abnormal operation in a four-stroke engine, such as a misfiring cylinder, can cause significant quantities of unburned fuel to pass into the exhaust. In lean mixtures, flame speeds may be too low for combustion to be completed during the power stroke, or combustion may not occur. These conditions also cause high hydrocarbon emissions.

Unburned hydrocarbons emitted from the cylinder continue to react in the exhaust only if the temperature is above 600°C and oxygen is present. So hydrocarbon emissions from the tailpipe may be significantly lower than the hydrocarbons leaving the cylinder. This effect is important at stoichiometric or near-stoichiometric conditions because of the higher exhaust temperatures experienced.

Particulate matter. Unlike diesel engines, PM emissions from spark-ignition engines are generally not regulated. They frequently are not measured or reported, leading many to assume they are negligible. This is not the case. Particulate matter emissions from four-stroke spark-ignition engines result from unburned lubricating oil in the exhaust and from ash-forming fuel and oil additives such as tetra-ethyl lead. Although PM emission rates for spark-ignition engines are low compared with diesels, emissions can be significant when poor maintenance or engine wear lead to high oil consumption, or when oil is mixed with the fuel, as in two-stroke engines.

Toxic pollutants. Toxic chemicals emitted by spark-ignition engines include lead compounds, benzene, 1,3 butadiene, and aldehydes. Lead emissions are caused by tetra-ethyl lead, used as an octane enhancer in leaded gasoline. Benzene is one of the many hydrocarbons in gasoline-engine exhaust, accounting for about 4 percent of total hydrocarbons. Benzene in the exhaust is caused by fuel coming through unburned and by dealkylation of other aromatic compounds. 1,3 butadiene is a product of partial hydrocarbon combustion. Both benzene and 1,3 butadiene are carcinogens, so exposure to them is cause for concern. Studies in the United States indicate that motor vehicles are responsible for most human exposure to benzene and 1,3 butadiene (U.S. EPA 1990).

Aldehydes are intermediate products of hydrocarbon combustion. Highly reactive, they form other products during combustion. The small amounts of aldehyde emissions found in gasoline, diesel, and other engines using hydrocarbon fuels are caused by the quenching of partially reacted mixture (because of contact with a cold surface, for instance). For fuels containing ethanol and methanol, the primary oxidation reactions proceed through formaldehyde and acetaldehyde, respectively, so these compounds are often found in significant concentrations in the exhaust. Both formaldehyde and acetaldehyde are irritants and suspected carcinogens.

Nitrous oxide. Uncontrolled auto engines emit a few milligrams per kilometer of nitrous oxide (N_2O). If the engine has a catalytic converter, increased nitrous oxide formation occurs because of the reaction of nitric oxide and ammonia with the platinum in the catalyst. No more than 5 to 10 percent of the nitric oxide in exhaust is converted into nitrous oxide in this way. The conversion in the catalyst is highly temperature dependent. As the catalyst warms up after a cold start, nitrous oxide levels increase to about 5.5 times the inlet level at 360°C. Emissions then decrease to the inlet level at 460°C. At higher temperatures, the catalyst destroys nitrous oxide instead of forming it (Prigent and Soete 1989). Nitrous oxide is thus formed primarily during cold starts of catalyst-equipped vehicles.

Crankcase Emissions

All piston engines experience leakage or blowby of compressed gas past the piston rings. In spark-ignition engines, this leakage consists of unburned or partly burned air-fuel mixture and contains unburned hydrocarbons. In older vehicles, blowby gases were vented to the atmosphere. Hydrocarbon emission levels from the crankcase were about half the level of uncontrolled exhaust emissions. These emissions are now controlled by venting the crankcase to the air intake system by means of a positive crankcase ventilation (PCV) valve. The unburned hydrocarbons in the blowby gas are recycled to the engine and burned. The flow of air through the crankcase also helps to prevent condensation of fuel and water in the blowby gases, thus reducing oil contamination and increasing engine life. One significant drawback of the PCV system is that the recycling of blowby gases tends to foul the intake manifold. An effective gasoline detergent can eliminate this problem.

Evaporative and Refueling Emissions

Gasoline-fueled vehicles emit a significant amount of hydrocarbons as evaporative emissions from their fuel system. Four main sources of evaporative emissions have been identified: *breathing* (diurnal) losses from fuel tanks caused by the expansion and contraction of gas in the tank with changes in air temperature; *hot-soak* emissions from the fuel system when a warm engine is turned off; *running losses* from the fuel system during vehicle operation; and *resting losses* from permeation of plastic and rubber materials in the fuel system. *Refueling emissions* consist of gasoline vapor displaced from the fuel tank when it is filled.

Evaporative and refueling emissions are strongly affected by fuel volatility. Diurnal emissions also vary depending on the daily temperature range and the amount of vapor space in the fuel tank. Hot-soak emissions result from the conduction of heat from the warm engine to the carburetor, which is open to the atmosphere. Hot-soak losses are greatly reduced in sealed fuel systems, such as those used with fuel injection systems. Such systems may have higher running losses, however, due to the recirculation of hot fuel from the engine back to the fuel tank.

Evaporative emissions can be reduced by venting the fuel tank and the carburetor to a canister containing activated carbon. This material adsorbs volatile emissions from the fuel system when the engine is not running. When the engine is running, intake air is drawn through the canister, purging it of hydrocarbons, which then form part of the fuel mixture fed to the engine. A larger canister can also be used to control refueling vapor emissions, or these emissions can be controlled by capturing them through the refueling nozzle and conducting them back to the service station tank.

Engine Design

Spark-ignition vehicle engines are either two stroke or four stroke. The distinction is important for emissions since two-stroke engines emit more hydrocarbons and particulate matter than four-stroke engines of similar size and power. Two-stroke engines are less fuel efficient than four strokes, but have higher power output, quicker acceleration, and lower manufacturing costs. Because of their performance and cost advantages, two-stroke engines are used extensively in motorcycles and in small power equipment such as chainsaws and outboard motors. They are particularly common in small motorcycles (50 to 150 cc engine displacement), where their poor fuel economy is of less importance to a consumer. They were also used in some small automobiles, particularly in Eastern Europe.

The piston and cylinder of a typical four-stroke engine are shown in figure A3.1.2. Engine operation takes place in four distinct steps: intake, compression, power, and exhaust, with each step corresponding to one *stroke* of the piston (180 degrees of crankshaft rotation). During intake, the intake valve admits a mixture of air and fuel, which is drawn into the cylinder by the vacuum created by the downward motion of the piston. Figure A3.1.2 shows the piston near the end of the intake stroke, approaching *bottom-dead-center*. During compression, the intake valve closes, and the upward motion of the piston compresses the air-fuel mixture into the combustion chamber between the top of the piston and the cylinder head.

The compression stroke ends when the piston reaches top-dead-center. Just before this point, the air-fuel mixture is ignited by a spark from the spark plug and begins to burn. Combustion of the air-fuel mixture takes place near top-dead-center, increasing the temperature and pressure of the trapped gases. During the power stroke, the pressure of the burned gases pushes the pis-

Figure A3.1.2 Piston and Cylinder Arrangement of a Typical Four-Stroke Engine

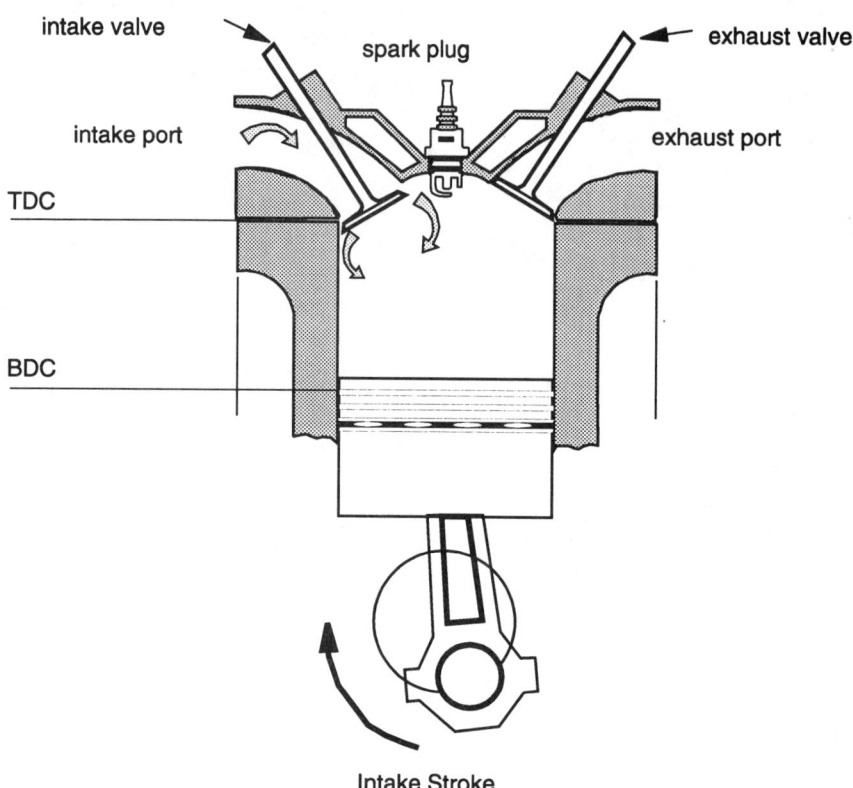

Intake Stroke

Source: Weaver and Chan 1995

ton down, turning the crankshaft and producing power. As the piston approaches bottom-dead-center again, the exhaust valve opens, releasing the burned gases. During the exhaust stroke the piston ascends toward top-dead-center, pushing the remaining burned gases out the open exhaust port. The exhaust valve then closes and the intake valve opens for the next intake stroke.

In a four-stroke engine, combustion and the resulting power stroke occur once every two revolutions of the crankshaft. In a two-stroke engine, combustion occurs in every revolution of the crankshaft. Two-stroke engines eliminate the intake and exhaust strokes, leaving only the compression and power strokes. Rather than occupying distinct phases of the cycle, exhaust and intake occur simultaneously (figure A3.1.3). As the piston approaches bottom-dead-center in the power stroke, it uncovers exhaust ports in the wall of the cylinder. The high-pressure combustion gases blow into the exhaust manifold. As the piston gets closer to the bottom of its stroke, intake ports are uncovered, and fresh air-fuel mixture is forced into the cylinder while the exhaust ports are still open. Exhaust gas is *scavenged* (forced) from the cylinder by the pressure of the incom-

ing charge. In the process, mixing between exhaust gas and the charge takes place, so some of the fresh charge is also emitted in the exhaust. If fuel is already mixed with the air in the charge (as in all current two-stroke motorcycle engines), this fuel will be lost in the exhaust. The loss of fuel during scavenging is one of the main causes of high hydrocarbon emissions from two-stroke motorcycle engines.

The other reason for high hydrocarbon emissions from two strokes is their tendency to misfire under low-load conditions. At low loads, the amount of fresh charge available to scavenge the burned gases from the cylinder is small, so a significant amount of the burned gas remains in the cylinder to dilute the incoming charge. The resulting mixture of air, fuel, and exhaust burns less readily and is more difficult to ignite than an air-fuel mixture alone. At light loads, the mixture sometimes fails to ignite, allowing the fuel vapor in the cylinder to pass unburned into the exhaust. These occasional misfires are the cause of the popping sound made by two-stroke engines under light-load conditions.

The mechanical systems required by four-stroke engines to open and close their intake and exhaust valves

Figure A3.1.3 Exhaust Scavenging in a Two-Stroke Gasoline Engine

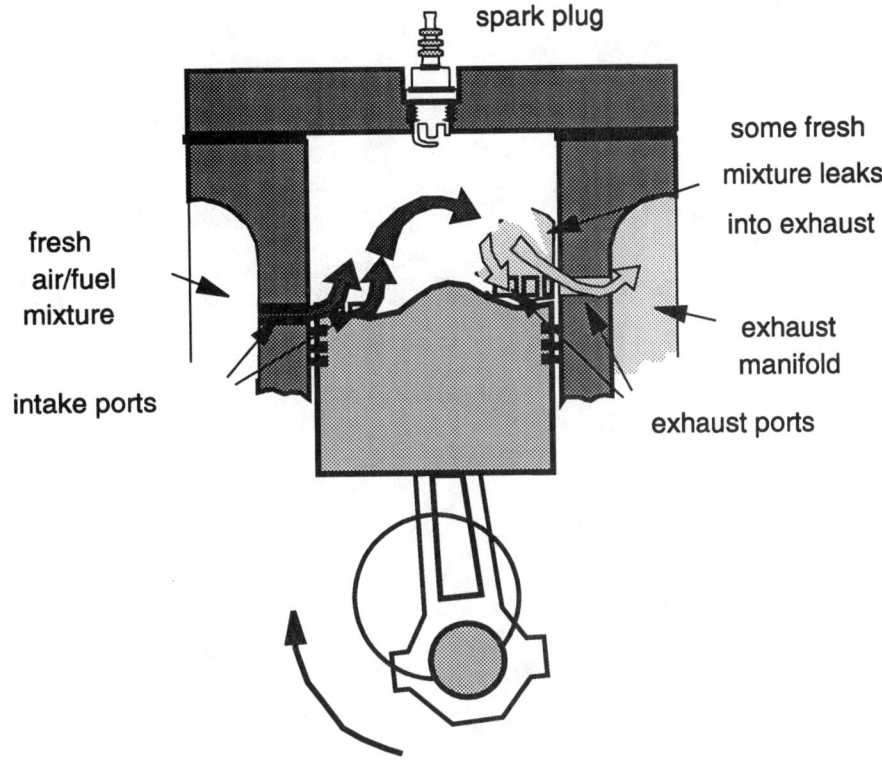

Source: Weaver and Chan 1995

at the right time make these engines relatively complex to manufacture. The mechanical layout of one common type of four-stroke engine, with an overhead camshaft, is shown in figure A3.1.4. The valves are opened by lobes on the camshaft, which is driven at one-half engine speed by a sprocket and chain arrangement from the crankshaft. The camshaft lobes press on the valve followers, pushing up the rocker arms, and make the valves open at the appropriate time in every second crankshaft revolution. The camshaft, valve linkage, crankshaft bearings, and pistons are lubricated by oil pumped from the oil sump at the bottom of the crankcase. The other common design, with the camshaft located lower in the engine and driving the valves through pushrods, is even more complicated.

A two-stroke engine is much simpler (figure A3.1.5). As the piston rises during the compression stroke, it creates a vacuum in the crankcase, which draws the air-fuel mixture from the carburetor into the crankcase. As the piston descends during the power stroke, the air-fuel mixture in the crankcase is prevented by the reed valve from going out the intake port, and instead is forced up into the cylinder, where it displaces the burned gases in the cylinder into the exhaust in the

scavenging process diagrammed in figure A3.1.3, as shown. The complex valve gear, camshaft, and related mechanisms of the four-stroke engine are not needed in the two stroke. For small motorcycles with two-stroke engines, the cost difference between two-stroke and four-stroke engines is U.S.$60 to U.S.$80 per unit (Weaver and Chan 1994).

Since the crankcase in a two-stroke engine serves as the pump for the scavenging process, it cannot be used as an oil sump; gasoline would mix with the oil and dilute it. Lubrication for two-stroke engines is provided by mixing two-stroke oil with the fuel. As the fuel-oil mixture is atomized and mixes with the air, the light hydrocarbons in the fuel evaporate, leaving a mist of oil droplets or particles. Some of this mist contacts the internal parts of the engine, lubricating them, but most of it passes into the cylinder with the fresh charge. A significant part of the frest charge passess unburned into the exhaust, carrying the oil droplets with it. Some oil may also escape unburned from the combustion process. Since the oil particles scatter light that strikes them, they are visible as white or blue smoke. This smoke is one of the distinguishing features of two-

Figure A3.1.4 Mechanical Layout of a Typical Four-Stroke Engine

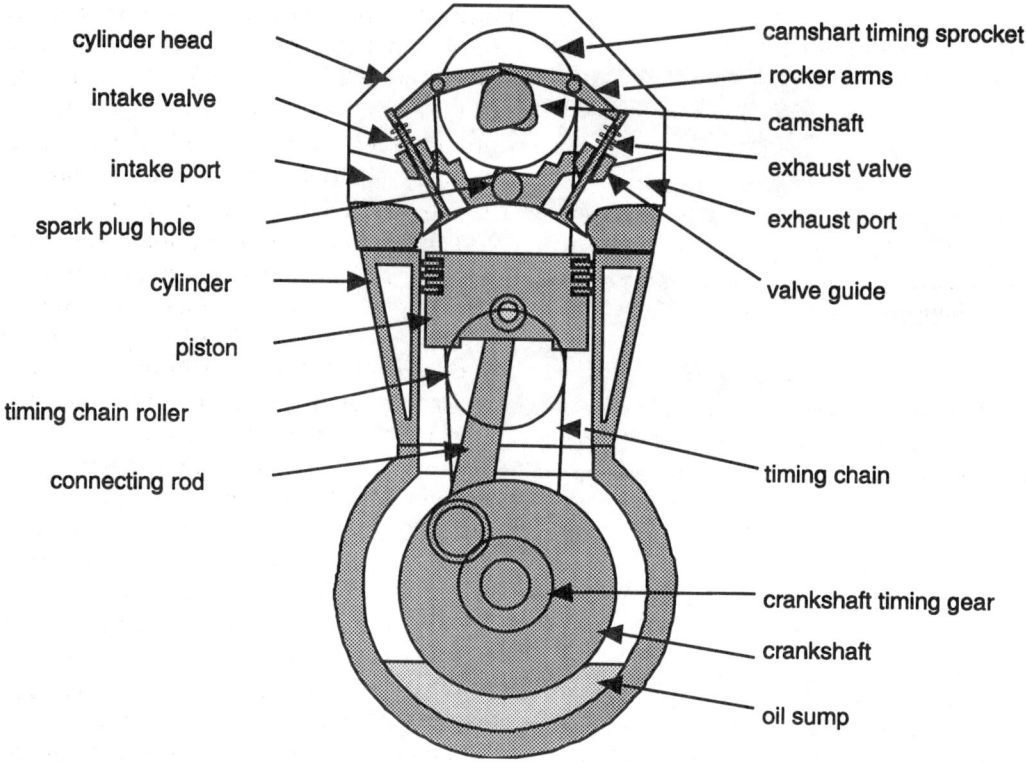

cylinder head

intake valve

intake port

spark plug hole

cylinder

piston

timing chain roller

connecting rod

camshart timing sprocket

rocker arms

camshaft

exhaust valve

exhaust port

valve guide

timing chain

crankshaft timing gear

crankshaft

oil sump

Source: Weaver and Chan 1995

Figure A3.1.5 Mechanical Layout of a Typical Two-Stroke Motorcycle Engine

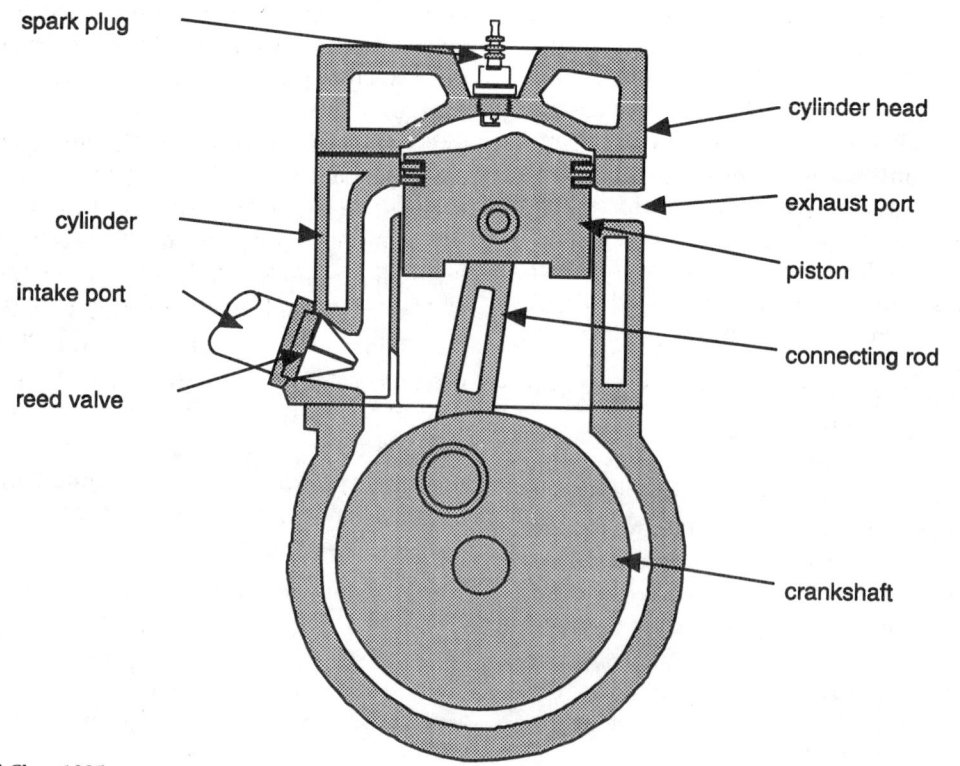

spark plug

cylinder

intake port

reed valve

cylinder head

exhaust port

piston

connecting rod

crankshaft

Source: Weaver and Chan 1995

stroke engines, and the reason for their high particulate matter emissions.

Emission Control Technology for Four-Stroke Engines

Fuel Metering Systems

Because of the importance of air-fuel ratio control for emissions, fuel metering systems are crucial in the emission control systems of Otto-cycle engines. Two types of metering systems are common: carburetors and fuel injection systems. Both are affected by altitude.

Carburetors. These systems were used almost universally before the advent of emission regulations and are still common where stringent emission controls are not in effect. In a carburetor, air going into the engine is accelerated in one or more venturis. The pressure differential in the venturi throat is used to draw fuel into the airstream and atomize it. In advanced carburetors, this is supplemented by devices that provide mixture enrichment at idle and light-load conditions, and during cold starts.

Carburetors cannot maintain precise air-fuel ratio control under all conditions and are subject to change over time. They are not suitable for applications that require precise and invariant air-fuel ratio control. To address these problems, carburetors with electronic air-fuel ratio adjustment were marketed in the early 1980s for use with three-way catalyst systems. Such systems are inferior to electronic fuel injection, however, and they are no longer produced in significant numbers.

Fuel injection systems. During the 1980s, air-fuel ratio control systems for light-duty gasoline vehicles evolved from mechanical systems such as carburetors to direct control of fuel quantity through electronic fuel injection. These systems provide rapid and precise control of the air-fuel ratio. There are two basic types of fuel injec-

tion systems: central (*throttle-body*) injection systems, with one or two centrally-located fuel injectors; and multi-port fuel injection systems, with one fuel injector located at the inlet to each cylinder. The multi-port systems reduce cylinder-to-cylinder variations in air-fuel ratio and simplify intake manifold design, since formation of fuel puddles in the intake manifold is no longer a problem. Because they have fewer parts, central fuel injection systems are cheaper, while multi-port systems have better emissions and performance.

The first widely used fuel injection systems were developed and marketed by Robert Bosch AG (Bosch 1986). In these systems, fuel is injected continuously through nozzles at each intake port. The rate of injection is controlled by varying the pressure supplied to the nozzles by an electric fuel pump. Fuel injection systems now use fully electronic control. Fuel is provided to the injectors at constant pressure by a pump and pressure regulating valve. The injectors are solenoid valves, which are controlled by the engine computer. The computer controls the quantity of fuel injected by varying the length of time the valve remains open during each revolution of the crankshaft.

Electronic multi-port fuel injection systems either fire all the fuel injectors at once, or each injector is fired sequentially at the optimal time during engine rotation. All-at-once systems are simpler and cheaper, while sequential fuel injection gives the most precise and flexible control over the injection process. Sequential systems allow better air-fuel mixing and thus better performance and emissions. As vehicle emission standards grow more demanding in industrialized countries, more vehicles are being equipped with multi-port, sequential fuel injection systems.

Altitude effects on air-fuel ratio. Atmospheric pressure, air temperature, and air density vary with changes in altitude (table A3.1.1). Reduced air density at high altitudes reduces the power of a naturally aspirated gasoline engine and upsets normal operating characteristics, resulting in excessive pollutant emissions. Fuel

Table A3.1.1   Effect of Altitude on Air Density and Power Output from Naturally Aspirated Gasoline Engines in Temperate Regions

Altitude (meters)	Air pressure (mm of mercury)	Air density (kilograms per cubic meter)	Air temperature (degrees Celsius)	Excess air coefficient	Engine power reduction (percent)
0	760.0	1.225	15.0	1.00	0
1,000	674.1	1.112	8.5	0.89	11.3
2,000	596.1	1.007	2.0	0.80	21.5
3,000	525.8	0.999	-4.5	0.71	30.8
4,000	452.3	0.819	-11.0	0.63	39.2
5,000	405.1	0.736	-17.5	0.56	46.7

Source: Babkov and Zamakhayev 1967

mixture composition is characterized by its *excess air coefficient*, which is the ratio of air supplied to the quantity required. Gasoline-fueled automobile engines usually operate with an excess air coefficient ranging from 0.8 to 1.2. Air density decreases with altitude, with the air admitted to the engine reducing by about 4 to 5 percent per 100 meters of altitude and, with it, engine power. As a result of the reduced air density the fuel-air mixture becomes richer, leading to a further decrease in engine power and an increase in emissions of carbon monoxide and hydrocarbons.

For example, at 2,000 meters (the altitude of Mexico City), air pressure is only 80 percent of that at sea level, so a given volume of air contains only 80 percent as much oxygen. For vehicles using carburetors or continuous fuel injection systems, the change in atmospheric pressure from one elevation to another affects the air-fuel ratio. If a carburetor is set for stoichiometric operation at sea level, it will mix the same mass of fuel with the same volume of air at 2,000 meters. But the same volume of air contains 20 percent less oxygen, so the resulting air-fuel mixture is 25 percent too rich. Conversely, a carburetor set for stoichiometric operation at 2,000 meters will produce too lean a mixture at sea level.

It is important that vehicles be adjusted for the altitude at which they operate. This is a problem at high altitudes, since vehicles are often preset for operation near sea level. When vehicles are driven to high-altitude areas, the resulting rich mixture has little effect on vehicle performance, though it increases fuel consumption and carbon monoxide emissions. Since performance is not affected, the vehicle owner may not find it necessary to have the engine adjusted. With proper adjustment, vehicle emissions at high altitudes are similar to those at sea level (although engine power falls in proportion to air pressure). Adjustments are not necessary in vehicles equipped with modern electronic fuel injection systems, since these systems automatically compensate for altitude changes.

Exhaust Gas Recirculation

Dilution of the incoming charge with spent exhaust affects pollutant formation, and can be useful in controlling nitrogen oxide emissions. In four-stroke engines, this dilution is achieved through exhaust gas recirculation. Exhaust gas recirculation can be used in two strokes, but a more convenient alternative is to reduce the scavenging ratio, thus increasing the amount of spent gas in the cylinder. The effect of exhaust gas dilution is similar to that of excess combustion air—diluting the combustion reactants reduces flame temperature and flame speed. Exhaust gas has a greater effect on flame speed and nitrogen oxide emissions than the same quantity of excess air (Heywood 1988). This is caused by the greater heat capacity of the carbon dioxide and water contained in the exhaust and the reduced oxygen content of the charge. As with excess combustion air, too much exhaust leads to unacceptable variation in combustion and increased hydrocarbon emissions. The degree of exhaust gas dilution that can be tolerated depends on the ignition system, combustion chamber design, and engine speed. Ignition systems and combustion chamber designs that improve performance with lean mixtures (*lean burn—fast burn combustion systems*) also improve performance with high levels of exhaust gas recirculation.

Combustion Timing

The relationship between the motion of the piston and the combustion of the charge has a major effect on pollutant emissions and engine efficiency. Combustion should be timed so that most of the combustible mixture burns near or slightly after the piston reaches top-dead-center. Mixtures that burn late in the expansion stroke do less work on the piston, decreasing fuel efficiency. Mixtures that burn before top-dead-center increase the compression work done by the piston, also decreasing efficiency. Since combustion takes time to complete, it is necessary to compromise between these two effects.

The timing of the combustion process is determined by the timing of the initial spark, the length of the ignition delay, and the rate of flame propagation through the mixture. Flame propagation is controlled by the geometry and turbulence level in the combustion chamber. Only the timing of the initial spark can be controlled without redesigning the engine. The greater the ignition delay and the slower the flame propagation rate, the earlier the initial spark must be to maintain optimal combustion timing. For typical gasoline engines, the optimal spark advance is 20 to 40 degrees crankshaft rotation before top-dead-center. The optimal spark advance is also a function of engine speed—at higher speeds, a larger angular advance is required, since the ignition delay time remains roughly constant.

The portion of the air-fuel mixture that burns at or before top-dead-center accounts for a disproportionate share of nitrogen oxide emissions, since the burned gases remain at high temperatures for long periods. To reduce nitrogen oxide emissions, it is common to retard the ignition timing somewhat in emission-controlled engines. Excessively retarded ignition timing can increase hydrocarbon emissions, reduce power output, and increase fuel consumption.

With lean mixtures, the ignition delay becomes longer and the flame speed becomes slower, so optimal spark timing is further advanced than for a stoichiometric mixture. For this reason, design features to increase flame speeds and reduce combustion time are important for lean-burn engines. Several combustion systems to achieve these goals have been developed.

Fast-burn techniques. To reduce knocking and improve efficiency, the time required for combustion should be minimized. This can be done by designing the combustion chamber to maximize flame speed and burning rate or minimize the distance the flame has to travel, or both.

The fraction of fuel burned and its derivative, the combustion rate, are shown for conventional and fast-burn combustion chambers in figure A3.1.6. In the conventional chamber, the flame spreads at a relatively slow rate, giving a long combustion time. To avoid having too much fuel burn late in the expansion stroke, it is necessary to advance the timing so that a significant part of the fuel burns before top-dead-center. The hot gases produced in this combustion, instead of expanding, must be further compressed. This subtracts from the net work output of the engine, and also increases nitrogen oxide emissions.

With the fast-burn chamber, the flame spreads rapidly because of turbulence, giving a higher combustion rate and shorter combustion duration. This means that the start of combustion can be later than in a conventional engine. Since very little combustion takes place until after top-dead-center, the burned gases are not recompressed, efficiency is increased, and nitrogen oxide emissions are lower. There is also less fuel burned during the later stages of the expansion stroke, which also contributes to better efficiency. Finally, reducing

the combustion time reduces the time available for the remaining unburned mixture to undergo pre-flame reactions and self-ignite, causing knock. Reducing the tendency to knock allows an increase in compression ratio, further increasing efficiency (box A3.1.1).

Ignition systems. The type of ignition system used and the amount of energy delivered have an important effect on the ignition delay and subsequent combustion. For any flammable mixture, there is a minimum spark energy required for ignition. Both the minimum ignition energy and the ignition delay are lowest for stoichiometric mixtures, and increase greatly as the mixture becomes leaner or more diluted. The minimum ignition energy also increases with increasing gas velocity past the spark plug. Increasing the spark energy beyond the minimum required for ignition helps reduce both the length and the variability of the ignition delay. Lean-burn engines and engines with high exhaust gas recirculation require much higher energy ignition systems than conventional stoichiometric engines.

High-ignition energies are attained by increasing the spark gap (and thus the breakdown voltage) or the stored energy available to supply the arc. The first approach is more effective. Transistorized coil and distributorless electronic ignition systems are increasingly used. Driven by a computer chip, these can provide flexible control of ignition timing.

Stratified charge. To obtain reliable ignition of lean or diluted air-fuel mixtures, it is advantageous to have a

Figure A3.1.6 Combustion Rate and Crank Angle for Conventional and Fast-Burn Combustion Chambers

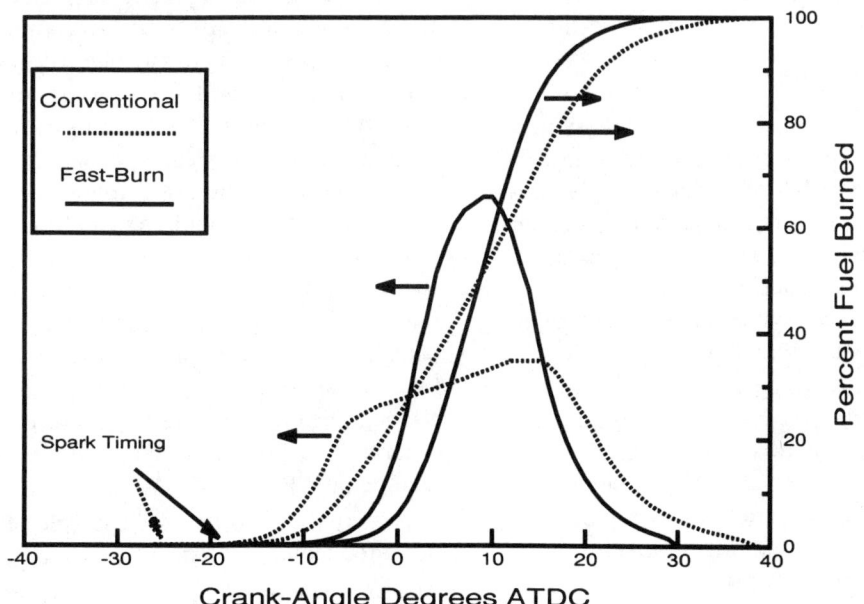

Source: Weaver and Chan 1995

Box A3.1.1 Compression Ratio, Octane, and Fuel Efficiency

The efficiency that is theoretically possible for an internal-combustion engine is a function of its *compression ratio*. A higher compression ratio increases efficiency, though improvement is less for ratios larger than 12:1. Frictional losses also increase with the compression ratio, so the optimal compression ratio for most engines is between 12:1 and 16:1. Compression ratios are limited in spark-ignition engines by the problem of *knock*. As unburned air-fuel mixture in the engine cylinder is compressed ahead of the flame front, its temperature increases, and it undergoes pre-flame reactions. Given sufficient time and temperature, the mixture self-ignites. The resulting shock waves can damage the engine or cause it to overheat.

The ability of a fuel to resist self-ignition is measured by its octane level. There are two octane measurements, the research octane number (RON) and the motor octane number (MON). Compared with other fuels for spark-ignition engines, gasoline has poor anti-knock performance. Typical RON for refined gasoline ranges from 91 to 97, and gasoline obtained directly from crude oil distillation can have a RON as low as 70. The RON for methanol is about 106, for propane 112, and more than 130 for natural gas.

Knocking increases with compression ratio. Fuel knock resistance therefore determines the compression ratio that can be used, and thus the fuel efficiency of the engine. Because diesel engines do not premix air and fuel in the cylinder, they do not suffer from knock, and can use higher compression ratios. This is one of the main reasons that diesel engines are more fuel-efficient than gasoline engines. Because of gasoline's poor knock resistance, gasoline engines are mostly limited to compression ratios of 8:1 or 9:1. More knock-resistant fuels, such as methanol and natural gas, use compression ratios of 11:1 or greater. Fast-burn techniques are especially important for engines using lean mixtures. Faster combustion and higher compression ratio allow lean mixtures to be burned, reducing emissions and improving efficiency. This is the basis of the lean-burn/fast-burn engine. These engines have a 15 to 20 percent advantage in fuel consumption compared with stoichiometric engines. Pre-catalyst emissions of carbon monoxide and nitrogen oxides are also lower, though hydrocarbon emissions from these engines may be higher than with a conventional stoichiometric engine.

Source: Weaver and Chan 1995

---

more easily ignitable mixture in the space near the spark plug. Combustion of this mixture ignites the more difficult-to-ignite mixture in the rest of the chamber. This charge stratification is achieved by placing the spark plug in a separate prechamber with its own fuel feed, or by stratifying the charge in the main combustion chamber through a combination of fluid dynamics and the timing of fuel injection. Because stratified charge systems are complex and costly, they are rarely used in automotive engines. They are common in large lean-burn engines used in stationary applications.

Cold-Start Emission Control

Engines using gasoline and other liquid fuels require special systems to start under cold conditions. In cold conditions, fuel does not fully evaporate. It is necessary to enrich the mixture, providing more fuel than normal so that even partial fuel vaporization produces an ignitable mixture. In carbureted vehicles, this is accomplished by the choke; in vehicles with fuel injection, extra fuel must be injected to provide the necessary enrichment.

The rich mixture and poor combustion under cold-start conditions cause high hydrocarbon and carbon monoxide emissions (table A3.1.2). This problem is aggravated in vehicles with catalytic converters because the catalyst has not reached operating temperature, and is thus ineffective in reducing these emissions. In modern emission-controlled vehicles, cold starting accounts

for more than 80 percent of total hydrocarbon and carbon monoxide emissions (Hellman and others 1989). It is important to reduce the time spent operating in this mode. Cold-start emission control devices include automatic chokes (under thermostatic or electronic control) and inlet air heaters. These heaters divert exhaust gas past the inlet manifold or pull engine intake air through a shroud surrounding the exhaust. This reduces warmup time but does not affect the need for initial enrichment to start the engine. A few recent engines include an electronically controlled electric heater grid in the air intake to assist in fuel evaporation.

To comply with the stringent low-emission vehicle (LEV) emission standards adopted in California, suppliers have focused on reducing cold-start hydrocarbon emissions. The simplest approach, and the most pursued, is to design the catalytic converter system to light-off quickly, which requires that either the main catalytic converter or a smaller "precatalyst" be positioned close to the engine, and that the heat capacity of the intervening exhaust system be minimized. This can make the catalyst more vulnerable to thermal damage under high loads. Emissions of hydrocarbons and nitrogen oxides are significantly lowered for close-coupled catalysts whereas carbon monoxide emissions remain unchanged (ACEA/EUROPEA 1995).

A number of other technologies have been developed to reduce cold start emissions. Of these, the electrically heated catalyst (EHC) system is the most effective. By preheating the catalytic converter, cold

Table A3.1.2 Cold-Start and Hot-Start Emissions with Different Emission Control Technologies
(grams per test)

Technology	Carbon monoxide			Hydrocarbons			Nitrogen oxides			Fuel consumption (liters per 100 kilometers)		
	Hot start	Cold start	CS:HS[a]	Hot start	Cold start	CS:HS[a]	Hot start	Cold start	CS:HS[a]	Hot start	Cold start	CS:HS[a]
Uncontrolled	49.7	100.6	2.0	32.2	25.1	0.8	12.2	11.8	0.97	10.1	12.9	1.3
Lean-burn	9.6	29.8	3.1	11.1	11.1	1.0	8.6	8.7	1.0	9.0	10.2	1.1
Three-way catalyst	11.1	32.0	2.9	1.8	3.8	2.1	3.4	4.8	1.4	9.8	11.3	1.2
Lean-burn and oxidation catalyst	1.8	22.9	12.7	2.5	6.2	2.5	7.2	8.2	1.1	8.9	10.4	1.2

Note: Based on ECE-15 testing procedure. All vehicles tested had a 1.6 liter engine.
a. Ratio of cold-start to hot-start emissions and fuel consumption.
Source: Pearce and Davies 1990

start hydrocarbon and carbon monoxide emissions can be reduced by more than two-thirds. Although effective in reducing cold-start emissions, the EHC has many drawbacks, including cost, weight, and energy consumption. In addition to the heated catalyst, an EHC system requires additional battery capacity, heavy cables, and expensive electronics to handle the amperages involved, an air pump and air injection system to supply oxygen to the catalyst, and most likely a larger alternator. Other approaches to preheating the catalyst include the use of separate burners igniting the exhaust itself. This last requires air injection combined with a rich air-fuel ratio to form an ignitable mixture. Another approach that has been developed has been to capture the hydrocarbon emissions temporarily by adsorbtion on specially designed materials when the engine is cold. As the exhaust temperature increases, the adsorbed hydrocarbons are released. The catalytic converter by then is warmed up and operational.

Idling Emission Control and Fuel Cut-Off Systems

Where the idling period is a large portion of the total driving period, as in many congested cities in developing countries (half of the driving time in Delhi is spent in the idling range of engine operation during peak traffic hours), total exhaust emissions of carbon monoxide and hydrocarbons may be reduced by lowering the idling speed (say, from 900 rpm to 600 rpm) and adjusting the ignition timing to achieve a stable idling operation. Similarly, where deceleration constitutes a significant portion of the driving period (20 to 25 percent), exhaust emissions may be reduced by cutting off the fuel supply during deceleration through a solenoid valve installed in the primary idling circuit of the carburetor. Use of such a fuel cut-off system during deceleration can reduce exhaust emissions by 90 percent, while vehicle handling in downhill driving is improved (Khatri and others 1994). Electronic fuel in-

jection systems incorporate fuel cutoffs as a matter of course.

Exhaust Aftertreatment

A useful alternative to controlling emissions within the engine cylinder is to reduce them instead with subsequent treatment of the exhaust gas. This allows the combustion process within the cylinder to be optimized (within some limits) for best power and fuel economy, rather than for lowest emissions. The two aftertreatment technologies that have seen wide use on spark-ignition vehicles are air injection and the various types of catalytic converters. In addition, thermal reactors have been used to promote oxidation of engine-out carbon monoxide and hydrocarbon emissions but systems of this type have faded from use (Faiz and others 1990).

Air injection. The exhaust gases expelled from the cylinder of an Otto-cycle engine contain significant amounts of unburned hydrocarbons and carbon monoxide. If sufficient oxygen is present, these gases will continue to react in the exhaust system, reducing the quantities of these pollutants that are ultimately emitted from the tailpipe. The reaction rate is extremely sensitive to temperature—a minimum of 600°C is needed to oxidize hydrocarbons significantly, and a minimum of 700°C for carbon monoxide. To provide the needed oxygen under rich or stoichiometric conditions, additional air is injected into the exhaust manifold. This air is provided either by a separate air pump or by a system of check valves that use the normal pressure pulsations in the exhaust manifold to draw in air from outside. The latter system, pulse air injection, is cheaper but provides a smaller quantity of air.

Air injection was first used as an emissions control technique in itself. It is still used for this purpose in heavy-duty gasoline engines and four-stroke motorcycles (both engine types operate under rich conditions at full load). Air injection is also used with oxidizing catalytic converters to ensure the mixture entering the catalyst has an air–fuel ratio greater than stoichiometer. In vehicles with three-way catalytic converters, air injection before the catalyst must be avoided for it to control nitrogen oxide emissions. So in three-way catalyst vehicles, air injection is used primarily during cold starts, and is cut off during normal operation.

Catalytic converters. To perform their function, catalyst materials in the catalytic converter must be in close contact with exhaust gas. The most common catalyst design is a single piece of ceramic (ceramic monolith), extruded to form numerous small parallel channels. Exhaust gases flow lengthwise through the parallel channels, which provide a high surface area. Other types include ceramic bead beds and cellular structures comprised of metal alloys. Alloys are mostly used in situations where mechanical or thermal shock would damage ceramic.

Chemical reactions in the catalytic converter take place at the catalyst surface. The average catalytic converter contains only a fraction of a gram of the catalytic metals. To make efficient use of this expensive material, it must have a high surface area readily accessible to the exhaust gases. The most common and damaging maintenance problem with catalytic converters is the reduction of the surface area of the exposed catalyst. For example, lead in fuel and phosphorus from fuel or engine oil can form deposits on the surface of the catalyst material, blocking the pores and destroying its efficiency. Excess temperatures can make the metal crystals sinter together, losing surface area, or even melting the ceramic monolith. This causes a loss of porosity and a drop in conversion efficiency. Such high temperatures are commonly caused by excessive combustible materials and oxygen in the exhaust (due to a misfiring cylinder, for instance). These materials react in the catalytic converter and can raise its temperature enough to cause permanent damage. Correcting the cause of the misfire will not restore emissions performance unless the catalyst is replaced. Reliable fuel and ignition systems are vital to the effective performance of catalyst systems.

Emission Control Technology for Two-Stroke Engines

Compared with four-stroke spark-ignition engines, two strokes exhibit vastly higher hydrocarbon and particulate matter emissions. Carbon monoxide emissions are comparable to four-stroke engines, while nitrogen oxide emissions are somewhat less. The major sources of hydrocarbon emissions in two strokes are the loss of unburned air-fuel mixture into the exhaust during scavenging and emissions caused by misfire or partial combustion at light loads. Studies indicate that fresh-charge losses into the exhaust can be as high as 30 percent for conventional two-stroke engines (Hare and others 1974; Batoni 1978; Nuti and Martorano 1985). Under light-load conditions, the flow of fresh charge is reduced, and substantial amounts of exhaust gas are retained in the cylinder. This residual gas leads to incomplete combustion or misfire. Misfiring causes the popping sound produced by two-stroke engines at idle and light loads and the problems these engines often have in maintaining stable idle. Unstable combustion is a major source of unburned hydrocarbon exhaust emissions at idle or light-load conditions (Tsuchiya and others 1980; Abraham and Prakash 1992; Aoyama and others 1977).

Technologies to reduce two-stroke engine emissions include advanced fuel metering systems, improved scavenging characteristics, combustion chamber modifications, improved ignition systems, exhaust aftertreatment technologies, and improvements in engine lubrication. The application of these technologies to small utility, moped, and motorcycle engines to reduce exhaust emissions has been reported in a number of studies, and there is real-world experience with some of these techniques, particularly in Taiwan (China).

Advanced Fuel Metering

Air and fuel metering can improve engine performance and fuel consumption and reduce exhaust emissions. Conventional carburetor systems for two-stroke motorcycle engines are designed to provide smooth and stable operation under a variety of speed and load conditions, but give little consideration to fuel consumption or exhaust emissions. The advantages of fuel injection in two-stroke engines are twofold—precise control of the air-fuel ratio, and the ability to reduce the loss of fresh charge into the exhaust by in-cylinder injection or properly timed injection into the transfer port.

Electronic fuel injection systems in current four-stroke automobiles provide a metered amount of fuel based on a measure of the air flow into the engine. The fuel supply system consists of a fuel pump, fuel filter, and pressure regulator. The fuel injector is a high-speed solenoid valve connecting the pressurized fuel supply to the engine air intake. By opening the valve, the electronic control unit permits pressurized fuel to spray into the air intake, where it mixes with air, vaporizes, and is inducted into the engine.

Similar fuel injection systems have been developed for use in advanced two-stroke engines. These systems typically inject fuel into the transfer port rather than directly into the cylinder. With precise injection timing, it is possible to reduce the hydrocarbon content of the air that short-circuits the combustion chamber during scavenging. Because of the need to assure a combustible mixture at the spark plug, it is not possible to completely eliminate short-circuiting by this means.

An alternative fuel injection approach eliminates short-circuiting entirely by injecting the fuel directly into the cylinder near or after the time the exhaust port closes. Because direct injection into the cylinder provides little time for fuel mixing and vaporization, direct fuel injection systems must inject quickly to achieve fuel atomization. This generally requires the use of air-blast injection, in contrast to the fuel-only atomization used in indirect injection systems. For direct-injection systems, an air pump is required to supply compressed air for the air-blast injection. A fuel injection system should deliver extremely small fuel droplets to control spray penetration and fuel distribution. It must mix fuel adequately with available air in the available time at the high engine speeds typical of two-stroke operation. Also, the fuel droplet size must be consistent throughout the fuel spray to ensure minimum coalescence of the droplets toward the end of the spray plume.

Atomization quality is based on the size of the droplets, measured in terms of the Sauter Mean Diameter (SMD). To vaporize the fuel spray quickly, a fuel droplet SMD of 10 to 20 microns is required for direct fuel injection. For indirect injection, a fuel droplet SMD of 100 microns is acceptable, since the fuel vaporizes in the intake port and during the compression stroke.

Because of the achievements reported for engines using the Orbital Combustion Process (OCP) and similar direct-injection approaches, two-stroke engines are currently a major area of automotive research and development. Some prototype two-stroke engines have reached emissions levels comparable to four-stroke engines (Duret and Moreau 1990; Huang and others 1991; Monnier and Duret 1991; Laimbock and Landerl 1990; Douglas and Blair 1982; Plohberger and others 1988; Blair and others 1991). Limited studies have been carried out on the application of direct fuel-injection systems in small two-stroke engines. In one such study, the Institute Francais du Petrole (IFP) designed and devel-

oped a marine outboard two-stroke engine based on a converted production engine, using a direct air-assisted fuel-injection system with compressed air supplied by the pumping action of the crankshaft (Monnier and Duret 1991). This system, named IAPAC, includes a surge tank that stores the compressed air from the crankcase. This surge tank serves as a reservoir to supply compressed air to the pneumatic fuel injection device.

Several versions of the IAPAC engine system were tested for performance and exhaust emissions (table A3.1.3). These included engines with the original and with a newly designed intake manifold, and an engine with an external compressor for supplying the compressed air. A 1.2 liter, 85-horsepower engine had 27 percent lower fuel consumption and reductions in hydrocarbon and carbon monoxide emissions of 84 percent and 79 percent, respectively. Although nitrogen oxide emissions increased by 133 percent, these levels are still low compared with four-stroke nitrogen oxide emissions.

Researchers at the Industrial Technology Research Institute (ITRI) of Taiwan (China) have demonstrated a low-pressure, air-assisted, direct fuel injection system in a two-stroke scooter engine (Huang and others 1991; 1993). Prior to this system, ITRI tested a cylinder-wall fuel injection system with and without air assist for fuel atomization. Exhaust emissions and fuel economy results for two 82 cc, two-stroke scooters with carbureted engines under best-tuned conditions are shown in table A3.1.4.

ITRI researchers achieved a fuel droplet SMD of 70 microns for the cylinder-wall injection system without air assist. This was done using an injector nozzle with spiral grooves. Emissions were reduced by 25 percent for hydrocarbons and 8 percent for carbon monoxide, and fuel economy increased 14 percent. As expected, nitrogen oxide emissions increased substantially because of leaner combustion.

With low-pressure air assist in the cylinder-wall injection system, fuel droplet SMD decreased to 35 microns. Compared with the carburetor system, this air-assisted

Table A3.1.3   Engine Performance and Exhaust Emissions for a Modified Marine Two-Stroke Engine
(grams per kilowatt hour)

| Configuration | Emissions | | | BSFC[a] | Power (kilowatts) |
	Carbon monoxide	Hydrocarbons	Nitrogen oxides		
Production	211.0	164	1.5	514	59.2
IAPAC with original intake system	69.9	55	3.5	396	56.8
IAPAC with new intake system	61.5	44	3.5	365	56.8
IAPAC with new intake system and external compressed air (HIPAC)	43.9	26	3.5	373	56.8

a. Break specific fuel consumption.
Source: Monnier and Duret 1991

Table A3.1.4 Exhaust Emissions and Fuel Economy for a Fuel-Injected Scooter
(grams per kilometer)

Configuration	Emissions			Fuel economy (kilometers per liter)
	Carbon monoxide	Hydrocarbons	Nitrogen oxides	
Carburetor	3.7	3.83	0.03	42.1
Cylinder wall injector	3.4	2.90	0.06	48.0
Air-assisted cylinder wall injector	3.0	2.58	0.09	50.4
Air-assisted cylinder head injector	2.5	2.12	0.16	52.8
Air-assisted cylinder head injector with skip-injection	1.6	1.62	0.17	55.1
Air-assisted cylinder head injector with skip-injection and catalytic converter	0.09	0.28	0.16	55.3
Air-assisted cylinder head injector with skip-injection, and catalytic converter with secondary air	0.08	0.25	0.16	55.0

Source: Huang and others 1993

fuel injection system reduced hydrocarbon and carbon monoxide emissions by 34 and 18 percent, respectively, and improved fuel economy by 20 percent. Nitrogen oxide emissions increased by 200 percent. Larger hydrocarbon and carbon monoxide reductions and further fuel economy improvement were achieved with the air-assisted fuel injection system in the cylinder head. This produced a 45 percent reduction in hydrocarbons and a 32 percent reduction in carbon monoxide emissions, with a 53 percent improvement in fuel economy. This is presumably due to the better charge stratification possible with the fuel injection system in the cylinder head. The fuel droplet SMD was reduced to 15 microns with this system.

Significant reductions in hydrocarbon and carbon monoxide emissions can be achieved with fuel- injection systems. Innovative designs are needed to develop an economical and efficient fuel injection system for two-stroke motorcycle engines.

Skip-Firing

Another advantage of electronic fuel injection systems is the ability to shut off fuel injection in some engine cycles. Fuel supply can be shut off for consecutive engine cycles under idle and light-load conditions. This allows exhaust gas to be purged from the combustion chamber, providing better combustion conditions for the next engine firing cycle. Irregular combustion under light-load conditions can thus be eliminated or minimized.

Researchers at ITRI have applied the skip-injection technique to a scooter engine to minimize unburned hydrocarbon emissions caused by irregular combustion under idle and light-load conditions (Huang and others 1991; 1993). They found that without skip-firing, the in-

dicated mean effective pressure (IMEP) varied significantly at idle, and many cycles could be identified as having incomplete combustion or complete misfire. This caused high concentrations of unburned hydrocarbon emissions—3,500 to 4,000 ppm of hexane equivalent in the exhaust. Several skip-injection modes were investigated, including fuel injection every two, three, four, and five cycles. Results indicated that IMEP variations decreased as skipped injections increased. In an engine dynamometer test with fuel injected every four cycles, hydrocarbon emissions and fuel flow at idle were reduced by 50 percent and 30 percent, respectively. The skip-injection mode was also applied and tested in a scooter engine. Hydrocarbon and carbon monoxide emissions were reduced by 58 percent and 57 percent, respectively, with a 31 percent improvement in fuel economy.

Scavenging Control Technologies

In a two-stroke engine, exhaust and intake events overlap. As the piston finishes its downward stroke and begins to move from the bottom of the cylinder to the top, exhaust ports in the walls of the cylinder are uncovered. High-pressure combustion gases blow into the exhaust manifold. As the piston nears the bottom of its stroke, intake ports are opened and fresh air or air-fuel mixture is blown into the cylinder while the exhaust ports are still open. Piston movement timing (measured in crank angle) and cylinder port configuration are the major factors for controlling scavenging. Ideally, the fresh charge would be retained in the cylinder (trapping efficiency) while the spent charge from the last cycle is exhausted (scavenging efficiency). The two goals conflict. Cylinder ports and timing are generally designed for scavenging efficiency, achieving maximum power output and smoother idle at the expense of increased short-circuiting and hydrocarbon emissions. It is possible to reconfigure intake and exhaust ports to fine-tune scavenging characteristics for lower emissions, but this involves performance tradeoffs. Another way to increase trapping efficiency, with minimum impact on performance, is applying exhaust charge control technology.

Exhaust charge control technology modifies exhaust flow by introducing control valves in the exhaust or by using the exhaust pressure pulse wave. Using the exhaust pressure pulse wave to control intake and exhaust flow requires a long exhaust pipe, and is effective only for a restricted range of engine speeds. Control valves are usually used to control the exhaust flow rate in small engines. The critical variable for exhaust charge control techniques is the contraction ratio, defined as the ratio of the unrestricted exhaust passage area to the restricted exhaust passage area regulated by the valve. The effectiveness of these techniques is measured by the delivery ratio, which is the ratio between the mass of air-fuel mixture actually delivered to the engine and the mass of air-fuel mixture contained by the engine displacement volume at ambient conditions.

Research has demonstrated the potential of exhaust charge control valves in small two-stroke engines (Hsieh and others 1992; Tsuchiya and others 1980; Duret and Moreau 1990). Significant reductions in hydrocarbon emissions and fuel consumption can be achieved, as can a reduction in unstable combustion at light load. The delivery ratio where rapid increase in irregular combustion occurs (defined as the critical delivery ratio) is 0.2 (Tsuchiya and others 1980). This is similar to a recent finding of 0.25 as the critical delivery ratio (Hsieh and others 1992). With exhaust charge control, the critical delivery ratio decreases from 0.25 to 0.20 and 0.15 at low and medium engine speeds (1,500 and 3,000 rpm), respectively (Hsieh and others 1992). The exhaust charge control technique reduces irregular combustion under light-load conditions. Hydrocarbon emissions and fuel consumption were reduced by 30 percent and 6 percent, respectively, when the exhaust charge control technique was used in a test engine. At the same delivery ratio, the engine with exhaust charge control produced higher power output. A 60 percent reduction in hydrocarbon emissions and a 20 percent reduction in fuel consumption can be achieved with an exhaust charge control valve (Duret and Moreau 1990).

Honda has incorporated a "Revolutionary Controlled Exhaust Valve (RC Valve)" in a 150 cc, two-stroke motorcycle equipped with a capacitive-discharge ignition, computerized controller, and servo motor to attain high power efficiency at low- and high-speed conditions (Tan and Chan 1993). Although the RC Valve is intended to improve engine performance, it can also serve as an emissions control device.

Other Engine Modifications

Other engine modifications and techniques include improved combustion chamber and piston configurations, improved lean–dilute combustion to prevent misfire, and changes in ignition timing and design.

Combustion chamber. Combustion chamber and piston configurations can be improved to induce more turbulent motions during the compression stroke and to control the direction of the fresh charge to minimize short-circuiting. Improved combustion chamber and piston configurations can also minimize the formation of pockets or dead zones in the cylinder. Researchers at Graz University of Technology (GUT) in Austria designed a combustion chamber that concentrated the squish area above the exhaust port (Laimbock and Landerl 1990). This design forces the fresh charge to overflow the spark plug, which improves cooling and allows the engine to run leaner without pre-ignition.

Lean–dilute combustion technology. Lean-burn engines can develop low nitrogen oxide and carbon monoxide emissions and good fuel economy, but they require a high-energy ignition system or stratified charge to avoid misfire and cyclic instability, resulting in high hydrocarbon emissions. Stratified-charge two-stroke engines with fuel injection systems include the OCP, PROCO, and DISC engines developed by Orbital, Ford, and General Motors, respectively. Another concept is the two-step, stratified-charge for a crankcase-scavenged, two-stroke engine (Blair and others 1991). In this system, air is introduced to scavenge the burned gas in the cylinder, then a rich mixture is delivered to the cylinder just after scavenging.

Ignition timing. Ignition timing on two-stroke motorcycle engines is usually chosen to optimize power output rather than fuel economy or exhaust emissions. The effects of ignition timing on combustion and emissions in two-stroke engines are similar to the effects in four-stroke engines. Retarding ignition timing reduces power and increases fuel consumption, but reduces nitrogen oxide and hydrocarbon emissions. Retarding ignition timing, especially at high loads, may recover the increased nitrogen oxide emissions that otherwise result from a lean mixture in low-emission, two-stroke engines.

The effect of ignition timing on direct-injected, spark-ignition engines is different from that in other spark-ignition engines. In direct-injection systems, advancing ignition timing at light load reduces hydrocarbon emissions by reducing the dispersion of the fuel cloud. This cloud is less likely to contact the walls of the combustion chamber. This reduces the hydrocarbons produced by the quenching effect at the combustion chamber walls, as well as the filling of crevice volumes with unburned mixture. Unburned hydrocarbons caused by flame quenching and crevice volumes are major sources of hydrocarbon exhaust emissions. With better combustion quality at advanced ignition timing, carbon monoxide emissions are also reduced. Nitrogen oxide emissions, however, are increased with advanced ignition timing.

Dual spark-plug ignition. Researchers at ITRI have used a dual spark-plug ignition in a scooter two-stroke engine to determine the effects on engine torque and unburned hydrocarbon emissions (Huang and others 1991). The engine with dual spark plugs had lower hydrocarbon emissions and better engine torque at low- and medium-load. Improvements were thought to be the result of increased combustion speed and decreased mixture bulk quenching when the dual-plug ignition was used. ITRI's findings also showed that additional spark plugs did not improve the hydrocarbon emissions under idling or light-load conditions.

Exhaust Aftertreatment Technologies

Aftertreatment technologies such as thermal oxidation and catalytic converters provide emission control beyond that achievable with engine and fuel-metering technologies alone. Catalytic converters are used extensively in automobiles and have also been demonstrated in two-stroke motorcycles and other small engines (Burrahm and others 1991; Laimbock and Landerl 1990; Hsien and others 1992).

Thermal oxidation. Thermal oxidation is used to reduce hydrocarbon and carbon monoxide emissions by promoting oxidation in the exhaust. This oxidation takes place in the exhaust port or pipe and may require additional air injection. Substantial reductions in hydrocarbon and carbon monoxide emissions can be achieved if the exhaust temperature is high enough for a long period. The temperatures required for hydrocarbon and carbon monoxide oxidation are 600 and 700°C, respectively. These requirements are difficult to meet in small engines with short exhaust systems, but they have been demonstrated with a four-stroke engine by introducing secondary air into the stock exhaust manifold upstream of the engine muffler. Air injection at low rates into the stock exhaust system was found to reduce hydrocarbon emissions by 77 percent and carbon monoxide emissions by 64 percent (White and others 1991). This was only effective under high-power operat-

ing conditions. The temperatures required to achieve oxidation substantially increase the external temperature of the exhaust pipe.

Oxidation catalyst. Like thermal oxidation, the oxidation catalyst is used to promote oxidation of hydrocarbon and carbon monoxide emissions in the exhaust stream. It also requires sufficient oxygen for the reaction to occur. Some of the requirements for a catalytic converter to be used in two-stroke engines include high hydrocarbon conversion efficiency, resistance to thermal damage, resistance to poisoning by sulfur and phosphorus compounds in the lubricating oil, and low light-off temperature. Additional requirements for catalysts in motorcycle engines include extreme vibration resistance, compactness, and light weight.

Using catalytic converters in two-stroke engines is difficult because of the high concentration of hydrocarbons and carbon monoxide in their exhaust. If combined with sufficient air, these concentrations cause catalyst temperatures that easily exceed the temperature limits of the catalyst, and also pose the risk of fire or injury to motorcycle users. Catalytic converters in two strokes require engine modifications to reduce pollutant concentrations in the exhaust, and may require that air supply be limited to the exhaust before the catalyst.

Oxidation catalytic converters have been applied to small engines. Researchers at GUT and ITRI have published data on the application of catalytic converters in small two-stroke engines. The Graz researchers focused on reducing exhaust emissions from two-stroke moped, motorcycle, and chainsaw engines by using catalytic converters, and on improving the thermodynamic characteristics of the engine, such as gas exchange and fuel handling systems, cylinder and piston geometry and configurations, and exhaust and cooling systems. The effects of catalytic converters on two moped engines tested under the ECE-15 driving cycle are shown in table A3.1.5. In the 1.2 horsepower engine, the catalytic converter reduced hydrocarbon and carbon monoxide emissions by 96.7 and 98.5 percent, respectively. There was no change in nitrogen oxides. For the 2.7 horsepower engine, hydrocarbon, carbon monoxide, and nitrogen oxide emissions were reduced by 96.4, 87.9, and 27.8 percent, respectively. Hydrocarbon and carbon monoxide emissions were reduced even further when an advanced engine with a watercooled cylinder and four scavenger ports was used with the catalytic converter. Data on catalyst temperatures and emissions durability are not available.

ITRI researchers retrofit a catalytic converter to a 125 cc, two-stroke motorcycle engine and demonstrated effective emissions control and durability

Table A3.1.5   Moped Exhaust Emissions
(grams per kilometer)

Engine configuration	Carbon monoxide	Hydrocarbons	Nitrogen oxides
1.2 horsepower			
Without catalyst	2.716	1.013	0.049
With catalyst	0.09	0.015	0.049
2.7 horsepower			
Without catalyst	3.205	0.771	0.09
With catalyst	0.116	0.093	0.065
Advanced engine with catalyst	0.037	0.022	0.067

Source: Douglas and Blair 1982

(Hsien and others 1992).The researchers evaluated the effects of catalyst composition and substrate, the cell density of the substrate, the converter size and installation location, and the use of secondary air injection on the catalytic effectiveness and engine performance. Their conclusions were: (i) metal substrate is superior to ceramic substrate for converter efficiency and engine performance, since the thin walls of the metal substrate have a larger effective area and lower back pressure; (ii) exhaust temperature profile, space availability, and the effects on engine exhaust tuning must be considered when installing the catalytic converter; (iii) additional catalyst would improve the carbon monoxide conversion efficiency in the rich air–fuel mixture typical of two-stroke motorcycle engines; (iv) the cell density of the substrate should be less than 200 cpsi to minimize pressure loss and maintain engine power; (v) hydrocarbon and carbon monoxide conversion efficiency increases significantly when secondary air is supplied; and (vi) exhaust smoke opacity was reduced with the use of the catalytic converter. Reduced opacity was probably the result of catalytic oxidation of the lubricating oil vapor. This has been observed in other engines.

ITRI also retrofit a catalytic converter to a two-stroke scooter along with fuel injection and skip-firing at idle. Adding the catalytic converter improved overall emissions control efficiency from 58.2 percent to 92.8 percent for hydrocarbons, and from 56.8 to 97.6 percent for carbon monoxide emissions. Efficiency improved only slightly with the use of secondary air. This is because the fuel-injected engine operated at lean conditions, so oxygen was available in the catalytic converter even without air injection.

Catalytic converters have been used for several years on two-stroke motorcycles and mopeds to meet emission standards in Austria, Switzerland, and Taiwan (China). In-use experience indicates they are acceptable, although special heat shielding is needed to protect passengers from contact with the catalyst housing,

which can reach 500°C. Conclusive data on the durability of these converter systems are not available.

Lubricating Oil Technologies

Lubricating oil is the major source of particulate matter emissions from two-stroke engines. Since the crankcase of a two-stroke engine is used for pumping air or the air-fuel mixture into the combustion chamber, it cannot be a lubricant-oil reservoir. Oil mist is therefore injected into the incoming air stream. As this stream passes through the crankcase, lubrication is provided for cylinder walls, crankshaft bearings, and connecting-rod bearings. Ball or roller bearings are typically used instead of the plain bearings of a four-stroke engine. The oil mist continues to the combustion chamber, where it is partly burned. Remaining unburned oil recondenses in the exhaust plume and creates blue or white smoke. Because phosphorus or other deposit-forming additives in the oil would poison a catalytic converter, two-stroke oils for catalyst-equipped motorcycles need to be formulated without these compounds.

Lubrication system. Three approaches are used to supply lubricating oil to two-stroke engines: pre-mixing with the fuel when it is added to the tank; line-mixing, in which oil is metered into the fuel between the fuel tank and the engine; and oil injection, in which the oil is metered directly into the intake manifold or other points using a pump controlled by engine speed or throttle setting. The second and third approaches are the most common in current motorcycles since they control the flow rate of the oil and provide reliable lubrication. Injection provides the best control. Orbital has designed an electronic lubrication system for their OCP two-stroke engines that reduces the amount of oil required by the engine. Several Yamaha two-stroke motorcycles marketed in Asia have also used an electronic oil metering system to alter the oil flow to the carburetor according to engine load. The Yamaha Computer-Controlled Lubrication System (YCLS) supplies required lubricating oil to the engine according to engine speed using an electronic control unit and three-way control valve (Tan and Chan 1993). In a fuel injected two-stroke engine, this function could be handled by the same electronic control unit as the fuel injection system.

Low-smoke oil. Conventional two-stroke lubricating oils are based on long-chain paraffin or naphthene molecules that break down slowly and are resistant to combustion. Using synthetic, long-chain polyolefin materials instead of naphthenes and paraffins can significantly reduce smoke opacity from two-stroke engines. Because of the double carbon bonds in the polyolefin chain, these chains break down more quickly and burn more completely. Studies have proven that substituting polyisobutylene for bright stock or other heavy lube-oil fractions in two-stroke lubricating oils can reduce engine smoke levels (Sugiura and Kagaya 1977; Kagaya and Ishimaru 1988; Eberan- Eberhorst and others 1979; Brown and others 1989). Lubricity is the same for polyisobutylene and bright stock. Low-smoke, polyisobuyltene-based lubricating oil is now required for two-stroke mopeds and motorcycles in some countries in southeast Asia, including Thailand.

Low-smoke oils will not solve the particulate problem for two strokes. Oil in the 20 to 30 percent of the fresh charge that short-circuits the cylinder will be unaffected by combustion. There is also a possibility that combustion of polyisobutene may increase emissions of toxic air contaminants, particularly 1,3 butadiene. Research is needed to assess the effect of these oils on particulate matter and other emissions.

References

Abraham, M. and S. Prakash. 1992. "Cyclic Variations in a Small Two-Stroke Cycle Spark-Ignited Engine—An Experimental Study." *SAE Paper* 920427. Society of Automotive Engineers, Warrendale, Pennsylvania.

ACEA/EUROPIA (European Automobile Manufacturer's Association). 1995. *European Program on Emissions, Fuels, and Engine Technologies.* EPEFE Report, Brussels.

Aoyama, T., M. Nakajima, M. Onishi, and S. Nakahiko. 1977. "Abnormal Combustion in Two-Stroke Motorcyle Engines." *SAE Paper* 770189, Society of Automotive Engineers, Warrendale, Pennsylvania.

Babkov, V., and M. Zamakhayev. 1967. *Highway Engineering.* MIR Publishers, Moscow.

Batoni, G. 1978. "Investigations into the Two-Stroke Motorcycle Engine." *SAE Paper* 780170, Society of Automotive Engineers, Warrendale, Pennsylvania.

Blair, G.P., R.J. Kee, C.E. Carson, and R. Douglas. 1991. "The Reduction of Emissions and Fuel Consump-

tion by Direct Air-Assisted Fuel Injection into a Two-Stroke Engine." Paper prepared for the 4th Graz Two-Wheeler Symposium, Technical University, Graz, Austria.

Bosch, Robert AG. 1986. *Automotive Handbook.* 2nd edition. Stuttgart.

Brown, R.M., D.A. Fog, and D.H. Garland. 1989. "Smoke Reduction In Two-Stroke Gasoline Engines." Paper prepared for the 5th International Pacific Conference on Automotive Engineering. Beijing, China.

Burrahm, R.W., J.J. White, and J.N. Carroll. 1991. "Small Utility Engine Emissions Reduction Using Automotive Technology." *SAE Paper* 911805, Society of Automotive Engineers, Warrendale, Pennsylvania.

Douglas, R. and G.P. Blair. 1982. "Fuel Injection of a Two-Stroke Cycle Spark Ignition Engine." *SAE Paper* 820952, Society of Automotive Engineers, Warrendale, Pennsylvania.

Duret, P. and J.F. Moreau. 1990. "Reduction of Pollutant Emissions of the IAPAC Two-Stroke Engine with Compressed Air Assisted Fuel Injection." *SAE Paper* 900801. Society of Automotive Engineers, Warrendale, Pennsylvania.

Eberan-Eberhorst, C.G., and H. Martin. 1979. "Performance Test Methods For Two-Stroke-Cycle Engine Lubricants Including Lean Fuel-Oil Ratio Conditions." *SAE Paper* 790078, Society of Automotive Engineers, Warrendale, Pennsylvania.

Faiz A., K. Sinha, M. Walsh, and A. Varma 1990. "Automotive Air Pollution – Issues and Options for Developing Countries." *Working Paper Series 492,* World Bank, Washington, D.C.

Hare, C.T., K.J. Springer, W.R. Houtman, and T. Huls, 1974. "Motorcycle Emissions, Their Impact, and Possible Control Techniques." *SAE Paper* 740267, Society of Automotive Engineers, Warrendale, Pennsylvania.

Hellman, K.H., R.I. Bruetsch, G.K. Piotrowski, and W.D. Tallen. 1989. "Resistive Materials Applied to Quick Lightoff Catalysts." *SAE Paper* 890799, Society of Automotive Engineers, Warrendale, Pennsylvania.

Heywood, J.B. 1988. *Internal Combustion Engine Fundamentals.* New York: McGraw Hill.

Hsieh, P.H., R.F. Horng, H.H. Huang, Y.Y. Peng, and J. Wang. 1992. "Effects of Exhaust Charge Control Valve on Combustion and Emissions of Two-Stroke Cycle Direct-Injection S.I. Engine." *SAE Paper* 922311. Society of Automotive Engineers, Warrendale, Pennsylvania.

Hsien, P.H., L.K. Hwang, and H.W. Wang. 1992. "Emission Reduction by Retrofitting a 125cc Two- Stroke Motorcycle with Catalytic Converter." *SAE Paper* 922175, Society of Automotive Engineers, Warrendale, Pennsylvania.

Huang, H.H., Y.Y. Peng, M.H. Jeng, and J.H. Wang. 1991. "Study of a Small Two-Stroke Engine with Low-Pressure Air-Assisted Direct-Injection System." *SAE Paper* 912350. Society of Automotive Engineers, Warrendale, Pennsylvania.

Huang, H.H., M.H. Jeng, N.T. Chang, Y.Y. Peng, J.H. Wang, and W.L. Chang. 1993. "Improvement of Exhaust Emissions on Two-Stroke Engine by Direct-Injection System." *SAE Paper* prepared for the March 1993 Society of Automotive Engineers Congress, Warrendale, Pennsylvania.

Kagaya, M., and M. Ishimaru. 1988. "A New Challenge For High Performance Two-Cycle Engine Oils." *SAE Paper* 881619. Society of Automotive Engineers, Warrendale, Pennsylvania.

Khatri, D.S., A. Ramesh, M.K.G. Babu, and F. Fallahzadeh. 1994. "Simple Techniques for Emission Control in a Passenger Car Engine." 3rd International Symposium on Transport and Air Pollution, *Poster Proceedings*, INRETS, Arcueil, France.

Laimbock, F.J. and C.J. Landerl. 1990. "50cc Two-Stroke Engines for Mopeds, Chainsaws, and Motorcycles with Catalysts." *SAE Paper* 901598, Society of Automotive Engineers, Warrendale, Pennsylvania.

Monnier, G., and P. Duret. 1991. "IAPAC Compressed Air Assisted Fuel Injection for High Efficiency Low Emissions Marine Outboard Two-Stroke Engines." *SAE Paper* 911849, Society of Automotive Engineers, Warrendale, Pennsylvania.

Nuti, M. and L. Martorano. 1985. "Short-Circuit Ratio Evaluation in the Scavenging of Two-Stroke S.I. Engines." *SAE Paper* 850177, Society of Automotive Engineers, Warrendale, Pennsylvania.

Pearce, T.C. and G.P. Davies. 1990. "The Efficiency of Automotive Exhaust Catalysts and the Effects of Component Failure." *Research Report 287*. TRRL. Crowthorne, Berkshire.

Plohberger and others. 1988. "Development of a Fuel-Injected Two-Stroke Gasoline Engine." *SAE Paper* 880170, Society of Automotive Engineers, Warrendale, Pennsylvania.

Prigent, M. and G. De Soete. 1989. "Nitrous Oxide N_2O in Engines Exhaust Gases - A First Appraisal of Catalyst Impact", *SAE Paper* 890492, SAE International, Warrendale, Pennsylvania.

Sugiura, K. and M. Kagaya. 1977. "A Study of Visible Smoke Reduction From a Small Two-Stroke Engine Using Various Engine Lubricants." *SAE Paper* 770623, Society of Automotive Engineers, Warrendale, Pennsylvania.

Tan, M. and T. Chan. 1993. "Technical Handout" Hong Leong Sdn. Bhd., Distributor of Yamaha Motorcycles in Malaysia.

Tsuchiya, K., S. Hirano, M. Okamura, and T. Gotoh. 1980. "Emission Control of Two-Stroke Motorcycle Engines by the Butterfly Exhaust Valve." *SAE Paper* 800973, Society of Automotive Engineers, Warrendale, Pennsylvania.

U.S. EPA (United States Environmental Protection Agency). 1990. "Volatile Organic Compounds from On-Road Vehicles: Sources and Control Options." Report prepared for the Long-range Transboundary Air Pollution Convention. United Nations Economic Commission for Europe. Washington, D.C. (Draft.)

Weaver, C.S. and L.M. Chan. 1994. "Motorcycle Emission Standards and Emission Control Technology." Report to the Royal Thai Ministry of Science, Technology, and the Environment and The World Bank. Engine, Fuel, and Emissions Engineering, Inc. Sacramento, California.

Weaver, C.S. and L.M. Chan. 1995. "Company Archives" Multiple Sources. Engine, Fuel and Emissions Engineering, Inc., Sacramento, California.

White, J.J., J.N. Carroll, C.T. Hare, and J.G. Lourenco. 1991. "Emission Control Strategies for Small Utility Engines." *SAE Paper* 911807, Society of Automotive Engineers, Warrendale, Pennsylvania.

Appendix 3.2

Emission Control Technology for Compression Ignition (Diesel) Engines

This appendix provides a more detailed technical discussion to supplement the general information on diesel engine emissions and control technologies in chapter 3.

Diesel Combustion and Pollutant Formation

Diesel engine emissions are determined largely by the combustion process. During the compression stroke, a diesel engine compresses only air. The process of compression heats the air to about 700–900°C, which is well above the self-ignition temperature of diesel fuel. Near the top of the compression stroke, liquid fuel is injected into the combustion chamber under tremendous pressure through a number of small orifices in the tip of the injection nozzle.

Figure A3.2.1 shows the stages of diesel injection and the subsequent combustion process. As fuel is injected into the combustion chamber just before the piston reaches top-dead-center, the periphery of each jet mixes with the hot compressed air already present. After a brief period known as the ignition delay, this fuel-air mixture ignites. In the *premixed combustion* phase, the fuel-air mixture formed during the ignition delay period burns very rapidly, causing a rapid rise in cylinder pressure. The subsequent rate of burning is controlled by the rate at which the remaining fuel and air can mix. Combustion always occurs at the interface between the air and the fuel. Most of the fuel injected is burned in this *mixing-controlled combustion* phase, except under very light loads, when the premixed combustion phase dominates.

Nitrogen Oxides

Diesel engines are a significant source of nitrogen oxide emissions, primarily in the form of nitric oxide (NO). This gas results from the reaction between nitrogen and free oxygen at the high temperatures close to the flame front. The rate of NO formation in diesels is a function of oxygen availability, and increases exponentially with flame temperature. Emissions of nitrogen oxides (NO_x) can be reduced effectively by actions which reduce the flame temperature. These include delaying combustion into the expansion phase, cooling the air charge going into the cylinder, and exhaust gas recirculation (EGR). Since combustion always occurs under near-stoichiometric conditions, reducing the flame temperature by lean-burn techniques, as in spark-ignition engines, is impractical.

Most of the NO_x emitted is formed early in the combustion process, when the piston is still near the top of its stroke. This is when the flame temperature is highest. Work by several researchers (Wade and others 1987; Cartellieri and Wachter 1987) indicates that a large share of diesel NO_x is formed during the premixed combustion phase. It has been found that reducing the amount of fuel burned in this phase can significantly reduce these emissions. An effective way of reducing NO_x emissions is to retard the injection timing such that burning occurs later in the cycle. However, this may increase smoke and fuel consumption. Since NO_x emissions are highest at high loads, a flexible timing plan could be introduced. This would vary the injection timing in accordance with the engine conditions to obtain a better tradeoff among different pollutant emissions. This type of system would require the use of electronic controls, which are considerably more expensive than conventional fuel injection equipment.

Charge cooling reduces NO_x emissions by reducing the charge temperature. This technique is particularly attractive owing to the additional benefit of increased power. In the absence of emission controls, the range of NO_x emissions from turbocharged charge-cooled DI engines (7-17 g/kWh) is lower than for naturally aspirated (11-23 g/kWh) and turbocharged (13-28 g/kWh) DI engines (figure A3.2.2).

Figure A3.2.1 Diesel Combustion Stages

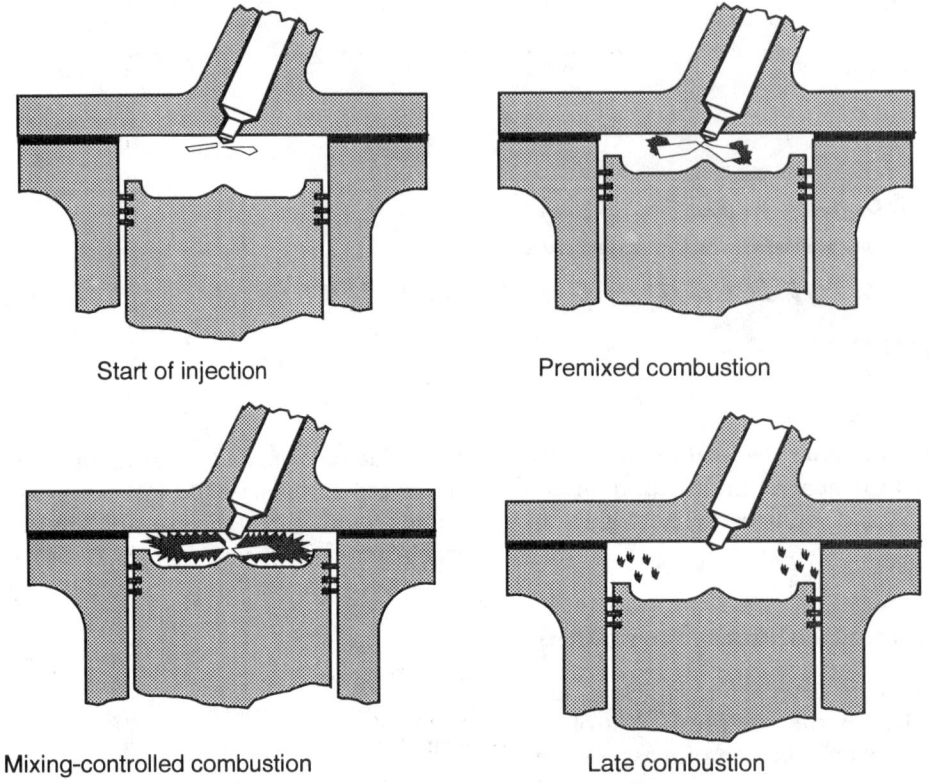

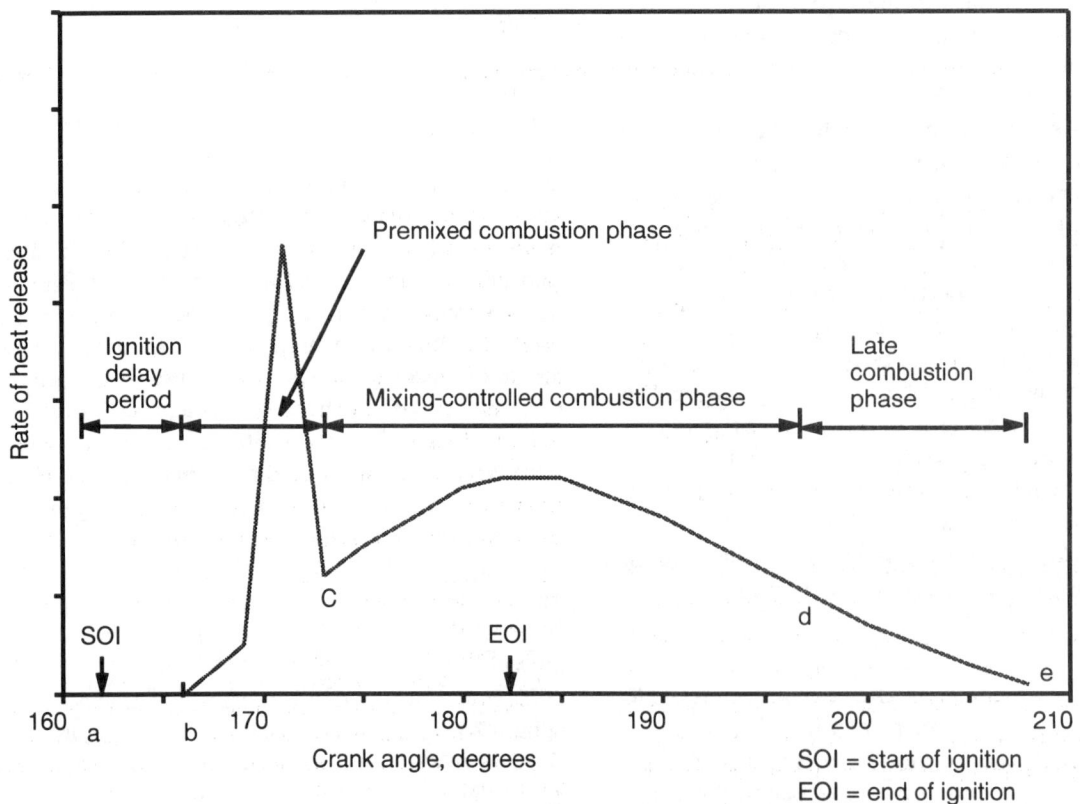

Source: Heywood 1988

Figure A3.2.2 Hydrocarbon and Nitrogen Oxide Emissions for Different Types of Diesel Engines
(European 13-mode test)

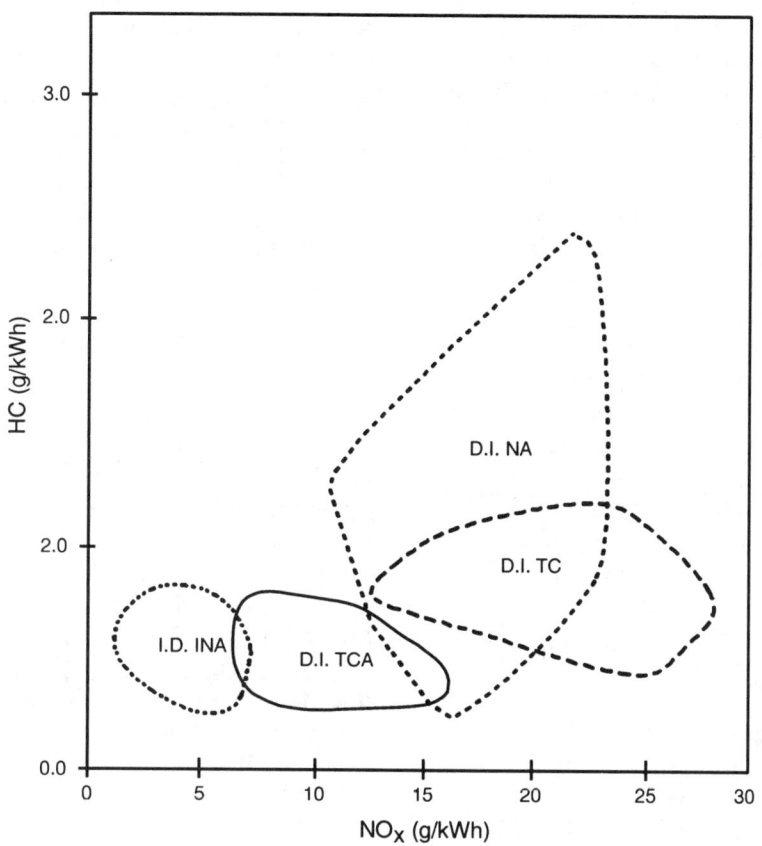

Note: D.I.NA = direct injection naturally aspirated; D.I.TC = direct injection turbocharged; D.I.TCA = direct injection turbocharged aftercooled, I.D.INA = indirect injection naturally aspirated.
Source: Latham and Tomkin 1988

A better tradeoff between NO_x emissions and fuel consumption can be achieved by optimizing the mixing process. The use of higher injection pressures reduces the heat release period, which allows the use of more retarded timings and leads to a better NO_x emission-fuel consumption tradeoff. Other methods of NO_x reduction include chemical removal, water injection, and exhaust gas recirculation. However, these techniques may introduce durability problems.

Particulate Matter

Diesel particulate matter[1] consists mostly of three components: soot formed during combustion, heavy hydrocarbons condensed or adsorbed on the soot, and sulfates. In older diesels soot was typically 40 to 80 percent of the total particulate mass, but soot emissions from modern emission-controlled engines are much lower. Most of the remaining particulate mass consists of heavy hydrocarbons adsorbed or condensed on the soot. This is referred to as the soluble organic fraction

(SOF) of the particulate matter. The SOF is derived partly from the lubricating oil, partly from unburned fuel, and partly from compounds formed during combustion.

Diesel soot is formed primarily during the mixing-controlled combustion phase. Primary soot particles are small spheres of graphitic carbon approximately 0.01 millimeter in diameter. These are formed by the reaction of gaseous hydrocarbons such as acetylene under conditions of moderately high temperature and oxygen deficiency. The primary particles then agglomerate to form chains and clusters of linked particles, giv-

---

1. The diesel combustion process results in fuel-rich zones in which carbon particles are formed. These particles adsorb organic compounds (primarily aliphatic and polycyclic aromatic hydrocarbons) and sulfuric acid during the exhaust gas cooling and dilution process. The hydrocarbons originating from the unburnt lubricating oil contribute significantly (up to 40 percent) to heavy-duty diesel PM emissions. Although PM emissions are related to smoke and hydrocarbon emissions, there is no direct correlation. Particulate matter emissions are assessed gravimetrically while smoke is measured by a light obscuration technique.

ing the soot its characteristic fluffy appearance. During mixing-controlled combustion, the local gas composition at the flame front is close to stoichiometric, with an oxygen-rich region on one side and a fuel-rich region on the other. The moderately high temperatures and oxygen deficiency required for soot formation are thus always present.

Most of the soot formed during diesel combustion is subsequently burned during the later portions of the expansion stroke. Typically, less than 10 percent of the soot formed in the cylinder survives to be emitted into the atmosphere; in modern emission-controlled diesel engines, less than one percent may be emitted. Soot oxidation is much slower than soot formation, however, and the amount of soot oxidized is dependent on the availability of high temperatures and adequate oxygen during the late stages of combustion. Conditions that reduce the availability of oxygen (such as poor mixing or operation at low air-fuel ratios), or which reduce the time available for soot oxidation (such as retarding the combustion timing) can cause a very large increase in soot emissions.

Hydrocarbons and Related Emissions

Diesel engines normally emit low levels of hydrocarbons. Diesel HC emissions are mainly due to incomplete combustion of the fuel or the lubricating oil. Hydrocarbon emissions can be controlled by the correct matching of the air-fuel mixing process and injection equipment. Further mechanical developments have reduced oil- derived hydrocarbon levels by reducing the amount of oil entering the cylinder. In extreme cases major structural changes may be required. Other reductions can be made by reducing the nozzle sac/hole volume and avoiding secondary injections. In general diesel engine hydrocarbon emissions are fairly easy to control.

Diesel hydrocarbon emissions (as well as the unburned-fuel portions of the particulate SOF) occur primarily at light loads. The major source of light-load hydrocarbon emissions is excessive air-fuel mixing, which results in air-fuel mixture(s) that are too lean to burn. Other hydrocarbon sources include fuel deposited on the combustion chamber walls or in combustion chamber crevices by the injection process, fuel retained in the orifices of the injector which vaporizes late in the combustion cycle, partly reacted mixture quenched by too-rapid mixing with air, and vaporized lubricating oil. Aldehydes and ketones (as partially-reacted hydrocarbons) and the small amount of carbon monoxide produced by diesels are probably formed in the same processes as the hydrocarbon emissions.

In addition to aldehydes and ketones, unregulated pollutants from diesel engines include polycyclic and nitro-polycyclic aromatic compounds (PAH and nitro-PAH), and small quantities of hydrogen cyanide, cyanogen, and ammonia. The presence of polycyclic aromatic hydrocarbons and their nitro-derivatives in diesel exhaust is of special concern, since these compounds include many known mutagens and suspected carcinogens. A significant portion of these compounds (especially the smaller two- and three-ring compounds) are apparently derived directly from diesel fuel. Typical diesel fuel contains a significant fraction of PAH by volume. Most of the larger and more dangerous PAH appear to form during the combustion process, possibly via the same acetylene polymerization reaction that produces soot (Williams and others 1987). Indeed, the soot particle itself can be viewed as a very large PAH molecule. PAH and nitro-PAH can be reduced by decreasing the aromatic content of the fuel.

Diesel engines emit significantly more aldehydes and ketones than comparable gasoline engines. Aldehyde emissions are largely responsible for diesel odor and cause irritation of the nasal passages and eyes. These emissions are strongly related to total hydrocarbon emissions and can be controlled by the same techniques. Sulfur dioxide emissions are the result of naturally occurring sulfur in the fuel and can only be reduced by lowering the fuel sulfur content. Hydrogen cyanide, cyanogen, and ammonia are present only in small quantities in diesel exhaust and are unlikely to become regulated.

Visible Smoke

Black smoke from diesel engines is due to the soot component of diesel particulate matter. Under most operating conditions, nearly all of this soot is oxidized in the cylinder, so the exhaust plume from a properly adjusted diesel engine is normally invisible—with a total opacity (absorbance plus reflectance) of 2 percent or less. Visible black smoke is generally due to injecting too much fuel for the amount of air available in the cylinder, or to poor mixing between fuel and air. These conditions can be prevented by proper design. The PM reductions required to comply with U.S. 1994 and Euro 1 emission standards have resulted in the virtual elimination of visible smoke emissions from properly functioning engines.

Under some conditions, diesel engines may also emit white, blue, or gray smoke. These are due to the presence of condensed hydrocarbon droplets in the exhaust. Unlike soot, these droplets scatter light, thus giving the smoke a bluish or whitish cast. Blue or gray smoke is generally due to vaporized lubricating oil and indicates an oil leak into the cylinder or exhaust system. White smoke is common when engines are first started in cold weather and usually goes away when the engine warms up.

Noise

Diesel engine noise is due mostly to the rapid combustion (and resulting rapid pressure rise) in the cylinder during the premixed burning phase. The greater the ignition delay, and the more fuel is premixed with air, the greater this pressure rise and the resulting noise emissions. Noise emissions and NO_x emissions thus tend to be related—reducing the amount of fuel burned in the premixed burning phase will tend to reduce both. Other noise sources include those common to all engines, such as mechanical vibration, fan noise, and so forth. These can be minimized by appropriate mechanical design.

Odor

The characteristic diesel odor is due primarily to partially oxygenated hydrocarbons (aldehydes and similar species) in the exhaust resulting from slow oxidation reactions in air-fuel mixtures too lean to burn normally. Unburned aromatic hydrocarbons may also play a significant role. The most significant aldehyde species are benzaldehyde, acetaldehyde, and formaldehyde, but other aldehydes such as acrolein (a powerful irritant) are significant as well. Aldehyde and odor emissions are closely linked to total hydrocarbon emissions—experience has shown that modifications that reduce total hydrocarbons also tend to reduce aldehydes and odor as well.

Toxic Air Contaminants

Diesel exhaust contains many organic species that are known or suspected of causing cancer or other health problems. These include formaldehyde, benzene, and polynuclear aromatic hydrocarbons, as well as other unidentified mutagenic compounds. Organic extracts of diesel particulate have repeatedly been shown to be mutagenic in the Ames test, a commonly used screening technique for potential carcinogens. California, based on a number of epidemiological studies, has proposed to regulate diesel exhaust itself as a suspected carcinogen. In general, measures that tend to reduce diesel hydrocarbon and particulate matter emissions also seem to reduce mutagenic (and presumed carcinogenic) activity.

Influence of Engine Variables on Emissions

The engine variables having the greatest effects on diesel emission rates are the air-fuel ratio, rate of air-fuel mixing, fuel injection timing, compression ratio, and the temperature and composition of the charge in the cylinder.

Air-Fuel Ratio

The ratio of air to fuel in the combustion chamber has an extremely important effect on emission rates for hydrocarbons and particulate matter. In diesel engines, the fact that fuel and air must mix before burning means that a substantial amount of excess air is needed to ensure complete combustion of the fuel within the limited time allowed by the power stroke. Diesel engines, therefore, operate with overall air-fuel ratios that are considerably lean of stoichiometric (λ greater than one). The minimum air-fuel ratio for complete combustion is about $\lambda = 1.5$. This ratio is known as the smoke limit, since smoke increases dramatically at air-fuel ratios lower than this limit. For a given amount of air in the cylinder, the smoke limit establishes the maximum amount of fuel that can be burned per stroke, and thus the maximum power output of the engine.

Figure A3.2.3 shows the typical relationship between air-fuel ratio and emissions in a diesel engine. At very high air-fuel ratios (corresponding to very light load), the temperature in the cylinder after combustion is too low to burn out residual hydrocarbons, so emissions of gaseous hydrocarbons and particulate SOF are high. At lower air-fuel ratios, less oxygen is available for soot oxidation, so soot emissions increase. As long as λ remains above 1.6, this increase is relatively gradual. Soot and visible smoke emissions show a strong non-linear increase below the smoke limit (at about $\lambda = 1.5$).

In naturally aspirated engines (those without a turbocharger), the amount of air in the cylinder is independent of the power output. Maximum power output for these engines is normally smoke-limited, that is, limited by the amount of fuel that can be injected without exceeding the smoke limit. Maximum fuel settings on naturally aspirated engines represent a compromise between smoke emissions and power output. Where diesel smoke is regulated, this compromise must result in smoke opacity below the regulated limit. Otherwise, opacity is limited by the manufacturer's judgment of the commercially acceptable smoke level, which varies significantly from country to country.

In conventional turbocharged engines, an increase in the amount of fuel injected per stroke increases the energy in the exhaust gas, causing the turbocharger to spin more rapidly and pump more air into the combustion chamber. For this reason, power output from turbocharged engines is not usually smoke-limited. Instead, it is limited by design limits such as the allowable turbocharger speed and mechanical and thermal loading of the engine components.

Turbocharged engines do not normally experience low air-fuel ratios during steady-state conditions, and therefore tend to exhibit low steady-state smoke even at full power. Low air-fuel ratios can occur in turbocharged engines under transient conditions, however. This is because the inertia of the turbocharger rotor keeps it from responding instantly to an increase in fuel input. Thus

Figure A3.2.3 Relationship between Air-Fuel Ratio and Emissions for a Diesel Engine

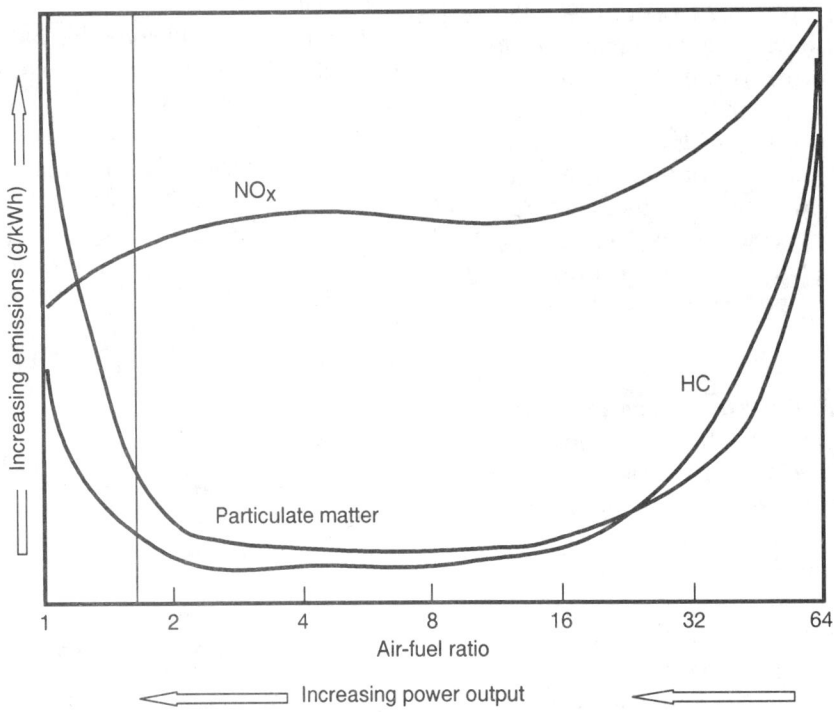

Source: Weaver and Chan 1995

the air supply during the first few seconds of full-power acceleration is less than would be supplied by the turbocharger in steady-state operation. Unless the fuel is similarly limited during this period, the air-fuel ratio will be below the smoke limit, resulting in a black 'puff' of dense smoke. The resulting excess particulate emissions can be significant under urban driving conditions. To overcome this problem, turbocharged engines in highway vehicles commonly incorporate an acceleration smoke limiter. This device modulates the fuel flow to the engine according to the boost pressure to allow the turbocharger time to respond.

The setting on the smoke limiter must compromise between acceleration performance (drivability) and low smoke emissions. In the United States, acceleration smoke emissions are regulated. ECE R24 also limits free-acceleration smoke opacity. Elsewhere, acceleration smoke opacity is limited by the manufacturer's judgment of commercial acceptability, or by the need to pass free-acceleration smoke tests.

The high transient smoke emissions observed from turbocharged engines not equipped with adequate acceleration smoke limiters have led some parties to conclude that turbocharged engines produce worse emissions than naturally aspirated engines. For example, an opacity test (ECE Regulation 24) carried out by

London Buses Limited (LBL) on 254 buses in its operational fleet showed that 68 percent of the naturally aspirated buses passed the test compared with a 36 percent pass rate for turbocharged buses (Gore 1991). When properly controlled, however, turbocharged engines tend to have lower emissions of particulate matter and oxides of nitrogen than naturally- aspirated engines of similar power output. Indeed, the strict emissions limits now effective in the United States have nearly eliminated naturally-aspirated engines from the U.S. market, and a similar trend in Europe with the advent of stricter diesel emission standards. High efficiency turbocharging such as variable geometry turbochargers (VGT) and charge cooling increase the air-fuel ratio and so lead to a further reduction in PM emissions under most engine conditions.

Air-Fuel Mixing

The rate of mixing between the compressed charge in the cylinder and the injected fuel is among the most important factors in determining diesel performance and emissions. The mixing rate during the ignition delay period determines how much fuel is burned in the premixed burning phase. The higher the mixing rate, the greater the amount of fuel burning in the premixed mode, and higher the noise and NO_x emissions. In the subsequent mixing-controlled combustion phase, the rate of combustion is limited by the mixing rate. The more rapid and complete this mixing, the greater the amount of fuel that burns near the top-dead-center, the higher the efficiency, and the lower the PM emissions.

In engine design practice, it is necessary to strike a balance between the rapid and complete mixing required for low soot emissions and best fuel economy, as the too-rapid mixing results in high NO_x emissions. The primary factors affecting the mixing rate are the fuel injection pressure, the number and size of injection orifices, any swirling motion imparted to the air as it enters the cylinder during the intake stroke, and air motions generated by combustion chamber geometry during compression. Much of the progress in in-cylinder emission control over the last decade has come through improved understanding of the interactions between these different variables and emissions, leading to improved design of fuel injection systems and combustion chambers.

Air-fuel mixing rates in current emission-controlled engines are the product of extensive optimization to assure rapid and complete mixing under nearly all operating conditions. Poor mixing may still occur during 'lug-down'–high-torque operation at low engine speeds. Turbocharger boost, air swirl level, and fuel injection pressure are typically poorer in these 'off-design' conditions. Significant improvements can be made in smoke emission and fuel consumption by optimizing the swirl ratio as a function of engine speed. Maintenance problems such as injector tip deposits can also degrade air-fuel mixing and result in greatly increased emissions. This is a common cause of high smoke emissions.

Fuel Injection and Combustion Timing

The timing relationship between the beginning of fuel injection and the top of the compression stroke has an important effect on diesel engine emissions and fuel economy. For best fuel economy, it is preferable that combustion begin at or somewhat before top-dead-center. Since there is a finite delay between the beginning of injection and the start of combustion, it is necessary to inject the fuel somewhat before this point (generally 5 to 15 degrees of crankshaft rotation before). The earlier fuel is injected, the less compression heating will have occurred, and the longer the ignition delay. The longer ignition delay provides more time for air and fuel to mix, increasing the amount of fuel that burns in the premixed combustion phase. More fuel burning at or just before top-dead-center also increases the maximum temperature and pressure attained in the cylinder. Both of these effects tend to increase NO_x emissions. On the other hand, earlier injection timing tends to reduce PM

and light-load hydrocarbon emissions. Fuel burning in premixed combustion forms little soot, while the soot formed in mixing-controlled combustion near top-dead-center experiences a relatively long period of high temperatures and intense mixing, and is thus mostly oxidized. The end-of-injection timing also has a major effect upon soot emissions—fuel injected more than a few degrees after TDC burns more slowly, and at a lower temperature, so that less of the resulting soot has time to oxidize during the power stroke. For a fixed injection pressure, orifice size, and fuel quantity, the end-of-injection timing is determined by the timing of the beginning of injection.

The result of these effects is that injection timing must compromise between PM emissions and fuel economy on the one hand and noise, NO_x emissions, and maximum cylinder pressure on the other. The terms of the compromise can be improved to a considerable extent by increasing injection pressure, which increases mixing and advances the end-of-injection timing. Another approach under development is split injection, in which a small amount of fuel is injected early in order to ignite the main body of fuel quantity injected near top-dead-center. Some fuel injection systems can now vary the injection rate, within limits, under electronic control.

Compared to uncontrolled diesel engines, modern diesel engines with emission controls generally have moderately retarded injection timing to reduce NO_x, in conjunction with high injection pressure to limit the effects of retarded timing on PM emissions and fuel economy. The response of fuel economy and PM emissions to retarded timing is not linear—up to a point, the effects are relatively small, but beyond that point deterioration is rapid. Great precision in injection timing thus becomes necessary—a change of one degree crank angle can have a significant impact on emissions. The optimal injection timing is a complex function of engine design, engine speed and load, and the relative stringency of emission standards for different pollutants. Attaining the required flexibility and precision of injection timing has posed a major challenge to fuel injection manufacturers.

Charge Temperature

The process of compressing the intake air in turbocharged engines increases its temperature. Reducing the temperature of the compressed air charge going into the cylinder has benefits for both PM and NO_x emissions. Reducing the charge temperature directly reduces the flame temperature during combustion, and thus helps to reduce NO_x emissions. In addition, the relatively colder air is denser, so that (at the same pressure) a greater mass of air can be contained in the same fixed cylinder volume.This increases the air-fuel ratio in the cylinder and thus helps to reduce soot emissions. By increasing the air available while decreasing piston temperatures, charge-air cooling can also make possible a significant increase in power output while remaining within the engine thermal limits. For this reason, many high-powered turbocharged engines incorporate charge-air coolers even in the absence of emission controls. Charge-air coolers are almost universal on low-emission engines, many of which use air-to-air aftercooling to achieve the lowest possible temperature. However, excessively cold charge air can reduce the burnout of hydrocarbons, and thus increase light-load hydrocarbon emissions.This can be counteracted by advancing injection timing, or by reducing charge air cooling at light loads.

Charge Composition

Nitrogen oxide emissions are heavily dependent on flame temperature. By altering the composition of the air charge to increase its specific heat and the concentration of inert gases, it is possible to decrease the flame temperature significantly.The most common way of accomplishing this is through exhaust gas recirculation (EGR).At moderate loads, EGR has been shown to be capable of reducing NO_x emissions by a factor of two or more with little effect on PM emissions.Although soot emissions are increased by the reduced oxygen concentration, the soluble organic portion of the PM and gaseous HC emissions are reduced due to the higher incylinder temperature caused by the hot exhaust gas. EGR has been used for some time in light-duty diesel engines in order to reduce NO_x. Preliminary results of a study of EGR effects in advanced heavy-duty diesel engines have shown the potential to achieve extremely low NO_x emissions (less than 3 g/kWh) combined with extremely low PM emissions (Needham and others 1991; Havenith and others 1993). EGR is considered one of the most promising NO_x control technologies. Extensive development, however, is needed with regard to combustion system designs, the recirculation and admission system, cooling at high load, rematching of turbo machinery, wear limitation, and transient control (OECD/IEA 1993).

Compression Ratio

Diesel engines rely on compression heating to ignite the fuel, so the engine's compression ratio has an important effect on combustion. A higher compression ratio results in a higher temperature for the compressed charge, and thus in a shorter ignition delay and higher flame temperature.The effect of a shorter ignition delay is to reduce NO_x emissions, while the higher flame temperature would be expected to increase them. In practice, these two effects nearly cancel each other out, so

that changes in compression ratio have little effect on NO_x.

Engine fuel economy, cold starting, and maximum cylinder pressures are also affected by the compression ratio. For an idealized diesel cycle, the thermodynamic efficiency is an increasing function of compression ratio. In a real engine, however, the increased thermodynamic efficiency is offset after some point by increasing friction so that a point of maximum efficiency is reached. With most heavy- duty diesel engine designs, this optimal compression ratio is about 12 to 15.To ensure adequate starting ability under cold conditions, however, most diesel engine designs require a somewhat higher compression ratio—in the range of 15–20 or more. Generally, higher-speed engines with smaller cylinders require higher compression ratios for adequate cold starting.

Emissions Tradeoffs

There is an inherent conflict between some of the most powerful diesel NO_x control techniques and PM emissions.This is the basis for the much-discussed "tradeoff" relationship between diesel NO_x and PM emissions.The tradeoff is not absolute—a combination of technical advances can make it possible to reduce both NO_x and particulate emissions (figure A3.2.4).These tradeoffs do place limits on the extent to which any one pollutant can be reduced, however. To minimize emissions of all pollutants simultaneously requires careful optimization of the fuel injection, fuel-air mixing, and combustion processes over the full range of engine operating conditions.

Improvements in Engine Technology

Diesel engine emissions are determined by the combustion process within the cylinder. This process is central to the operation of the diesel engine. Virtually every characteristic of the engine affects combustion in some way, and thus has some direct or indirect effect on emissions. Some of the engine systems affecting diesel emissions are the fuel injection system, the engine control system, air intake and combustion chamber, and the air charging system. Actions to reduce lubricating oil consumption can also impact hydrocarbon and PM emissions. Finally, some advanced technologies now under development may give even lower pollutant emissions in the future (Walsh and Bradlow 1991).

Combustion System Types

The geometries of the combustion chamber and the air intake port control the air motion in the diesel combus-

Figure A3.2.4 Estimated PM-NO$_x$ Trade-Off Over Transient Test Cycle for Heavy-Duty Diesel Engines

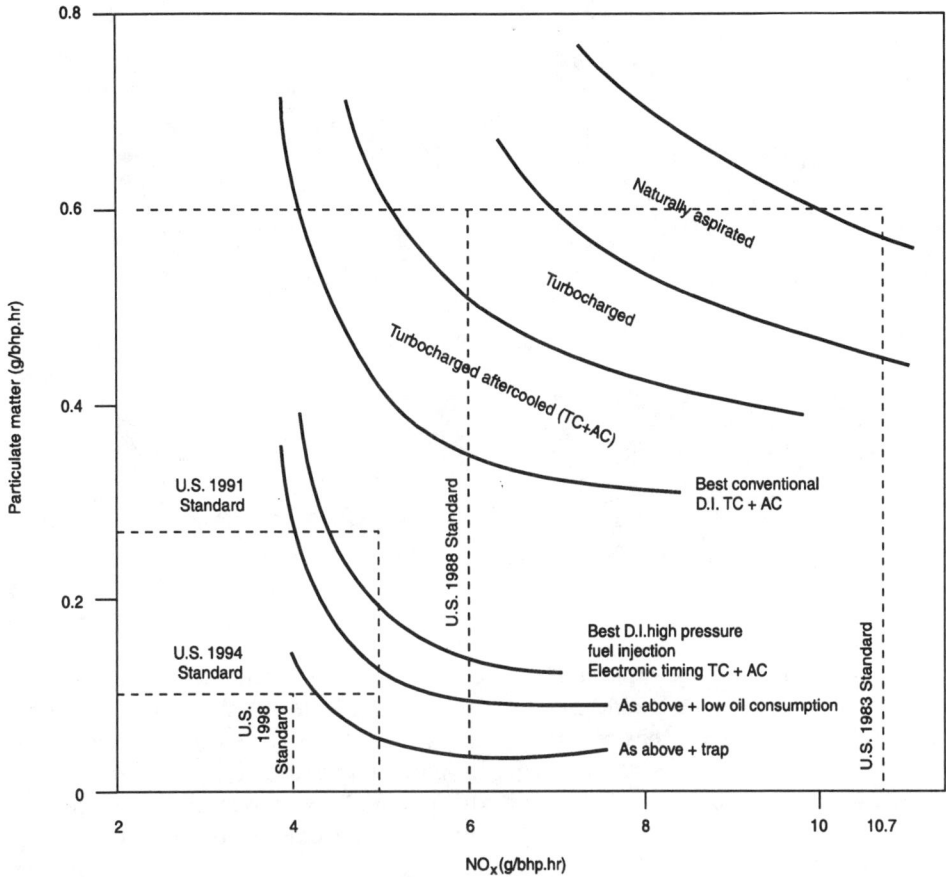

Note: Current engines improve significantly on these trends meeting U.S. 1998 standards without a trap.
Source: Adapted from Latham and Tomkin 1988

tion chamber, and thus play an important role in air-fuel mixing and emissions. A number of different combustion chamber designs are used in diesel engines. Virtually all commercial diesel engines make use of one of the common combustion chamber designs shown in figure A3.2.5. The most fundamental difference between the combustion chambers is between the indirect-injection (IDI) and direct-injection (DI) designs. In an indirect-injection engine, fuel is injected into a separate 'prechamber,' where it mixes and partly burns before jetting into the main combustion chamber above the piston. In the more common direct-injection engine, fuel is injected directly into a combustion chamber hollowed out of the top of the piston. DI engines can be further divided into high-swirl, low-swirl (quiescent chamber), and wall-wetting designs. The latter has many characteristics in common with indirect-injection systems.

Fuel-air mixing in the direct-injection engine is limited by the fuel injection pressure and any motion imparted to the air entering the chamber. In high-swirl DI engines, a strong swirling motion is imparted to the air entering the combustion chamber by the design of the

intake port. These engines typically use moderate-to-high injection pressures, and three to five spray holes per nozzle. Low-swirl engines rely primarily on the fuel injection process to supply the mixing. They typically have very high fuel injection pressures and six to nine spray holes per nozzle. Wall-wetting DI engines also have fairly high swirl, but the injection system is designed to deposit the fuel on the combustion chamber wall, where it vaporizes and burns relatively slowly.

In the IDI engine, the mixing between air and fuel is driven primarily by air swirl induced in the prechamber as air is forced into it during compression, and by the turbulence induced by the expansion out of the prechamber during combustion. These engines typically have better high-speed performance than DI engines, and can use cheaper fuel injection systems. For this reason, they have been used extensively in passenger car and light truck applications. Historically, IDI diesel engines have also exhibited lower emission levels than DI engines. With recent developments in DI engine emission controls, however, this is no longer the case. Disadvantages of the IDI engine are the extra heat and frictional losses

Figure A3.2.5 Diesel Engine Combustion Chamber Types

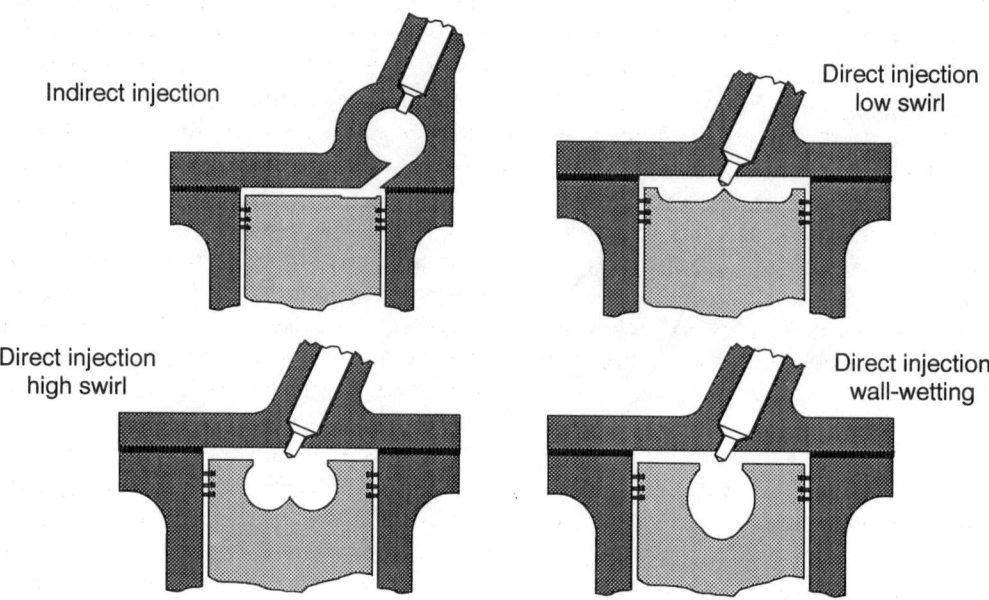

Source: Weaver and Chan 1995

due to the prechamber. These result in a 5 to 15 percent reduction in fuel efficiency compared to a DI engine. Because of this, nearly all heavy-duty truck engines are of the DI type, and there is an increasing trend toward DI engines in passenger cars and light trucks.

DI Combustion Chamber Design

Changes in the engine combustion chamber and related areas have demonstrated a major potential for emission control. Design changes to reduce the crevice volume in DI diesel cylinders increase the amount of air available in the combustion chamber. Changes in combustion chamber geometry —such as the use of a reentrant lip on the piston bowl—can markedly reduce emissions by improving air-fuel mixing and minimizing wall impingement by the fuel jet. Optimizing the intake port shape for best swirl characteristics has also yielded significant benefits. A number of light- and medium-duty engines now incorporate variable air intake systems to optimize swirl characteristics across a broader range of engine speeds.

Crevice volume. The crevice volume includes the clearance between the top of the piston and the cylinder head, and the 'top land'—the space between the side of the piston and the cylinder wall above the top compression ring. The air in these spaces contributes little to the combustion process. The smaller the crevice volume, the larger the combustion chamber volume can be for a given compression ratio. Thus, reducing the crevice volume increases the amount of air available for

combustion. To reduce these volumes, engine makers have reduced the clearance between the piston and cylinder head through tighter production tolerances, and have moved the top compression ring toward the top of the piston. This increases the temperature of the top ring and poses design problems for the piston top and cooling system. These problems have been addressed through redesign and the use of more expensive materials. The higher piston ring temperature also requires higher quality engine oil to avoid formation of damaging deposits.

Combustion chamber shape. Numerous test results indicate that, for high swirl DI engines, a reentrant combustion chamber shape (in which the lip of the combustion chamber protrudes beyond the walls of the bowl) provides a substantial improvement in performance and emissions over the previous straight-sided bowl designs. Nearly all of the manufacturers of high-swirl engines are developing or using this approach. Similar improvements in the performance of low-swirl DI engines have come primarily through refinements to the classic 'mexican hat' combustion chamber shape.

Intake air swirl. Optimal matching of intake air swirl with combustion chamber shape and other variables is critical for emissions control in high-swirl engines. Swirl is determined mostly by the design of the air intake port. Unfortunately, the selection of a fixed swirl level involves some tradeoffs between low-speed and high-speed performance. At low speeds, higher swirl provides better mixing, permitting more fuel to be injected and thus greater torque output at the same smoke level. However, this can result in too much swirl at higher speeds, impairing the airflow to the cylinder.

Attaining an optimal swirl level is most difficult in smaller light-duty and medium-duty DI engines, as these experience a wider range of engine speeds than do heavy-duty engines. One solution to this problem is to vary the swirl ratio as a function of engine speed. A number of production light- and medium-duty engines now use this approach with a noticeable reduction in PM and NO_x emissions (Shimada, Sakai, and Kurihara 1986).

Fuel Injection

The fuel injection system in a diesel engine includes the machinery by which the fuel is transferred from the fuel tank to the engine, then injected into the cylinders at the right time for optimal combustion, and in the correct amount to provide the desired power output. The quality, quantity, and timing of fuel injection determine the engine's power, fuel economy, and emission characteristics so that the fuel injection system is one of the most important components of the engine.

The fuel injection system normally consists of a low-pressure pump to transfer fuel from the tank to the system, one or more high-pressure fuel pumps to create the pressure pulses that actually send the fuel into the cylinder, the injection nozzles through which fuel is injected into the cylinder, and a governor and fuel metering system. These determine how much fuel is to be injected on each stroke, and thus the power output of the engine.

The major areas of concentration in fuel injection system development have been on increased injection pressure, increasingly flexible control of injection timing, and more precise governing of the fuel quantity injected. Systems offering electronic control of these quantities, as well as fuel injection rate, have been introduced. Some manufacturers are also pursuing technology to vary the rate of fuel injection over the injection period, in order to reduce the amount of fuel burning in the premixed combustion phase. Reductions in NO_x and noise emissions and maximum cylinder pressures have been demonstrated using this approach (Gill 1988). Other changes have been made to the injection nozzles themselves, to reduce or eliminate sac volume and to optimize the nozzle hole size and shape, number of holes, and spray angle for minimum emissions.

Injection system types. The most common diesel fuel injection system consists of a single fuel pump (typically mounted at the side of the engine) driven by gears from the crankshaft and connected to individual injection nozzles at the top of each cylinder by special high-pressure fuel lines. These pump-line-nozzle injection systems can be further divided into two subclasses: "distributor" fuel pumps, in which a single pumping element is mechanically switched to connect to the high-pressure fuel lines for each cylinder in turn, and "in-line" pumps having one pumping element per cylinder, each connected to its own high-pressure fuel line. Distributor-type systems are less costly and are commonly used in light- duty engines. In-line fuel injection pumps have better durability and can reach higher injection pressures. They are much more common in engines used for heavy-duty vehicles.

The most common alternative to the pump-line nozzle injection systems are systems using unit injectors, in which the individual fuel metering and pumping element for each cylinder is combined in the same unit with the injection nozzle at the top of the cylinder. The pumping elements in a unit injector system are generally driven by the engine camshaft.

Worldwide, many more engines are made with pump-line-nozzle injection systems than with unit injectors. This is primarily due to the higher cost of unit injector systems. Due to the absence of high- pressure fuel lines, however, unit injectors are capable of higher injection pressures than pump-line-nozzle systems. With improvements in electronic control, these systems offer better fuel economy at low emission levels than the pump-line-nozzle systems. For this reason, most of the new heavy-duty engines produced in the United States are now equipped with unit injectors (Balek and Heitzman 1993).

A new type of fuel injection system has recently been introduced. This system uses electro-hydraulic actuators supplied with fuel from a "common rail" instead of mechanically driven pumping elements. These systems allow great flexibility in control of fuel injection rate and timing.

Fuel injection pressure and injection rate. High fuel injection pressures are desirable to improve fuel atomization and fuel-air mixing, and to offset the effects of retarded injection timing by increasing the injection rate. A number of studies have been published on the effects of higher injection pressures on PM or smoke emissions. All show marked reductions as injection pressure is increased. High injection pressures are most important in low-swirl, direct-injection engines, since the fuel injection system is responsible for most of the fuel-air mixing in these systems. For this reason, low-swirl engines tend to use unit injector systems, which can achieve peak injection pressures in excess of 1,500 bar.

The injection pressures achievable in pump-line-nozzle fuel injection systems are limited by the mechanical strength of the pumps and fuel lines, as well as by pressure wave effects, to about 800 bar. Improvements in system design to minimize pressure wave effects, and increases in the size and mechanical strength of the lines and pumping elements have increased the injection pressures achievable in pump-line-nozzle systems substantially from those achievable a few years ago.

The pumping elements in nearly all present fuel injection systems are driven by a fixed mechanical linkage from the engine crankshaft. This means that the pumping rate, and thus the injection pressure, are strong functions of engine speed. At high speeds, the pumping element moves rapidly, and injection pressures and injection rates are high. At lower speeds, however, the injection rate is proportionately lower, and injection pressure drops off rapidly. This can result in poor atomization and mixing at low speeds, and is a major cause of high smoke emissions during lugdown. Increasing the pump rate to provide adequate pressure at low speeds is impractical, as this would exceed the system pressure limits at high speed.

A new type of in-line injection pump provides a partial solution to this problem (Ishida, Kanamoto, and Kurihara 1986). The cam driving the pumping elements in this pump has a non-uniform rise rate, so that pumping rate at any given time is a function of the cam angle. By electronically adjusting a spill sleeve, it is possible to select the portion of the cam's rotation during which fuel is injected, and thus to vary the injection rate. Injection timing varies at the same time, but the system is designed so that desired injection rate and injection timing correspond fairly well. A 25 percent reduction in PM emissions and a 10 percent reduction in HC emissions has been obtained using this system, with virtually no increase in NO_x emissions. Fuel injection pumps incorporating this technology are now used in a number of production engines. The same approach has also been applied to unit injector systems, using an electronically controlled spill valve.

Another approach to increasing injection pressure at low engine speeds is the use of electro-hydraulic "common rail" injection systems. Through appropriate design and control schemes, such systems can control and maintain fuel injection pressures nearly independently of engine speed (Hower and others 1993).

Initial injection rate and premixed burning. Reducing the amount of fuel burned in the premixed combustion phase can significantly reduce total NO_x emissions. This can be achieved by reducing the initial rate of injection, while keeping the subsequent rate of injection high to avoid high PM emissions due to late burning. This re-

quires varying the rate of injection during the injection stroke. This represents a difficult design problem for mechanical injection systems, but should be possible using electro-hydraulic injectors. Another approach to the same end is split injection, in which a small amount of fuel is injected in a separate event ahead of the main fuel injection period. Experimental data show a marked beneficial effect from reducing the initial rate of injection (Gill 1988). A substantial reduction in NO_x emissions could be achieved through this technique with no significant adverse impacts on HC or PM emissions. As an additional benefit, engine noise and maximum cylinder pressures for a given power output are reduced.

Low sac/sacless nozzles. The nozzle sac is a small internal space in the tip of the injection nozzle. The nozzle orifices open into the sac so that fuel flowing past the needle valve first enters the sac and then sprays out of the orifices. The small amount of fuel remaining in the sac tends to burn or evaporate late in the combustion cycle, resulting in significant PM and HC emissions. The sac volume can be minimized or eliminated by redesigning the injector nozzle. One manufacturer reported nearly a 30 percent reduction in PM emissions through elimination of the nozzle sac. It is also possible to retain some of the sac while designing the injector nozzle so that the tip of the needle valve covers the injection orifices when it is closed. This valve-covers-orifice (VCO) injector design is now common on production engines designed for North American emissions standards.

Engine Control Systems

Traditionally, diesel engine control systems have been closely integrated with the fuel injection system, and the two systems are often discussed together. These earlier control systems (still in use on many engines) are entirely mechanical. In recent years an increasing number of computerized electronic control systems have been introduced in diesel engines. With the introduction of these systems, the scope of the engine control system has been greatly expanded.

Mechanical controls. Most current diesel engines still rely on mechanical engine control systems. The basic functions of these systems include basic fuel metering, engine speed governing, maximum power limitation, torque curve "shaping", limiting smoke emissions during transient acceleration, and (sometimes) limited control of fuel injection timing. Engine speed governing is accomplished through a spring and flyweight system which progressively (and quickly) reduces the maximum fuel quantity as engine speed exceeds the rated value. The maximum fuel quantity itself is generally set through a simple mechanical stop on the rack controlling injection quantity. More sophisticated systems allow some "shaping" of the torque curve to change the maximum fuel quantity as a function of engine speed.

Acceleration smoke limiters are needed to prevent excessive black smoke emissions during transient acceleration of turbocharged engines. Most are designed to limit the maximum fuel quantity injected as a function of turbocharger boost, so that full engine power is developed only after the turbocharger comes up to speed.

Many pump-line-nozzle fuel injection systems incorporate mechanical injection timing controls. Since the injection pump is driven by a special shaft geared to the crankshaft, injection timing can be adjusted within a limited range by varying the phase angle between the two shafts, using a sliding spline coupling. A mechanical or hydraulic linkage slides the coupling back and forth in response to engine speed and load signals.

In mechanical unit injector systems, the injectors are driven by a direct mechanical linkage from the camshaft, making it very difficult to vary the injection timing. Formerly, some California engines with unit injectors used a mechanical timing control that operated by moving the injector cam followers back and forth with respect to the cam. Although effective in limiting light load hydrocarbon and PM emissions, these systems have proved to be troublesome and unpopular among users.

Electronic controls. The advent of computerized electronic engine control systems has greatly increased the potential flexibility and precision of fuel metering and injection timing controls. In addition, it has made possible whole new classes of control function, such as road speed governing, alterations in control strategy during transients, synchronous idle speed control, and adaptive learning, including strategies to identify and compensate for the effects of wear and component-to-component variation in the fuel injection system.

By continuously adjusting the fuel injection timing to match a stored "map" of optimal timing vs. speed and load, an electronic timing control system can significantly improve on the NO_x/PM emissions and NO_x emission/fuel consumption tradeoffs possible with static or mechanically-variable injection timing. Most electronic control systems also incorporate the functions of the engine governor and the transient smoke limiter. This helps to reduce excess PM emissions due to mechanical friction and lag time during engine transients, while simultaneously improving engine performance. Potential reductions in PM emissions of up to 40 percent have been demonstrated with this approach (Wade and others 1987).

Other electronic control features, such as cruise control, upshift indication, and communication with an electronically controlled transmission also help to reduce fuel consumption, and will thus likely reduce in-use emissions. Since the effect of these technologies is to reduce the amount of engine work necessary per kilometer, rather than the amount of pollution per unit of work, these effects are not necessarily reflected in dynamometer emissions test results.

Turbocharging and Intercooling

A turbocharger consists of a centrifugal air compressor mounted on the same shaft as an exhaust gas turbine. By increasing the mass of air in the cylinder prior to compression, turbocharging correspondingly increases the amount of fuel that can be burned without excessive smoke, and thus increases the potential maximum power output. The fuel efficiency of the engine is improved as well. The process of compressing the air, however, increases its temperature, increasing the thermal load on critical engine components. By cooling the compressed air in an intercooler before it enters the cylinder, the adverse thermal effects can be reduced. This also increases the density of the air, allowing an even greater mass of air to be confined within the cylinder, and thus further increasing the maximum power potential. Turbocharging and intercooling offer an inexpensive means to simultaneously improve power-weight ratios, fuel economy, and control of NO_x and PM emissions (OECD/IEA 1993).

Increasing the air mass in the cylinder and reducing its temperature can reduce both NO_x and PM emissions as well as increase fuel economy and power from a given engine displacement. Most heavy-duty diesel engines are equipped with turbochargers, and most of these have intercoolers. In the United States virtually all engines were equipped with these systems by 1991, and in Europe turbocharging and aftercooling are expected to become standard feature on all heavy-duty engines. Recent developments in air charging systems for diesel engines have been primarily concerned with increasing the turbocharger efficiency, operating range, and transient response characteristics; and with improved intercoolers to further reduce the temperature of the intake charge. Tuned intake air manifolds (including some with variable tuning) have also been developed to maximize air intake efficiency in a given speed range.

Turbocharger refinements. Turbochargers for heavy-duty diesel engines are already highly developed, but efforts to improve their performance continue. The major areas of emphasis are improved matching of turbocharger response characteristics to engine requirements, improved transient response, and higher efficiencies. Engine/turbocharger matching is especially critical because of the inherent conflict between the response characteristics of the two types of machines. Engine boost pressure requirements are greatest near the maximum torque speed, and most turbochargers are matched to give near optimal performance at that point. At higher speeds, however, the exhaust flow rate is greater, and the turbine power output is correspondingly higher. Boost pressure under these circumstances can exceed the engine's design limits, and the excessive turbine backpressure increases fuel consumption. Thus some compromise between adequate low speed boost and excessive high speed boost must be made.

Variable geometry turbocharger. Because of the inherent mismatch between engine response characteristics and those of a fixed geometry turbocharger, a number of engine manufacturers are considering the use of variable geometry turbines instead (Wallace and others 1986). In these systems the turbine nozzles can be adjusted to vary the turbine pressure drop and power level in order to match the engine's boost pressure requirements. Thus, high boost pressures can be achieved at low engine speeds without wasteful overboosting at high speed. The result is a substantial improvement in low speed torque, transient response, and fuel economy, and a reduction in smoke, NO_x, and PM emissions.

Prototype variable geometry turbochargers (VGT) have been available for some time, but they have not been used in production vehicles until recently. The major reasons for this are their cost (which could be 50 percent more than a comparable fixed geometry turbocharger), reliability concerns, and the need for a sophisticated electronic control system to manage them. With the recent deployment of electronic engine controls on most new engines in North America, these arguments have lost much of their force, and the fuel economy and performance advantages of the VGT are great enough to outweigh the costs in many applications. As a result, variable geometry turbochargers are beginning to enter the market in limited numbers.

Other superchargers. A number of alternative forms of supercharging have been considered, with a view to overcome the mismatch between turbocharger and engine response characteristics. The gas dynamic supercharger, for example, has the major advantages of superior low-speed performance and improved transient response.

Aftercoolers. Most aftercoolers rely on the engine cooling water as a heat sink, since this minimizes the components required. The relatively high temperature of this water (about 90°C) limits the benefits available, however. For this reason, many heavy-duty diesel engines are now being equipped with low-temperature charge-air cooling systems.

The most common type of low-temperature charge-air cooler rejects heat directly to the atmosphere through an air-to-air heat exchanger mounted on the truck chassis in front of the radiator. Although bulky and expensive these charge air coolers are able to achieve the lowest charge air temperatures—in many cases, only 10 or 15°C above ambient. An alternative approach is low- temperature air to water intercooling, in which the basic water-air intercooler is retained, but with drastically reduced radiator flow rates to lower the water temperature. This water is then passed through the intercooler before it is used for cooling the rest of the engine.

Intake manifold tuning. Tuned intake manifolds have been used for many years to enhance airflow rates on high performance gasoline engines, and are now used on some diesel engines as well A tuned manifold provides improved airflow and volumetric efficiency at speeds near its resonant frequency at the cost of reduced volumetric efficiency at other speeds. A variable-resonance manifold has been shown to improve airflow characteristics at both low and high speeds for light-duty engines (Thoma and Fausten 1993).

Lubricating Oil Control

A significant fraction of diesel particulate matter consists of oil-derived hydrocarbons and related solid matter; estimates range from 10 to 50 percent. Reduced oil consumption has been a design goal of heavy-duty diesel engine manufacturers for some time, and the current generation of diesel engines already uses relatively little oil compared to their predecessors. Further reductions in oil consumption are possible through careful attention to cylinder bore roundness and surface finish, optimization of piston ring tension and shape, and attention to valve stem seals, turbocharger oil seals, and other possible sources of oil loss. Some oil consumption in the cylinder is required with present technology, however, for the oil to perform its lubricating and corrosion retarding functions.

The reduction in diesel fuel sulfur content has reduced the need for corrosion protection, and has opened the way for still greater reductions in lubricating oil consumption. Changes in oil formulation can also help to reduce PM emissions by 10 to 20 percent (Dowling 1992).

Key Elements in Controlling Engine-Out Emissions

In the last decade, engine manufacturers have made enormous progress in reducing emissions of NO_x and particulate matter from diesel engines. Most engine manufacturers have followed a broadly similar approach to reducing diesel emissions, although the specific techniques used differ considerably from one manufacturer to the next. This typical approach includes the following major elements:

- Minimize parasitic hydrocarbon and PM emissions (those not directly related to the combustion process) by minimizing injection nozzle sac volume and reducing oil consumption to the extent possible.
- Reduce PM emissions, improve fuel-efficiency, and increase power and torque output by turbocharging naturally-aspirated engines, and increasing the boost pressure and improving the match between turbocharger and engine in already-turbocharged engines.
- Reduce PM and NO_x (with some penalty in HC) by cooling the compressed charge air as much as possible, via air-air or low temperature air-water aftercoolers.
- Further reduce NO_x to meet regulatory targets by severely retarding fuel injection timing over most of the speed/load range. Minimize the adverse effects of retarded timing on smoke, starting, and light-load hydrocarbon emissions by means of a flexible timing system to advance the timing under these conditions.
- In light-duty diesels only, further reduce NO_x by the use of exhaust gas recirculation under light-load conditions. EGR is not yet used in commercial heavy-duty diesel engines, but likely will be used to meet future NO_x standards less than 5 g/kWh.
- Recover the PM increase due to retarded timing by increasing the fuel injection pressure and controlling the fuel injection rate.
- Improve air utilization (and reduce hydrocarbon and PM emissions) by minimizing parasitic volumes in the combustion chamber, such as the clearance between piston and cylinder head clearance and between the piston and the walls of the cylinder.
- Optimize in-cylinder air motion through changes in combustion chamber geometry and intake air swirl to provide adequate mixing at low speeds (to minimize smoke and PM) without over-rapid mixing at high speeds (which would increase hydrocarbon and NO_x, emissions, and fuel consumption).
- Control smoke and PM emissions in full-power operation and transient accelerations through improved control of fuel injection, under both steady-state and (especially) transient conditions, frequently through electronic control of the fuel injection system.

Taken together, the changes made to diesel engines used in North America to limit their emissions have re-duced PM emissions by more than 80 percent and NO_x emissions by 50 to 70 percent compared to uncontrolled levels. Similar changes are now being made in European engines to meet the new EU emission standards. At the same time that emissions have been reduced, engine power output per unit of weight or engine displacement has increased by more than 30 percent. Fuel efficiency has also improved by 20 percent compared to the engines of two decades ago. Achieving these emission reductions and fuel economy improvements has required the complete redesign of large portions of the engine and combustion system, with significant costs to manufacturers and engine purchasers. A detailed technical review of control measures involving multi-valve technology, improved turbocharging and exhaust gas recirculation, and electronic ignition control is provided in TUV Rheinland/OPET 1991 and OECD/IEA 1993.

Vertical Vehicle Exhausts for Dispersing Emissions

The exhaust pipes on heavy-duty vehicles may be either vertical (so that the exhaust is emitted above the vehicle) or horizontal. Although the choice of exhaust location does not affect overall pollutant emissions, it can significantly affect local concentrations of pollutants. Use of a vertical exhaust reduces the concentration of exhaust pollutants at breathing level. thus reducing human exposure to high local concentrations.

In the United States nearly all heavy-heavy trucks and many medium-heavy trucks are outfitted with vertical exhausts. Where horizontal exhausts are fitted, this is often due to some feature of the design that rules out a vertical exhaust. For instance, many medium-duty vans have too little clearance between the van body and the cab door for an exhaust pipe to be fitted. Trucks used for automobile transportation usually have horizontal exhausts to prevent smoke from blowing back over the cars being transported and damaging their finish. Other cases where truck design or use rule out vertical exhausts include garbage trucks with hydraulic lifts for "dumpsters" and specialized construction vehicles such as truck cranes. There appears to be no engineering reason for preferring horizontal over vertical exhausts in transit buses, except for front-engined buses or articulated models with the engine located in the middle. Some bus transit operators prefer the low stack since the exhaust plume then is less visible to the public.

Colucci and Barnes (1986) investigated the effects of vertical and horizontal exhausts on pollutant concentrations in a bus-stop situation and found that the pollutant concentration in the "breathing zone" near a bus stop averaged about eight times higher with a horizontal

Figure A3.2.6 Bus Plume Volume for Concentration Comparison between Vertical and Horizontal Exhausts

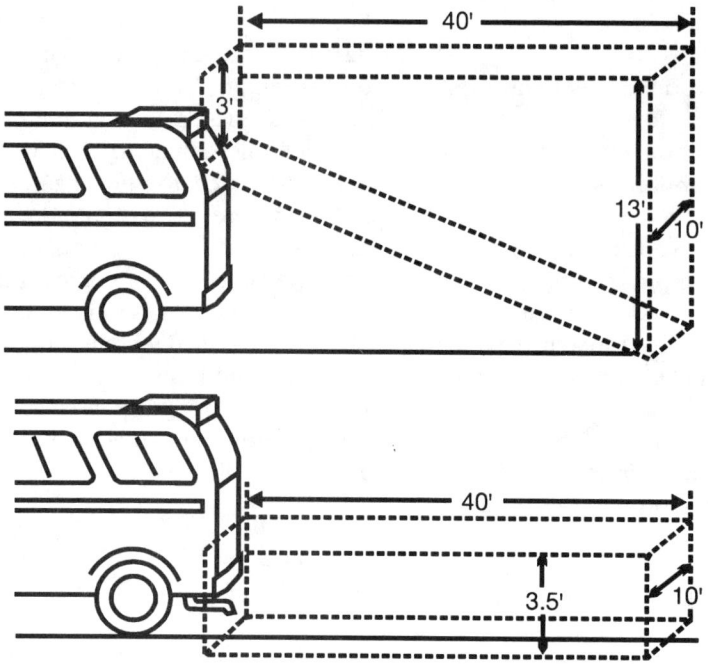

Source: Weaver and others 1986

Figure A3.2.7 Truck Plume Volume for Concentration Comparison between Vertical and Horizontal Exhausts

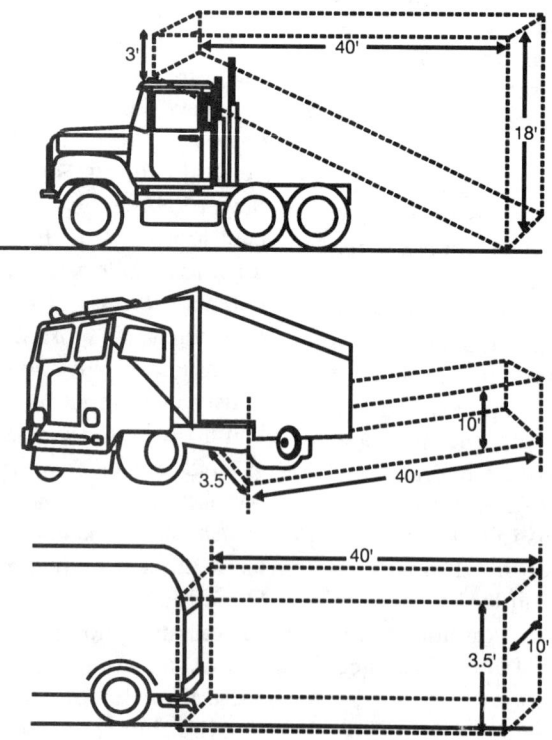

Source: Weaver and others 1986

than with a vertical exhaust. In order to estimate the effects of exhaust geometry on pollutant concentrations in other situations, Weaver and others (1986) conducted field observations of plume dispersion behind a number of smoky trucks and buses. Based on these "eyeball" measurements and simple geometric assumptions, rough estimates of the relative pollutant concentrations behind vehicles of each type moving on a crowded roadway were made. The geometric assumptions used in this calculation are shown in figures A3.2.6 and A3.2.7. These assumptions tend to understate the improvements due to the vertical exhaust. Despite this understatement, the calculated reduction in exposure to high local concentrations of pollutants due to vertical exhaust ranged from 65 to 87 percent.

Retrofitting vertical exhausts to trucks originally equipped with horizontal exhausts is feasible in most cases, with an estimated cost of about U.S.$300— U.S.$500 in an easy installation and up to U.S.$1,000 in a more difficult one. Retrofitting vertical exhausts to buses is also feasible in many cases. For the front-engine buses commonly used in developing countries, the modifications are similar to those required for a truck. A mandatory program of retrofitting existing buses was undertaken in Santiago, Chile in 1987, with considerable success. For rear engine buses, vertical exhaust retrofits may be more difficult. The back of a rear-engine bus is usually crowded with plumbing, air ducts, and other obstructions, many of which would need to be relocated in order to route the exhaust upward. These modifications are best undertaken when the bus is designed. It is recommended that rear engine buses be specified with vertical exhausts when they are ordered. Most bus manufacturers offer a choice of exhaust locations.

References

Balek, S.J. and R.E. Heitzman. 1993. "Caterpillar 3406E Heavy-Duty Diesel Engines." *SAE Paper* 932969, SAE International, Warrendale, Pennsylvania.

Cartellieri, W.P., and W.F. Wachter. 1987. "Status Report on a Preliminary Survey of Strategies to Meet US-1991 HD Diesel Emission Standards Without Exhaust Gas Aftertreatment." *SAE Paper* 870342, SAE International, Warrendale, Pennsylvania.

Colucci, J.M. and G.J. Barnes. 1986. "The Effect of Exhaust System Geometry on Exhaust Dilution and Odor Intensity." *SAE Paper* 710219, Warrendale, Pennsylvania.

Dowling, M. 1992. "The Impact of Oil Formulation on Emissions from Diesel Engines", *SAE Paper* 922198, SAE International, Warrendale, Pennsylvania.

Gill, A.P. 1988. "Design Choices for 1990s Low Emission Diesel Engines." *SAE Paper* 880350, SAE International, Warrendale, Pennsylvania.

Gore, B.M. 1991. "Vehicle Exhaust Emission Evaluations." Office of the Group of Engineers, London Buses Limited, London (Draft).

Havenith, C. J.R. Needham, A.J. Nicol, C.H. Such. 1993. "Low Emission Heavy Duty Diesel Engine for Europe." *SAE Paper* 932959, Society of Automotive Engineers, Warrendale, Pennsylvania.

Heywood, J.B. 1988. *Internal Combustion Engine Fundamentals*. New York: McGraw Hill.

Hower, M.J., R.A. Mueller, D.A. Oehlerking, and M.R. Zielke. 1993. "The New Navistart T 444E Direct-Injection Turbocharged Diesel Engine." *SAE Paper* 930269, SAE International, Warrendale, Pennsylvania.

Ishida, A. T. Kanamoto, and S. Kurihara. 1986. "Improvements of Exhaust Gas Emissions and Cold Startability of Heavy Duty Diesel Engines by New Injection-Rate-Control Pump". *SAE Paper* 861236, in *Combustion, Heat Transfer, and Analysis*. SAE International, Warrendale, Pennsylvania.

Latham, S. and P.R. Tomkin. 1988. "A Study of the Feasibility and Possible Impact of Reduced Emissions Levels from Diesel-Engined Vehicles" *Research Report No. 158*, Transport and Road Research Laboratory, Crowthorne, Berkshire, U.K.

Needham, J.R., D.M. Doyle, and A.J. Nicol. 1991. "The Low Nitrogen Oxide Truck Engine." *SAE Paper* 910731, Society of Automotive Engineers, Warrendale, Pennsylvania.

OECD (Organization for Economic Cooperation and Development)/IEA (International Energy Agency). 1993. *Cars and Climate Change*. Paris.

Shimada, T.K. K. Sakai, and S. Kurihara. 1986. "Variable Swirl Inlet System and Its Effect on Diesel Performance and Emissions", *SAE Paper* 861185 in *Combustion, Heat Transfer and Analysis*. Society of Automotive Engineers, Warrendale, Pennsylvania.

Thomas, F. and H. Fausten. 1993. "The New 4-Valve 6 Cylinder 8.0 Liters Mercedes-Benz Diesel Engine for the Executive Class Passenger Vehicle." *SAE Paper* 932875, SAE International, Warrendale, Pennsylvania.

TUV Rheinland-OPET. 1991. *The Diesel Engine Energy Stake and Environment Constraints*, Proceedings of an International Seminar held in Cologne, 22-23 May 1991, Germany.

Wade, W.R., C.E. Hunter, F.H. Trinker, and H.A. Cikanek. 1987. "The Reduction of NO_x and Particulate Emissions in the Diesel Combustion Process." *ASME Paper* 87-ICE-37 American Society of Mechanical Engineers, New York, NY.

Wallace, F.J., D. Howard, E.W. Roberts, and U. Anderson. 1986. "Variable Geometry Turbocharging of a Large

Truck Diesel Engine". *SAE Paper* 860452, SAE International, Warrendale, Pennsylvania.

Walsh M.P. and R. Bradlow. 1991. "Diesel Particulate Control Around the World." *SAE Paper* 910130, Society of Automotive Engineers, Warrendale, Pennsylvania.

Weaver, C.S., R.J. Klausmeier, L.M. Erickson, J. Gallagher, and T. Hollman. 1986. "Feasibility of Retrofit Technologies for Diesel Emissions Control", *SAE Paper* 860452, SAE International, Warrendale, Pennsylvania.

Weaver, C.S. and L.M. Chan. 1995. "Company Archives" Multiple Sources. Engine, Fuel And Emissions Engineering, Sacramento, California.

Williams, P.T., G.E. Andrews, and K.D. Bartle. 1987. "Diesel Particulate Emissions: The Role of Unburnt Fuel in the Organic Fraction Composition," *SAE Paper* 870554, SAE International, Warrendale, Pennsylvania.

Appendix 3.3

The Potential for Improved Fuel Economy

This appendix provides a general overview of possible fuel economy improvements in the automotive sector in developing countries. It is based mostly on information in several reports on energy use in developing countries, prepared by the Office of Technology Assessment (OTA) of the U.S. Congress.

Background

In most developing countries the transportation sector accounts for one-third of total commercial energy consumption and more than one-half of total oil consumption. China and India are exceptions; in these countries transportation accounts for less than 10 percent of commercial energy consumption, though this share is growing rapidly because of economic growth and the concomitant increase in motorization (OTA 1991a; Faiz 1993). Application of available technologies could significantly improve transport energy efficiency in developing countries. Retrofits—including turbochargers, radial tires, rebuilt motors using diesel fuel injection systems, and cab-mounted air deflectors to reduce wind resistance—could yield substantial energy savings. The main vehicle design factors affecting fuel consumption are weight, rolling resistance and aerodynamic drag (OTA 1995). Figure A3.3.1 shows how the aerodynamic shape of a heavy-duty truck can be improved with side-skirts, fill-in panels, and a roof-mounted deflector. The associated reduction in drag can produce fuel savings of about 10-14 percent (Waters 1990).

In passenger transport, automobiles in developing countries would benefit from such technologies as radial tires, improved aerodynamics, fuel injection, and electronic control of spark timing. Many other efficiency improvements are at various stages of development and commercialization (OTA 1991b). However, most improvements in vehicle fuel economy and performance over the next decade will come from the diffusion of technology already existing in prototypes and some new vehicles.

Opportunities for fuel economy improvements in light-duty vehicles are shown in figure A3.3.2. The two most important areas for improvements involve increasing the efficiency of the drivetrain (engine, transmission, and axles) and reducing the work needed to move the vehicle (tractive effort). The efficiency of the rapidly growing fleet of two- and three-wheelers could also be improved by using improved carburetors, electronic ignition, and four-stroke rather than two-stroke engines (OTA 1992).

These technologies are highly cost-effective, and could result in energy savings of 20 to 50 percent over current levels. Improved carburetors and electronic ignition in two-wheelers, for example, can increase energy efficiency by 10 to 15 percent, while using four-stroke rather than two-stroke engines increases energy efficiency by 25 percent.

Because of the small size of their markets, however, developing countries have little influence over the development and commercialization of these technologies. And several factors impede the adoption of these technologies or the full harnessing of their potential for improved fuel economy, including:

- *Poor infrastructure.* Aerodynamic and turbocharging benefits accrue only at higher speeds, which often are not possible on the rough, congested roads in developing countries. Similarly, poor and deteriorated pavements deter the use of larger and more energy efficient trucks. Many new technologies depend on high quality fuels and are incompatible with the variable quality of fuels often found in developing countries.

- *Maintenance and training.* Several technologies require maintenance skills that may not be available in developing countries. Forcing truck owners to

119

Figure A3.3.1 Aerodynamic Shape Improvements for an Articulated Heavy-Duty Truck

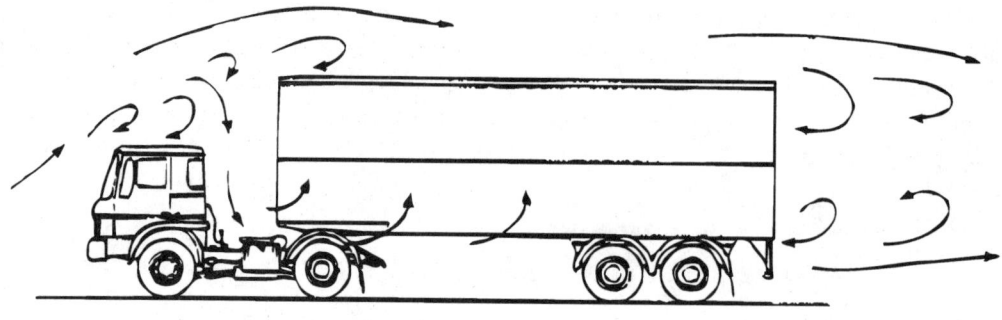

(a) poor aerodynamic shape

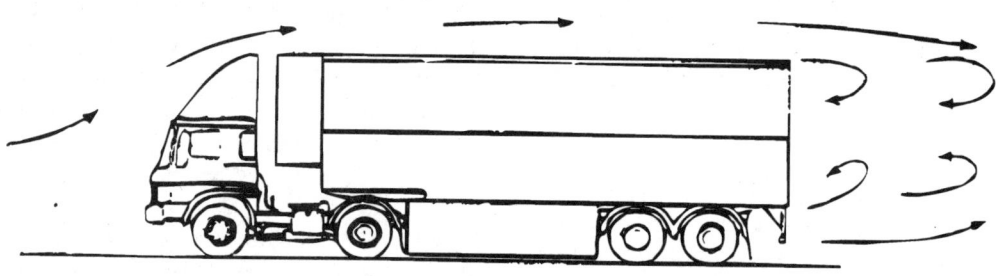

(b) improved aerodynamic shape

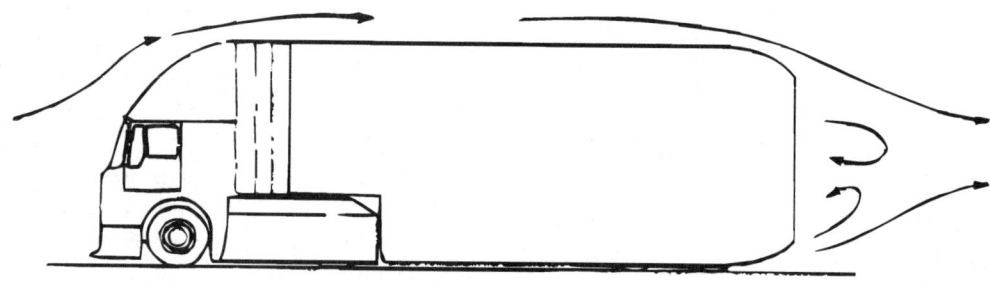

(c) good aerodynamic shape

Source: Waters 1990

seek out specialized firms for routine maintenance reduces the benefits of improved energy efficiency. Poor driving habits can also reduce efficiency gains.

- *High costs.* As in other sectors, potential users are deterred by the additional cost of improved technology. For many users, energy costs constitute a small part of total operating costs, so the efficiency benefit may be small in relation to the effort involved in finding and maintaining a more efficient vehicle and the various attendant uncertainties. En-

ergy efficient technologies are adopted more rapidly in transportation modes where fuel is a large share of total cost (such as in air and maritime transport) and where financial decisions are made on the basis of discount rates closer to commercial bank rates.

- *Fuel costs.* The willingness to pay for new technologies is also closely related to fuel costs. In oil-exporting developing countries, fuel prices are often lower than international prices, offering little incentive to economize on their use. In the oil-importing countries, gasoline prices are generally higher than international costs. Diesel prices, however, are often considerably lower than gasoline and international prices, again discouraging conservation.

- *Import duties.* High import duties on retrofit equipment and on vehicles with higher initial costs due to more efficient equipment deter the diffusion of such technologies, despite their potential to reduce oil (and refinery equipment) imports.

- *Low scrappage rates.* The vehicle fleet is older, less efficient, and less likely to be scrapped in developing countries because road vehicles are in great demand, maintenance is cheap, and replacement is expensive. While old vehicles can be retrofit to improve fuel efficiency, their owners are likely to be unable to afford the additional cost. Replacing old vehicles with new, while contributing to the improved efficiency of the fleet, can incur large capital outlays.

Figure A3.3.2 Technical Approaches to Reducing Fuel Economy of Light-Duty Vehicles

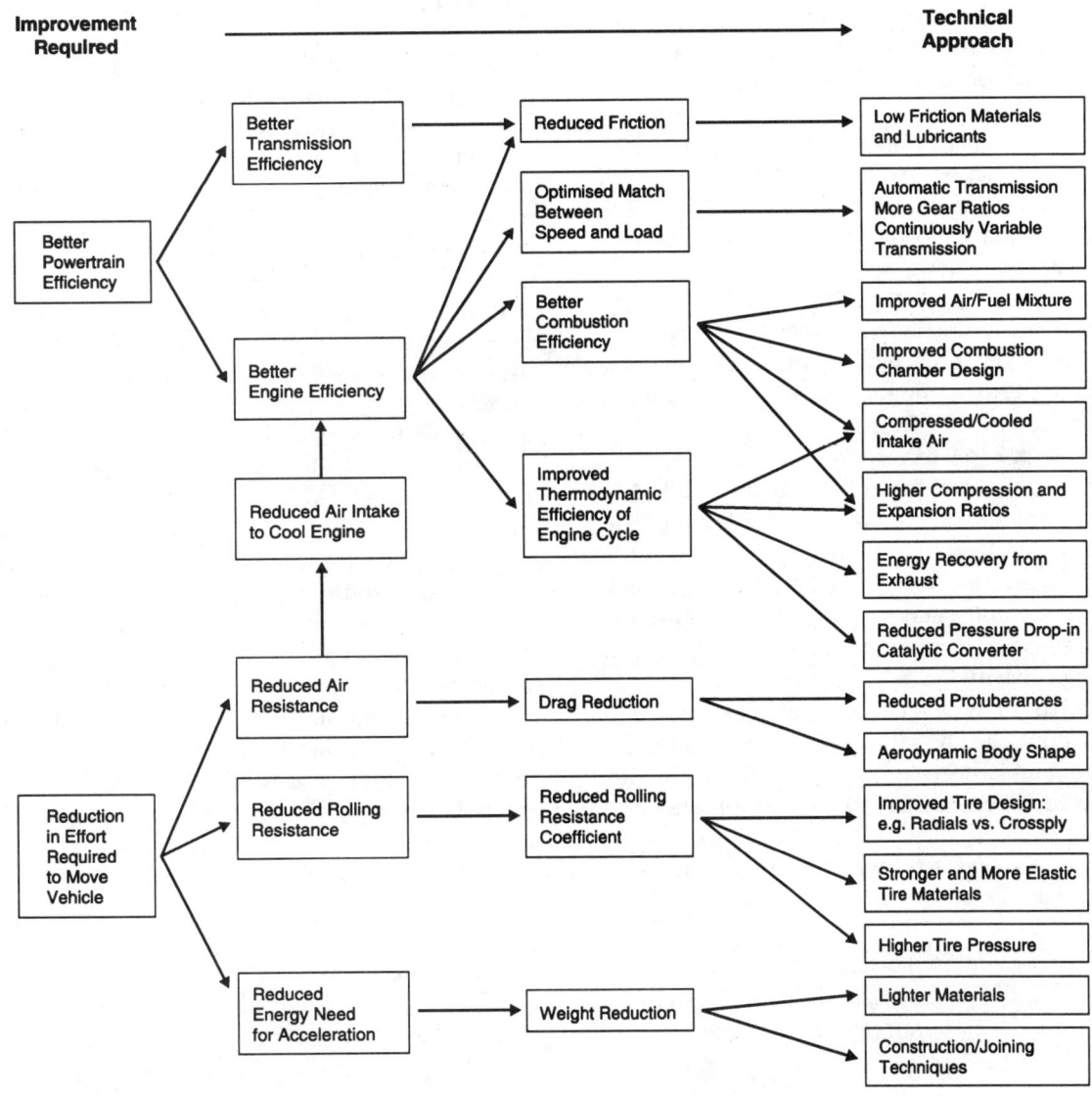

Source: OECD/IEA 1993

Trucks

The potential for improving the energy efficiency of truck fleets in developing countries is considerable. The truck fleets in developing countries are generally older, smaller, and less technologically sophisticated—and therefore less energy efficient—than those found in the industrialized countries (table A3.3.1)

Furthermore, truck fleet scrappage rates are much lower than in industrialized countries. Since repairs are relatively inexpensive, it is usually cheaper and easier to repair a vehicle than to replace it. It is generally not easy to retrofit trucks with efficiency improvements once they are in operation. One exception is periodic engine rebuilding. If done with modern technology, such as improved fuel injectors and injection pumps, rebuilding can improve engine efficiency by up to 5 percent.

However, most repair shops in developing countries lack the technical expertise to rebuild engines with improved technologies. If engines have to be sent to the factory for rebuilding, downtime and transport costs increase, making energy-saving investments less attractive.

The size of a truck affects its overall energy efficiency. Small trucks generally require more energy than large trucks to perform the same task, Chinese trucks, for example, average 4 to 5 tons, and the largest Indian trucks are rated at 8 to 9 tons. In the United States, mostly 40-ton trucks are used for freight haulage while 10–20 ton trucks are used for short-haul distribution. The poor highway infrastructure in many developing countries constrains the use of larger trucks. Improved truck carrying capacity and energy efficiency cannot be implemented without corresponding improvements in road carrying capacity.

As noted, trucks in developing countries are generally less technologically sophisticated. For example, most truck manufacturers in India build diesel engines comparable in technological sophistication and efficiency to those built in industrialized countries in the 1960s. This situation is slowly changing. Some Indian truck manufacturers now offer Japanese engines with improved fuel efficiency. Brazil assembles trucks based on a Western European design and exports them throughout Latin America. And many countries in Africa import trucks from Europe and Japan. Still, given the local conditions in developing countries (road conditions, maintenance challenges, variable fuel quality, and energy pricing policies), total diesel fuel economy stays quite low. As the share of developing countries in markets for trucks increases, they may be able to increase their leverage on industrial manufacturers to provide technologies more suitable for their conditions.

Technology alone may thus have a limited role in improving truck energy efficiency in developing countries. Proven efficiency improvements such as turbochargers, larger vehicles, and aerodynamic improvements, require smooth, uncongested, heavy-duty roads. Diesel vehicle efficiency drops sharply under conditions of varying load and speed such as are often found in developing countries.

Road construction and maintenance, however, are expensive, and the energy efficiency benefits are often difficult to measure. In the short term, improved roads increase energy efficiency by allowing for higher sustained speeds. Truck operating costs drop 15 to 40 percent when a road is paved. In the longer term, improved roads allow for turbochargers, aerodynamic improvements, and larger, heavier trucks. Improved roads may increase traffic (and therefore energy use), but additional traffic may contribute to overall economic development in an efficient manner.

Operational improvements also have the potential to improve efficiency. Since load factors are often low, improved communications, route scheduling, and overall coordination of freight transport can improve energy efficiency by ensuring full loads and reducing waste. Low loads, however, are often dictated by reg-

Table A3.3.1 Energy Efficiency of Trucks in Selected Countries

Country/region	Truck name	Capacity (metric tons)	Energy consumption (megajoules per metric ton per kilometer)
OECD	Mercedes Benz 1217 (1979)	7.0	1.0
OECD	Man-VW9136 (1980)	5.9	1.0
India	TATA 1201 SE/42	5.0	2.1
India	Ashok Leyland Beaver	7.5	1.6
China	Jiefang CA-10B	4.0	2.3
China	Dongeng EQ140	5.0	1.8

Note: OECD and Indian trucks use diesel, Chinese trucks use gasoline.
Source: Yenny and Uy 1985

ulatory and institutional factors. Many enterprises in developing countries do their own shipping rather than rely on commercial fleets that may be less reliable. These "own account" shippers are usually licensed to carry their own products only, resulting in empty backhauls and low average load factors. Entry and price regulations also contribute to low-load factors and high fuel consumption per unit of cargo hauled.

Automobiles

Automobile ownership levels are increasing rapidly. As the most energy intensive form of passenger transport, automobile energy efficiency is of critical importance. As with trucks, the scrappage rate of the automobile fleet in many developing countries is quite low. The reasons are much the same—low repair costs, minimal quality requirements for annual registration, and the high cost and limited availability of new vehicles. Government measures to increase scrappage rates—by purchasing old cars, establishing emission standards, or imposing registration fees that are inversely proportional to age—would increase the average energy efficiency of the auto fleet, but at a cost to users. In addition, the long useful life of the average car in developing countries indicates that new cars have sustainably high standards of energy efficiency.

The energy efficiency of vehicles sold in developing countries varies widely, mainly as a result of whether the vehicle is manufactured domestically or imported. The most popular car in India until recently had fuel consumption about twice that of a comparably sized Japanese or German car. The energy efficiency of cars produced in China is similarly low. But new automotive technologies have been introduced in both countries in recent years, increasing the fuel efficiency of domestically manufactured cars. The fuel efficiency of cars produced in other developing countries—Brazil, Mexico, and Republic of Korea—meet international standards after accounting for their size, accessories, and other factors.

Most developing countries, however, import their autos from the industrialized world, either in finished form (as in small African and Central American nations) or in the form of kits from which cars and light-duty trucks are assembled. Automobiles produced in industrialized countries and exported to developing countries are similar but not identical to those sold in the industrialized countries. Models sent to developing countries have smaller engines, fewer luxury accessories (such as air conditioning), lower compressions ratios (to allow for lower octane gasoline), and often do not use technologies such as fuel injection and electron-

ic engine controls. The lack of luxury accessories increases efficiency, but the lack of electronic engine controls and other technologies decreases it. The net effect is that autos exported to developing countries are of comparable efficiency to similar models produced and sold in the industrialized countries.

The energy efficiency of these vehicles would benefit from readily available technologies, though at an increased cost. The electronic control of spark timing and idle speed found in almost all industrialized country vehicles, for example, increases fuel efficiency by 4 to 5 percent at a cost of about U.S.\$75–100. Radial tires, improved aerodynamics, and fuel injection offer similar energy savings.

Though such calculations do not account for environmental and safety benefits, they do illustrate that investments in increased efficiency can provide a reasonable return. Evidence from industrialized countries, however, indicates that consumers demand a payback period of two years or less from efficiency investments; consumers in developing countries could be expected to demand even shorter payback period. In oil-importing developing countries gasoline prices tend to be above international levels, and there is a limit to how much gasoline prices can be increased without unduly widening the gap between gasoline and diesel prices (table A3.3.2). Purchase taxes in most countries are already structured to discourage the purchase of large autos, suggesting a role for auto efficiency standards.

Efficiency features such as fuel injection are considerably more complex than what is currently in place (carburetors) and require skilled labor for repair. But some efficiency features offer benefits in addition to fuel savings. Fuel injection is more reliable, does not require adjustment, and results in lower emissions than carburetors. Radial tires improve handling and safety and increase tire life.

Developing countries depend on industrialized countries for vehicle design. Further improvement in fuel efficiency thus depends on automotive technology advances in the industrialized countries. Vehicles currently sold in both the developing and the industrialized world are not nearly as efficient as is technically possible. Several manufacturers, including General Motors, Volkswagen, and Volvo have built prototype automobiles that achieve from 66 miles per gallon (3.5 liters/100 km) to 70 miles per gallon (3.3 liters/100 km). A prototype four-seater automobile introduced by Toyota in 1985 achieved 98 miles per gallon (2.4 liters/100 km). This vehicle uses a direct-injection diesel engine, a continuously variable transmission (CVT), and plastics and aluminum to reduce weight. While these vehicles are not in production, the long-term technical potential for efficiency improvements is large, and improvements

Table A3.3.2 International Gasoline and Diesel Prices
(U.S. cents per liter)

		Gasoline	Diesel				Gasoline	Diesel
North America (first half 1996 unless noted)					**Europe (first half 1996 unless noted)**			
Canada		51	37					
Mexico		39	26		*Western*			
United States		39	35		Austria	Dec '95	115	87
					Belgium		119	80
Latin America and the Caribbean					Denmark		113	89
(first quarter 1996 unless noted)					England		92	85
Antigua & Barbuda	1994	56	53		Finland		120	85
Argentina		84	28		France		118	82
Barbados	Dec '95	77	64		Germany		111	74
Bolivia		60	40		Greece		93	62
Brazil		65	37		Iceland	Dec '95	105	35
Chile		55	34		Ireland		104	90
Colombia		34	23		Italy		122	91
Costa Rica		41	28		Luxembourg		95	67
Curacao		66	25		Netherlands		116	80
Dominica		40	28		Norway		134	116
Ecuador		33	28		Portugal		103	70
El Salvador		47	27		Spain		91	68
Grenada		71	57		Sweden	Dec '95	117	101
Guatemala		45	35		Switzerland	Dec '95	102	101
Honduras		40	27					
Jamaica		27	24		*Central and Eastern (last quarter 1995 unless noted)*			
Nicaragua		89	32		Bulgaria		46	26
Panama		49	33		Croatia		75	64
Paraguay	Dec '95	44	28		Czech Republic		85	60
Peru		77	39		Estonia		33	33
Puerto Rico		36	31		Hungary		74	65
Trinidad & Tobago		48	22		Latvia		41	34
Uruguay		95	44		Lithuania		62	30
Venezuela		12	9		Macedonia		93	59
					Poland		55	42
Africa (last quarter 1995 unless noted)					Romania		29	19
Algeria		40	23		Russia	Jan. '96	40	22
Benin		36	28		Slovakia		66	40
Botswana		38	35		Slovenia		59	50
Burkina Faso		81	62		Turkey	Jul '96	63	37
Burundi		52	48		Yugoslavia/Serbia		76	84
Cameroon		68	50					
Central African Republic		82	64		**Asia (1994 unless noted)**			
Chad		80	70		Bangladesh	Mar. '96	36	31
Cote d' Ivoire		83	56		China		28	25
Egypt		29	12		Hong Kong		105	71
Ethiopia		32	24		India	Mar. '96	54	22
Ghana		38	33		Israel		62	16
Kenya	Apr '96	56	37		Japan	Jul '96	108	69
Madagascar		47	32		Kuwait	Oct '96	15	13
Mali		82	57		Nepal	Mar. '96	52	22
Morocco		94	47		Pakistan	Jul '96	45	20
Mozambique		53	32		Saudi Arabia	Sep '96	16	10
Niger		79	55		Sri Lanka	Mar. '96	74	22
Nigeria		13	3		Republic of Korea	Jan. '96	79	33
Sengal		94	62		Taiwan (China)		61	43
South Africa	Apr '96	48	48		Thailand		31	29
Sudan		50	25		Vietnam		34	34
Tanzania		56	44					
Togo		47	40		**Other (1995 unless noted)**			
Uganda		98	85		Australia		54	52
Zambia		60	57		New Zealand		63	35

Source: World Bank 1996

made by the major producers could quickly become the global standard.

Buses

Buses are the backbone of urban passenger transport in the developing world, providing essential low-cost transport, particularly for low-income groups. In many cities buses account for more than half of all motorized trips.

As with other road vehicles, a variety of technologies are available to improve the energy efficiency of bus transport, including turbochargers, smaller engines, automatic timing advances, and lighter frames (using aluminum or plastic rather than steel). However, fuel conservation technologies in buses are constrained by their operating environment. Most urban buses operate on congested streets, and low speeds and frequent speed changes are associated with low operating efficiency. To the extent that inefficient bus service has contributed to the use of automobiles and two- and three-wheelers, more efficient bus service could slow the increase in private vehicle ownership.

Some cities have introduced bus priority lanes in an effort to improve bus service. The most successful priority busways are in Brazil, notably in Curitiba, Porto Alegre, and São Paulo. These busways have increased bus speeds by more than 20 percent, created smoother traffic flows, and reduced fuel consumption.

Two- and Three-Wheelers

Two-wheelers are widely used in many Asian and other developing nations as an inexpensive mode of personal transportation for a growing urban middle class. Two- and three-wheelers vastly outnumber automobiles in many Asian cities and consume a large fraction of total gasoline (table A3.3.3). Two- wheelers fall between automobiles and buses in terms of energy efficiency per passenger-mile.

Two-wheeler engines are designed as either two-stroke or four-stroke. In the past, all but the largest motorcycles had two-stroke engines, since manufacturing these engines is simple and inexpensive. They also produce more power for a given displacement and require little maintenance. However, emissions from two-stroke engines (largely unburned gasoline) are ten times those of four-stroke engines of equal power, and fuel efficiency is 20 to 25 percent lower.

The problems of the two-stroke engine are made even worse when used on a three-wheeler vehicle. Such vehicles are underpowered, so the engine is usu-

Table A3.3.3   Gasoline Consumption by Two- and Three-Wheelers

Country	Percentage of total consumption
Bolivia	8
Brazil	2
Cameroon	12
China	12
India	45
Japan	14
Kenya	Less than 2
Taiwan (China)	50
Thailand	30
United States	Less than 1

Source: OTA 1992

ally operated near wide-open throttle. Two-stroke engines operating under these conditions produce high emissions and have poor fuel economy, with fuel consumption comparable to that of some small cars. Yet there is a widespread perception in India and parts of Asia that three-wheelers are efficient modes of public transport.

Improved technologies are available that can drastically reduce emissions and fuel consumption, though at some cost to the user. Improved carburetors and electronic ignition can improve fuel efficiency by 10 to 15 percent. Replacing two-stroke with four-stroke engines increases fuel efficiency by 25 percent. This technology costs about U.S.$100. Based on a gasoline price of U.S.$1.50/gallon (U.S.$0.40/liter), this investment is recouped in about 1.6 years, even without considering environmental benefits (OTA 1992).

References

Faiz, A. 1993. "Automotive Emissions in Developing Countries—Relative Implications for Global Warming, Audification and Urban Air Quality." *Transport Research, 27 (3): 167–86.*

OECD (Organization for Economic Cooperation and Development)/IEA (International Energy Agency). 1993. *Cars and Climate Change.* Paris.

OTA (Office of Technology Assessment). 1991a. *Energy in Developing Countries.* Report No. OTA-E-486, Washington, D.C. U.S. Government Printing Office.

———. 1991b. *Improving Automobile Fuel Economy: New Standards, New Approaches.* Report No. OTA-E-504, Washington, D.C.: U.S. Government Printing Office.

_____. 1992. *Fueling Development: Energy Technologies for Developing Countries*. Report No. OTA-E-516, Washington, D.C. U.S. Government Printing Office, Washington, D.C.

_____. 1995. *Advanced Automotive Technology, Visions of a Super-Efficient Family Car*. OTA-ETI-638, U.S. Government Printing Office, Washington, D.C.

Waters, M.H.L. 1990. "UK Road Transport's Contribution to Greenhouse Gases: A Review of TRRL and other Research." Transport Road Research Laboratory (TRRL) Contract Report No. 223, Crowthorne, Berkshire, U.K.

World Bank. 1996. International Gas Prices, Internal Memo Transportation, Water and Urban Division, The World Bank, Washington, D.C.

Yenny, J. and L. Uy. 1985. "Transport in China." *Working Paper 723*, World Bank, Washington, D.C.

4

Controlling Emissions from In-Use Vehicles

Inspection and maintenance (I/M) measures to control emissions from in-use vehicles are an essential complement to emission standards for new vehicles. Although difficult to implement, an effective inspection and maintenance program can significantly reduce emissions from uncontrolled vehicles. I/M programs are also needed to ensure that the benefits of new-vehicle control technologies are not lost through poor maintenance and tampering with emission controls.

I/M programs for gasoline vehicles commonly include measurement of hydrocarbon and carbon monoxide concentrations in the exhaust. These have limited effectiveness but can identify gross malfunctions in emission control systems. Newer programs such as the IM240 procedure developed in the United States use dynamometers and constant volume sampling to measure emissions in grams per kilometer over a realistic driving cycle. Inspection and maintenance of high-technology, computer-controlled vehicles can be enhanced substantially with on-board diagnostic systems. For diesel vehicles, smoke opacity measurements in free acceleration are the most common inspection method. This approach also has limited effectiveness but can identify serious emission failures. Opacity measurements can also be used to control white smoke emissions from two-stroke motorcycles.

On-road emission checks can improve the effectiveness of periodic I/M programs. Checks for smoke emissions from two-stroke and diesel vehicles can be made more effective by visual prescreening. The effectiveness of on-road checks for hydrocarbons and carbon monoxide can be enhanced by remote sensing the concentrations of these pollutants in the vehicle exhaust.

There are two main types of I/M programs: centralized programs, in which all inspections are done in high-volume test facilities operated by the government or contracted to competitively-selected private operators, and decentralized programs, in which both emissions testing and repairs are done in private garages.

The decentralized arrangement is generally less effective because of fraud and improper inspections (box 4.1). Centralized programs operated by private contractors yield better results (box 4.2) and are recommended for most developing countries.

Inspection and maintenance programs are an important part of motor vehicle emissions control efforts. When I/M programs identify non-complying vehicles, this information can be fed back to recall and assembly line test programs, allowing regulatory authorities to focus investigations and test orders on vehicles with consistently high emissions. Inspection and maintenance programs help identify equipment defects and failures covered by vehicle warranty schemes. These programs also discourage tampering with emission controls or misfueling; the threat of failing inspection is considered a strong deterrent. Without effective I/M programs, compliance with standards is significantly weakened.

Where emission standards have recently been introduced, measures are needed to address existing uncontrolled and high-emitting vehicles. Retrofit programs for some vehicles may reduce in-use emissions. Policies that accelerate the retirement or relocation of worn-out and inefficient vehicles and vehicles not equipped with emission controls can also be of value, particularly in developing countries where the high cost of vehicle replacement combined with the low cost of labor for repairs results in large numbers of older vehicles continuing in service well beyond their normal lifespan.

Inspection and Maintenance Programs

To ensure a reasonable level of maintenance and proper functioning of emission controls, many jurisdictions in Europe, Japan, the United States, and an increasing number of developing countries have established peri-

---

Box 4.1 Effectiveness of California's Decentralized "Smog Check" Program

The first wide-ranging test and repair I/M program was implemented in California in 1984, requiring gasoline-fueled light-duty vehicles to pass inspection every two years. Inspections were performed at decentralized, state-licensed private garages. California's Smog Check Program demonstrates the inherent weakness of decentralized test and repair I/M, and also the political difficulty of changing such a program once adopted. Despite rigorous enforcement efforts by the California Bureau of Automotive Repair (BAR), an independent audit of the Smog Check Program found that improper inspection and repairs were widespread and seriously reduced its potential effectiveness. Failure to carry out complete visual and functional checks of emission control systems, and falsification of after-repair emission tests to show that vehicles had been repaired to meet the emission standards, were identified as major problems. Covert audits carried out by BAR have shown that only about half of the vehicles tested receive proper visual and functional checks.

According to a 1995 review of the program by U.S. EPA, falsification of retest data had compromised the program's effectiveness to achieve reductions in HC emissions by at least 30 percent and CO emissions by about 39 percent. Such falsification is an inherent problem in test-and-repair systems, due to the conflict of interest in having a repair facility check its own work. Several reasons are attributed to the less than expected effectiveness of the California Smog Check Program: inspection stations do not identify problem vehicles; the stations engage in corruption and fraud; mechanics have difficulty in fixing broken cars—indeed, about half the cars repaired following an inspection have increased emissions; and motorists tamper with their cars to make them clean on inspection day.

The experience with California's Smog Check Program convinced the U.S. EPA to require test-only inspections for enhanced I/M programs in the U.S. But in California, the implementation of a test-only program has been blocked because of political resistance by the existing Smog Check inspection facilities. Instead, California is seeking to implement a complex "hybrid" program that would retain the costly and ineffective test-and-repair system for most vehicles and make increased use of remote sensing to identify possibly high-emitting vehicles on the road. The technical and political feasibility of this hybrid program is yet to be established.

Source: U.S. EPA 1995

---

odic inspection and maintenance programs for light-duty cars and trucks. Some jurisdictions have extended these programs to include heavy-duty trucks and, in a few cases, motorcycles. These programs often are supplemented by programs that identify and cite polluting vehicles on the road.

Inspection and maintenance serve two purposes in a vehicle emission control program. First, they help identify vehicles in which maladjustments or other mechanical problems are causing high emissions. In populations of modern, emission-controlled vehicles, a large fraction of total emissions is due to a minority of vehicles with malfunctioning emission control systems. Various researches have shown that 5 percent of the vehicle fleet causes 25 percent of all emissions, that 15 percent of the fleet is responsible for 43 percent of total emissions, and that 20 percent of the vehicles are responsible for 60 percent of emissions (Guensler 1994). The problem is aggravated by super emitters – vehicles with emission rates greater than five times the certification standard. Among uncontrolled vehicles, the difference in hydrocarbon and carbon monoxide emissions between a properly adjusted and maintained engine and one that is poorly adjusted can amount to a factor of four or more. Carbon monoxide emissions can increase by up to 400 percent through normal drift of engine settings between routine services (Potter and Savage 1986).

The second important role of I/M programs is to identify malfunctions and discourage tampering with emission control equipment, so that the emission controls continue to be effective over the useful life of the vehicle. A damaged catalytic converter or malfunctioning oxygen sensor can increase hydrocarbon and carbon monoxide emissions from modern emission-controlled vehicles by a factor of 20 or more, often without seriously affecting drivability. Similar malfunctions can increase nitrogen oxide emissions three to five fold as well. In diesel vehicles, a worn or damaged fuel injection system can increase emissions of particulate matter at least twenty fold. It has been shown that by identifying vehicles that have maintenance problems and requiring that they be repaired, an effective I/M program should be able to reduce average vehicle emissions by 30 to 50 percent. There are major political and institutional constraints to establishing such programs, however, and few I/M programs now in place are achieving their full potential.

A comprehensive I/M program requires the following major elements:

- A suitable test procedure, supplemented by inspection of emission control systems where necessary.
- Effective enforcement of vehicle compliance (for example, through the vehicle registration process).
- Adequate attention to repair procedures and mechanic training.

Box 4.2 Experience with British Columbia's AirCare I/M Program

In 1992, the Canadian province of British Columbia implemented a centralized emissions inspection and maintenance (I/M) program in the Lower Fraser Valley to control excess emissions from in-use cars and light duty trucks. The program, known as AirCare, utilized state-of-the-art inspection procedures. It was the first I/M program to measure HC, CO and NO_x emissions using the acceleration simulation mode (ASM), a steady-state loaded mode test that simulates vehicle acceleration. The inspection also included an idle test and an anti-tampering check to assure high emitting vehicles were identified and repaired. Shown below are average exhaust emission test results, before and after repair for HC, CO, and NO_x for the indicated model year groups.

Model year	Carbon monoxide (g/km)		Hydrocarbons (g/km)		Nitrogen oxides (g/km)	
	Before repair	*After repair*	*Before repair*	*After repair*	*Before repair*	*After repair*
Pre-1981	33	17	3.5	1.9	3.3	1.4
1981-1987	29	12	2.2	1.2	2.8	2.1
Post-1987	8.6	2.9	0.49	0.24	3.0	1.7

Repairs significantly reduced HC, CO, and NO_x emissions from failed vehicles in all model year groups; overall, about 88 percent of the repairs were effective in reducing emissions. Based on the audit results, it was estimated that emissions were reduced approximately 20 percent for HC, 24 percent for CO, and 2.7 percent for NO_x, results that correlate well with the reductions predicted by the U.S. EPA's MOBILE5a model—20 percent for HC, 20 percent for CO, and one percent for NO_x. In addition, the audit program found that fuel economy for the failed vehicles improved by approximately 5.5 percent corresponding to an estimated annual savings of C$72 per year per vehicle.

An audit program has demonstrated that the centralized AirCare program has resulted in very high quality inspections. For example, after reviewing over 2 million tests, the auditor concluded that incorrect emission standards were applied in only one percent of the inspections. Not one instance was found where a vehicle was given a conditional pass or waiver inappropriately. Waivers are given after at least C$200 has been spent on repairs at a certified repair facility. About one percent of the failed vehicles were found to be receiving waivers even though their emissions were excessive, i.e. they exceeded either 10 percent CO, 2,000 ppm HC or 4,000 ppm NO_x. Available data also indicated that vehicles are repaired sufficiently well so that they remain low emitting. For example, almost 53,000 vehicles which failed the test the first year were repaired adequately to pass the following year.

In summary, the British Columbia experience confirms that well-designed I/M programs, properly implemented at centralized facilities using a loaded mode test can achieve a substantial reduction in emissions. These reductions are approximately equal to those predicted by the MOBILE5a model. An added benefit is the substantial fuel savings. Improvements to the program such as including evaporative testing, reducing or eliminating cost waivers, adding the IM240 test or tightening the standards could further increase program effectiveness and benefits.

Source: Radian 1994

- Routine quality control.
- Enforcement of program requirements for inspectors and mechanics, especially in decentralized programs, through such means as undercover vehicles containing known defects.
- Periodic evaluation and review to identify problem areas and develop solutions.
- Comprehensive vehicle model year coverage that includes older vehicles.
- Minimization of repair cost waivers and other waivers and exemptions.

These requirements call for a properly designed program that is well funded, politically supported, and staffed with technically competent personnel. In most cases, it has proven difficult to sustain these elements over the long term, resulting in less effective programs.

Vehicle Types Covered

The tendency in industrialized countries such as the United States has been to design I/M programs around private passenger cars, with other vehicle classes included as an afterthought, if at all. This is inappropriate even in industrialized countries where passenger cars account for most pollutant emissions, and is even less appropriate in developing nations. Because of their larger share of the vehicle fleet and relatively high utilization, commercial vehicles such as buses, minibuses, taxis, and trucks account for a large share of motor vehicle emissions in developing countries. More frequent and more stringent inspections may therefore be justified for these vehicles, and such measures are likely to meet less political resistance than inspections of passenger cars. Concentrating on these vehicles may also be justified where governments, because of limited institutional capacity or political concerns, are unable to implement a more extensive I/M program. As these vehicles have impacts on public safety, it may also be appropriate to include safety inspections (brakes, tires, and so on) at the same time as the emissions inspection. Where safety inspections are already in effect, emissions inspection can be added to the program.

Inspection and maintenance programs and test procedures should target the vehicle types and pollutants of greatest concern. For example, if particulate emissions from diesel vehicles and two-stroke motorcycles are the most important vehicle-related air quality concern (as they are in much of the developing world), I/M programs initially should focus on controlling smoke emissions from such vehicles. There has been a tendency among developing countries to model their I/M programs after those in the industrialized countries, such as concentrating on hydrocarbon and carbon monoxide emissions from passenger cars. This is often inappropriate.

The potential emission benefits of improving maintenance for diesel vehicles are shown in figure 4.1, which compares the emissions and fuel economy from several buses tested as part of a Chilean field study. Among the

diesel buses, four had their engines rebuilt to manufacturers' specifications. One was tested in that condition, one was maintained in accordance with manufacturer's specifications, one was maintained according to average maintenance standards, and one received no maintenance. Additional tests were carried out on a poorly maintained in-service bus without rebuilding the engine. Emissions, especially of particulate matter, were many times higher for the engine that received no maintenance, and even higher for the poorly maintained in-service bus. The potential reduction in diesel emissions through adequate maintenance remains substantial in most developing countries. Further, the improvement in fuel economy from adequate maintenance should more than recoup the added costs.

Inspection Procedures for Vehicles with Spark-Ignition Engines

For vehicles with spark-ignition engines, the inspection element of an I/M program typically includes a measurement of tailpipe hydrocarbon and carbon monoxide concentrations. Some jurisdictions perform a visual check of the emission control systems, and some areas conduct a functional check of certain systems, such as the exhaust gas recirculation (EGR) system and ignition timing. Checks of evaporative control systems have recently been introduced in the United States. Ve-

Figure 4.1 Effect of Maintenance on Emissions and Fuel Economy of Buses in Santiago, Chile

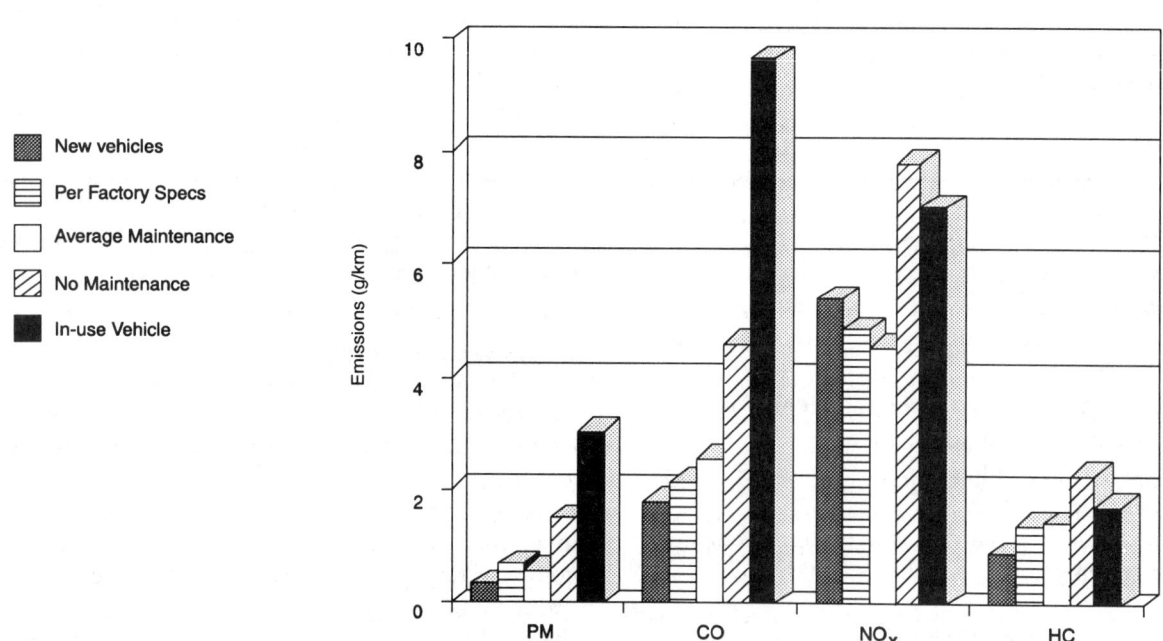

Source: Escudero 1991

hicles with malfunctioning or disabled emission controls or excessive exhaust pollution must be repaired and retested. Often there is provision for a cost waiver—a limit on the amount the vehicle owner is required to spend for repairs, except in the case of deliberate tampering. About 75 percent of the I/M programs in the United States include provision for waivers. Under the U.S. Clean Air Act, repair cost limits may not be less than U.S.$450 for vehicles subject to enhanced inspection and maintenance. For vehicles in less-polluted areas subject only to basic inspection and maintenance, the repair cost limits are U.S.$75 for pre-1981 vehicles and U.S.$200 for 1981 and later vehicles, adjusted for inflation.

Exhaust Emissions

Idle tests. The most common emission test in I/M programs is a measurement of hydrocarbon and carbon monoxide concentrations in the exhaust while the vehicle is idling. Many inspections in Finland, Germany, Sweden, and the United States supplement this measurement with a second measurement carried out with the engine running at 2,500 rpm with no load. Austria has adopted a 3,000 rpm high idle (Laurikko 1994). These measurements do not require a dynamometer or other expensive equipment.

The idle test was originally developed for vehicles with little or no emission control, and for these vehicles it can detect a large proportion of malfunctioning or maladjusted engines. These vehicles are normally equipped with mechanical carburetors or fuel injection systems, where the air-fuel ratio at idle is related to the air-fuel ratio under load. Thus measurements of hydrocarbon and carbon monoxide emissions at idle and 2,500 rpm provide a reasonable indication of the emissions under normal operating conditions for vehicles with mechanical air-fuel ratio control systems. These vehicles include older technology cars and light trucks, and most heavy-duty gasoline vehicles commonly used in developing countries.

The *idle/2,500 rpm test* procedure does not give satisfactory results for vehicles using electronic air-fuel ratio control systems. Despite the widespread use of this test procedure in I/M programs, the correlation is poor between idle/2,500 rpm test results and emissions measured under more realistic conditions. For example, Laurikko (1994) found that a reasonably good correlation existed between U.S. FTP results and the idle/2,500 rpm test for carbon monoxide emissions, but not for hydrocarbons. The test was also ineffective in detecting high NO_x emissions, as might be caused by a defective or disconnected oxygen sensor. Meaningful measurements of NO_x emissions require that the engine be under load.

The idle/2,500 rpm procedure passes many vehicles that have high emissions when measured in a more comprehensive emissions test. Passing a vehicle that has high emissions is referred to as an error of omission. The idle/2,500 rpm test sometimes fails vehicles that do not have excessive emissions when measured on the FTP and that are not malfunctioning. This is known as an error of commission. Errors of omission reduce the effectiveness of the I/M program, while errors of commission unnecessarily increase costs to the consumer and may even increase emissions, as mechanics try to fix an emissions control system that is not broken.

In the past a number of car models suffered from chronic errors of commission in the idle/2,500 rpm test procedure, usually because of the interaction between catalyst protection strategies (such as turning off air injection after a specified period at idle) and the I/M test procedure. These failures were a significant issue in the late 1980s, but most vehicles are now designed to avoid this problem. Regulations adopted by the U.S. Environmental Protection Agency require new vehicles to pass all applicable I/M tests with a significant margin for deterioration.

To help reduce the false failures associated with the idle test, some I/M programs require *preconditioning* at 2,500 rpm with no load for three minutes before a final idle test failure determination is made. This preconditioning helps ensure that the control system is in normal closed-loop operation and the catalytic converter is adequately warmed up. Tests in Finland have shown that 95 percent of the vehicles equipped with three-way catalytic converters reach stabilized readings within three minutes, and in Austria three-minute conditioning prior to measurement is mandatory (Laurikko 1994). Because of the increased noise resulting from running the engine at high speed, the Finnish testing procedure requires that measurements start at 2,500 rpm. If the readings stabilize before the three minutes have elapsed, the vehicle passes the high idle portion of the test. After passing the high-idle test, the conditioning is normally sufficient for measurement at normal idle. More than one-third of the vehicles that fail the initial test pass after extended preconditioning (Tierney 1991). Even though adequate preconditioning minimizes false failures, the correlation between idle/2,500 rpm results and FTP emissions remains poor.

Loaded dynamometer tests. Studies in California (Austin and Sherwood 1989) and Germany (OECD 1989) indicate that more representative test procedures, requiring dynamometer loading of the vehicles, give substantially better results, especially for NO_x emissions. An acceleration simulation mode test called the

ASM5015 (15 mph, with horsepower loading equal to that required to achieve 50 percent of the maximum acceleration rate on the FTP) gives reasonable correlation with the FTP and is effective in identifying vehicles with high NO_x emissions, related to EGR valve malfunction (U.S. EPA 1995). The ASM 5015 mode has been combined with the ASM2525[1] mode to create the *ASM2 test* procedure, which is a dynamometer loaded, steady state, raw exhaust concentration test that approximates mass emissions by relating emission concentrations to the displacement of the engine being tested. The ASM procedure has been adopted in New Jersey in the United States as part of an enhanced I/M program. British Columbia in Canada carried out pilot testing of ASM 5015 and ASM 2525 in 1992 as part of its regular I/M program and decided to drop the ASM 5015 mode from its official test procedure because of operational problems (U.S. EPA 1995).

---

1.   In the ASM2525 mode, the vehicle is operated at 25 mph with a dynamometer setting to achieve 25 percent of the maximum acceleration rate on FTP. The length of the modes used in the test procedure has not been fully specified by U.S. EPA or by any state. Based on preliminary analysis of the California results, each mode will need to last a minimum of 30 seconds to ensure that exhaust concentrations are reasonably stable and vehicles are adequately conditioned. As with other steady-state tests, preconditioning and/or second chance testing will also need to play a role to minimize false failures. For some vehicles, a minimum of 180 seconds of loaded operations will be needed to trigger operation of the purge mechanism. There are six cut points (ppm for HC and NO_x, percent for CO) applied in the ASM2 test, one for each of the three pollutants on each of the two modes. These six cut points must vary with vehicle weight, to ensure that large and small cars are held to comparable targets for total mass emissions.

Advanced "short" tests require a vehicle to simulate more rigorous driving conditions on a dynamometer under a loading similar to that which the vehicle would experience in actual driving. These procedures involve testing the vehicle under simulated transient driving conditions, using equipment similar to that for emissions certification but with an abbreviated driving schedule (see chapter 2). The most advanced of these tests and the one currently recommended for use in I/M programs in the United States is the *IM240*, so named because it is based on the first 240 seconds of the FTP. This 240-second test involves operating the vehicle over a transient driving cycle, simulating the stop-and-go of real driving in a manner similar to that experienced in the FTP. As shown in figure 4.2, the simulation requires fairly elaborate test equipment, including a dynamometer with flywheels to simulate the effect of the vehicle's inertia, and a more expensive sampling and measurement system (U.S. EPA 1993). The IM240 test procedure has shown very good correlation with the full FTP and is more effective than steady-state loaded tests in identifying high-emitting vehicles. A study (Martino, Wakim and Westand 1994) has attempted to correlate the FTP with the IM240 test. The correlation coefficient between the IM240 test and the FTP was shown to be 0.86 for CO and 0.90 for HC. The study concluded that this test is about three times more accurate in identifying vehicles exceeding emission standards than the idle test techniques. Test data for California vehicles (Weaver and Chan 1994) showed correlation coefficients for HC, CO, and NO_x of 0.95, 0.71, and 0.96, respectively, between the FTP and the IM240; this was significantly

Figure 4.2 Schematic Illustration of the IM240 Test Equipment

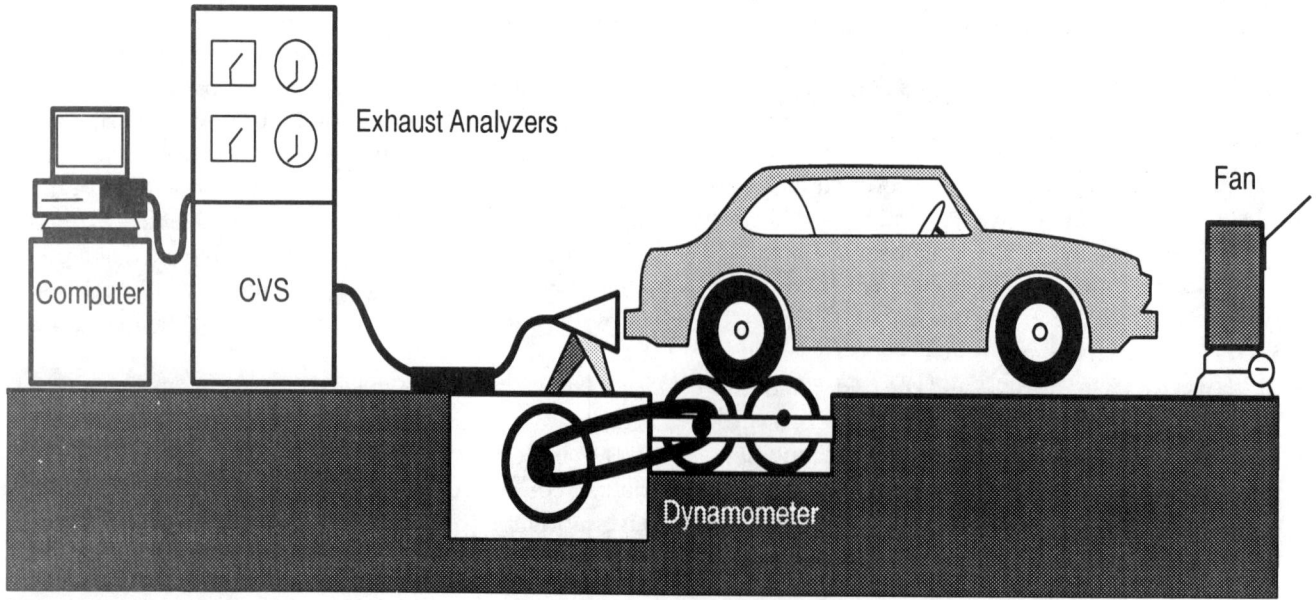

Source: U.S. EPA 1995

better than the correlation with either of the ASM test procedures.

Compared to earlier tests, the IM240 and similar short test cycles have many advantages. Since they are based on realistic driving cycles, they correlate better with the results of complete emissions tests but are considerably shorter and less complex to perform. Since the emissions measured in the test procedure are representative of those produced in actual driving, errors of omission and commission are greatly reduced. The low likelihood of errors of commission makes it possible to establish more stringent standards, thereby reducing errors of omission. The test procedure also provides realistic measurement of NO_x, HC, and CO emissions, making visual and functional checks of EGR, spark timing, and catalyst presence unnecessary. This reduces the opportunities for fraud and error in the inspection program.

The disadvantages of the IM240 test are the longer testing time (which reduces throughput) and the cost (U.S.$80,000–120,000) and complexity of the equipment required to perform it. A dynamometer with inertial simulation used for the IM240 test costs about U.S.$20,000, and the cost of the associated sampling and measuring equipment can exceed U.S.$100,000. By comparison, the cost of equipment to perform the ASM5015 test is about U.S.$40,000, while the simple idle/2,500 rpm test requires an analyzer costing U.S.$5,000-12,000. Competition and economies of mass production may reduce the cost of IM240 systems in the future by as much as 50 percent (U.S. EPA 1995).

To increase the effectiveness of I/M testing, especially for nitrogen oxides, many jurisdictions require supplementary visual and functional checks of the emission controls. For instance, the mechanic is required to confirm the presence of the catalytic converter, check that the fillpipe restrictor is in place, check for lead in the tailpipe (indicative of misfueling), confirm operation of the EGR valve, check ignition timing, ensure that there are no leaks in the exhaust system, and so on. Because these checks are time-consuming, many mechanics in decentralized I/M programs do not perform all of them.

Evaporative Emissions

Evaporative emissions account for about half the total hydrocarbon emissions from current vehicles. Two common failures of evaporative systems are vapor leaking from the fuel system and failure of the canister purge system, which results in an ineffective, saturated canister. U.S. EPA specifications for enhanced I/M programs require a pressure test for fuel vapor leakage and a functional check of the canister purge system (U.S. EPA 1993). Few vehicles are programmed to purge under idle or 2,500 rpm/no-load conditions, so there is no

way to check the proper operation of the purge system in the idle tests. Evaporative purge systems are operative in both the IM240 and ASM test procedures.

Motorcycle White Smoke Emissions

Measuring white smoke on two-stroke vehicles is even more problematic than measuring black smoke from diesels. This is because motorcycles exhibit greater variation in exhaust pipe configuration than trucks. Where white smoke and particulate emissions from two-stroke motorcycles are a significant problem (as in much of Asia), white smoke testing programs should be established to reduce these emissions.

Appropriate procedures for white smoke testing have been developed for application in Thailand (McGregor and others 1994). A similar procedure is applied for new vehicle certification in Taiwan (China). Smoke opacity measurements are carried out using an in-line smoke opacity meter similar to that used to measure diesel smoke opacity. The opacity meter is connected to an adapter attached to the motorcycle exhaust pipe. The adapter standardizes the path length through which the opacity measurement is taken. The motorcycle is 'revved' repeatedly by suddenly opening and closing the throttle with the engine at idle. The maximum smoke opacity measured is taken as the emission measurement.

Inspection Procedures for Vehicles with Diesel Engines

Emissions inspection procedures for diesel vehicles usually focus on smoke emissions. Smoke can be measured in a number of ways, including the Bosch and Hartridge methods or with a full-flow opacity meter. In the *Bosch* method, a spring-loaded sampler pulls a fixed volume of smoke through filter paper, depositing the smoke particles on the paper (Bosch 1986). The paper is then read by a photoelectric device, which produces a number indicating the degree of blackness of the collected particulate matter. The higher the number, the darker the smoke.

The Bosch method provides an accurate measure of soot and other dark material in the smoke, but it responds poorly, if at all, to smoke particles that are not black. Lubricating oil droplets, for instance, produce a bluish or grayish smoke (because of light scattering) but have little color in themselves. When collected on the Bosch filter, these droplets make the filter wet, but not black, and are not detected. This is a major drawback, since excessive oil smoke is a major contributor to total particulate emissions from diesel vehicles in use, particularly in developing countries. Because it fails to detect this smoke, the Bosch method correlates poorly with actual particulate emissions from in-use vehicles (figure

Figure 4.3 Bosch Number Compared with Measured Particulate Emissions for Buses in Santiago, Chile

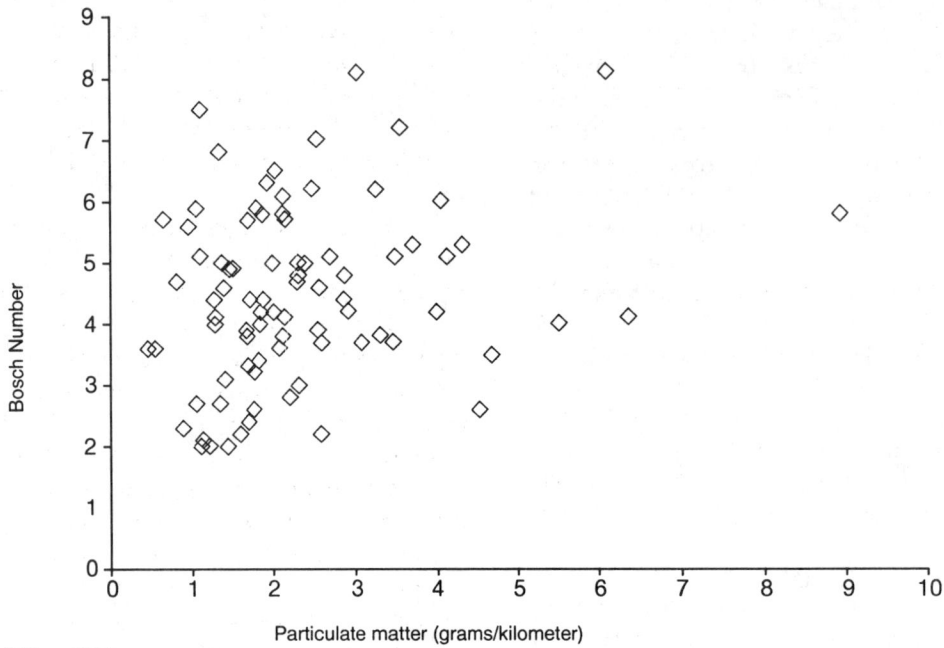

Source: Weaver and Chan 1995

4.3). Another problem with the Bosch method is that great skill is needed by the operator to trigger the sampling system at exactly the right moment to catch the results of the snap acceleration. Triggering the sampler too soon or too late (the common case) will give an erroneously low value.

For purposes of inspection and maintenance, a better measure of excessive particulate emissions is given by a light-transmission opacity meter (opacimeter). This measures the attenuation of a beam of light shining across the smoke plume in percent opacity. Since these opacimetric measurements include both the effects of light absorption (by soot) and light scattering (by oil or fuel droplets), they provide a better indicator of emissions of both types of smoke. Light-transmission opacity meters include full-flow, end-of-pipe opacity meters, and partial-flow, sampling meters such as the commonly used Hartridge meter. The *Hartridge* and other partial-flow meters draw a continuous sample of the exhaust into a chamber and measure the attenuation of a beam of light shining through the chamber. Exposure to heavy oil smoke may contaminate the Hartridge meter, requiring that it be cleaned.

For light-transmission opacity meters, the opacity measurement depends on the density of the smoke and the width of the plume of smoke that the light beam must pass through. This width is known as the path length. The path length is related to the light attenuation by the Beer-Lambert law:

$$N = 100 \, (1 - e^{-kl}) \qquad (4.1)$$

where N is the opacity (in percent), e is the base of natural logarithms, k is the absorption coefficient (smoke density), and l is the effective path length through the smoke (in meters). This equation can be used to correct smoke opacity measurements made at different path lengths to a single comparable basis.

This relationship applies both to full-flow and partial-flow opacity meters. For the full-flow meter, the path length can usually be approximated as the diameter of the exhaust pipe, which is typically in the range from 50 to 150 mm. For partial-flow meters, it is the length of the measuring chamber built into the instrument. This length varies depending on the meter: for the Hartridge meter, it is 457 mm (18 inches). This relatively long path length makes the Hartridge meter sensitive to low levels of smoke but impairs its ability to distinguish between moderate and extreme smoke levels. Other partial-flow meters use shorter measuring chambers.

In 1989, Chile switched from a Bosch-type filter method for measuring diesel smoke emissions to a test procedure utilizing a full-flow, end-of-pipe opacity meter. The Chilean experience has confirmed the greater effectiveness of the opacimeter. This technique is used in most I/M programs in the United States that include diesel smoke measurement. The most common meter is a portable, battery-operated system made up of a light transmitter/receiver assembly that fits over the engine exhaust pipe and is connected by a cable to a hand-held digital readout unit. This system can also be connected to a strip-chart recorder to produce a permanent record. Other opacimeter designs are available for use in fixed installations.

To be meaningful, diesel smoke measurements should be taken with the engine under load. Smoke levels under light-load conditions and at idle are normally very low, so that it is not possible to distinguish between clean vehicles and most smoky vehicles operating under light-load conditions. Only a small minority of seriously malfunctioning diesel vehicles exhibit measurable smoke opacity under idle or light-load conditions.

One commonly used technique for loading the engine to achieve a meaningful smoke measurement is the so-called *snap acceleration* or *free-acceleration* test. This test loads the engine with its own inertia, as the engine is rapidly accelerated from idle to full (governor-limited) speed. Test results are very sensitive to proper conduct of the test. The test is fast and convenient, however, and is therefore widely used for on-road enforcement.

In order to reduce the variability and improve the accuracy of the snap acceleration procedure, a committee organized by the Society of Automotive Engineers (SAE) has been working to develop a standardized Recommended Practice for this test. The core of the procedure (SAE J1667) is a *snap-idle* sequence that is administered by the inspector in the course of the opacity measurement. SAE J1667 differs from earlier versions of the snap-acceleration test procedure mostly in its calculation of the final result. The snap-acceleration test procedure, however, is very sensitive to operator style. It is important that the procedure conform exactly to the written directions in order to obtain the most accurate results (McGregor and others 1994).

Although simple, convenient, and widely used, this test does not always give a reliable measure of smoke emissions in actual use, since the operating condition involved (rapid acceleration under no load) is not characteristic of normal engine operation. Tests in California showed poor correlation between this test and more representative test procedures such as full-load acceleration (Weaver and Klausmeier 1988). This lack of correlation makes it necessary to set somewhat looser standards to avoid failing properly functioning engines. Like the idle test for gasoline vehicles, the snap acceleration is most useful as a crude indicator of serious emission malfunctions.

A more representative smoke value can be obtained by measuring opacity under stabilized, full-load conditions. These can be achieved either by using a dynamometer or by placing the drive wheels on a set of free rollers and loading the engine by using the vehicle's service brake. The *free-roller approach* is an ISO standard. It has an advantage in safety and convenience over the dynamometer, since a vehicle that slips off the rollers is restrained by its own brake and does not keep moving.

A disadvantage of the free-roller technique is that it may be difficult to control engine loading, especially in trucks equipped with air brakes. Another disadvantage is that drivers are often reluctant to apply the brake while running the engine at full power. Although a vehicle's brakes should be able to absorb the full power output of the engine for a considerable time before overheating, drivers often fear the vehicle will be damaged.

To control transient smoke emissions (especially a problem with turbocharged engines) it is desirable to include a *transient acceleration test* mode in the I/M procedure. Various technical options exist for performing such a test. For most engines, however, the snap-acceleration test provides a reasonable indicator of acceleration smoke performance.

Heavy-duty diesel vehicles (HDDV) are expected to be able to meet stricter exhaust standards through improvements in engine design, electronic controls, and the use of low-sulfur fuel. As commercial fleets are replaced with newer models, HDDV emissions are likely to decrease substantially. This does not imply that there is no need for I/M programs for heavy-duty vehicles. In California, CARB has estimated that about 42 percent of HC, and 56 percent of PM, and 6 percent of NO_x emissions emitted by HDDVs in 1990 resulted from engine tampering and maladjustment (Jacobs and others 1991). The primary causes of excess smoke emissions were found to be improper air-fuel ratio control settings, incorrect fuel injection timing, and inadequate intake air (e.g. from dirty air filters). Excess emissions from HDDVs occur to a large extent as a result of improper maintenance and deliberate engine tampering to coax more power out of an engine beyond its rated capacity. This tampering results in overfuelling and large amounts of excess hydrocarbons and smoke emissions. Reduction of these excess emissions has been the main target of I/M programs for HDDVs. Several such programs have been implemented in the United States, especially in the western states (table 4.1) and in some European and developing countries. The effectiveness of these programs has not been fully established although experience with such programs in Santiago and Singapore appears quite promising.

Institutional Setting for Inspection and Maintenance

The cost-effectiveness of I/M programs depends on the testing procedure and institutional setting. I/M programs are commonly either centralized (see box 4.1) or decentralized (see box 4.2), although some hybrid programs have also been adopted.

Table 4.1 Characteristics of Existing I/M Programs for Heavy-Duty Diesel Vehicles in the United States, 1994

State/area	Program type	Vehicles subject	Test procedure	Enforcement	Repair cost limits
Arizona	Centralized; based on annual registration	All locally registered HDDVs	Dynamometer lug down	Registration	Waiver granted for repairs over U.S.$300
California	Random roadside program for fleet vehicles	All HDDVs	Snap-idle[a]	Citation (ticket) with mandatory penalties	None
Southern California	Smoke patrol	All HDDVs	Visual opacity	Citation (ticket)	Not applicable
	Smoking vehicle reporting program	All HDDVs	None	Public reports of smoking vehicles result in letters sent to operators requesting that they repair their vehicles	None
Colorado	Decentralized; with annual fleet self testing	Locally registered HDDVs and other HDDVs which generally operate in program areas	Dynamometer lug down, on-road acceleration, on-road lug down and stall idle	Certificate of emissions compliance required, decal-based enforcement, snap-idle testing performed for data collection purposes not enforcement	Waivers granted after repair costs exceed U.S.$1,500
Nevada	Smoke patrol	All HDDVs	Visual opacity	Citation (ticket)	None
Washington	Centralized; based on annual registration with fleet self testing	Locally registered 1968 and later models HDDVs	Snap-idle	Registration	Waivers granted after repair costs exceed U.S.$50 for 1980 and older vehicles, and U.S.$150 for 1981 and newer vehicles

a. Smoke emission standard—all trucks, 55 percent opacity; 1991 and later models, 40 percent opacity.
Source: GVRD 1994

Centralized I/M

In centralized programs, vehicles must be inspected at one of a small number of high-volume inspection facilities. These facilities are regulated by the government, and run either by public employees or by independent contractors. In a typical program, the government franchises a single contractor to build and operate the inspection centers in a given area, while charging a set fee to the public. These franchises are normally awarded on a competitive basis, taking into account both the technical competence and experience of the bidders and the proposed fee. The fee is normally set at a level that allows the contractor to recover its capital and operating costs and make a profit over the time period covered by the franchise. In some cases, governments have issued multiple franchises for the same area. In Mexico City, for instance, franchises have been awarded to five different consortia, reducing the government's dependence on any one organization.

Centralized, fixed-lane inspection centers designed specifically for high-volume inspection of vehicles permit the cost-effective use of more sophisticated and comprehensive inspection procedures and equipment. A typical layout of a combined safety and emissions inspection station is shown in figure 4.4. These centers use automated and computerized inspection equipment and procedures and include consumer-oriented features, such as:

Figure 4.4 Schematic Illustration of a Typical Combined Safety and Emissions Inspection Station: Layout and Equipment

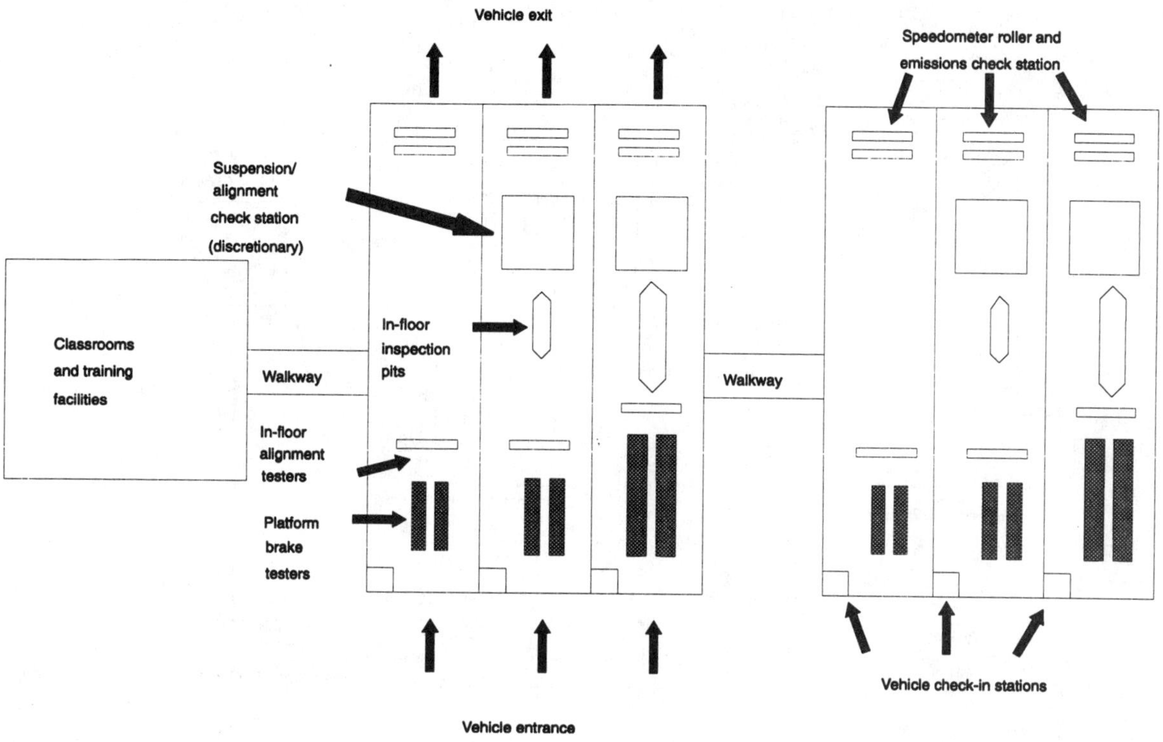

Source: Weaver and Chan 1995

- Subjective decisions are automated, minimizing the potential for human error or tampering with the test results.
- Inspection standards, test results, and pass-fail decisions can be computer-stored and printed out for the consumer.
- Diagnostic information suggesting the probable cause of failure can also be printed—helping the motorist and the mechanic, and lowering repair costs.
- Independent reinspection and evaluation of the effectiveness of vehicle repairs and costs allow the government to determine the effectiveness of the program and to upgrade repair industry performance.

In general, inspections in centralized programs are more effective because of better oversight, standardized training, and more experienced inspection personnel.

Decentralized I/M

In a decentralized program, vehicles are inspected at private service stations and garages, which also make repairs on vehicles that fail the emissions test. While these inspection and repair stations are generally licensed and authorized by regional or local government, they are not operated under direct government control.

This presents opportunities for fraud, both against the consumer (failing vehicles that should have passed in order to "repair" them) and against the system (passing vehicles that should have failed in return for a bribe or to keep the customer happy). To deter fraud, decentralized I/M programs in the United States require the use of emission analyzers, such as BAR90, which incorporate extensive automation of the inspection process.[2] A schematic illustration of an automated inspection process utilizing a four-gas (carbon monoxide, hydrocarbons, oxygen, and carbon dioxide) analyzer is shown in figure 4.5. Automation is intended to make fraud more difficult. In addition, well-run decentralized programs incorporate extensive overt and covert audits of inspec-

---

2. Specifications for the automated emission analyzers were first developed by the California Bureau of Automotive Repair (BAR) in 1984 to support that state's Smog Check Program. These BAR84 analyzer specifications were widely adopted in other U.S. jurisdictions and in several other countries. In response to studies showing continued widespread fraud, BAR in 1990 adopted more stringent automation requirements, including the printing of the inspection certificates on a remote printer not accessible to the inspector. Use of analyzers meeting this specification became mandatory in California in 1990. Because of their extensive automation, data recording, and other features, the BAR90 analyzers are fairly expensive—about U.S.$12,000 to U.S.$15,000 for a basic machine, compared with less than U.S.$5,000 for a simple CO and HC measurement system.

Figure 4.5 Schematic Illustration of an Automated Inspection Process

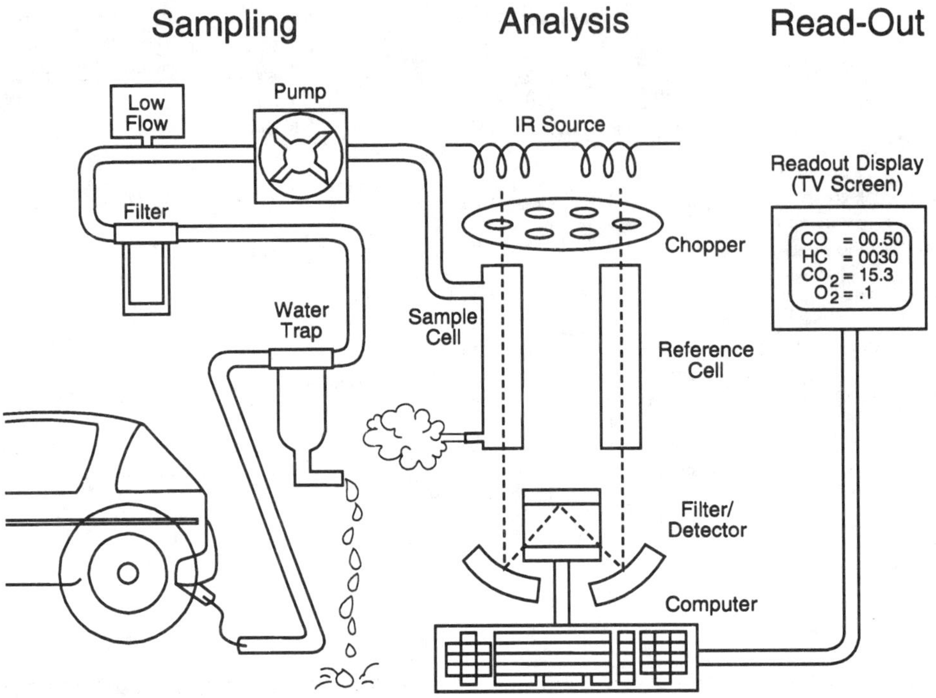

Source: University of Central Florida 1991

tion stations to make fraud more risky. This implies that a large surveillance team is needed in the regulatory agency to supervise inspection programs and ensure quality control. Fraud and poor performance are, nonetheless, still possible and common. Although the BAR90 analyzer itself reads the pollutant concentrations and makes the pass-fail decision, there is no way to ensure that the analyzer is inserted in the tailpipe correctly. Similarly, there is no way for the analyzer to detect whether mechanics have carried out the visual and functional checks they are prompted to perform, and many mechanics do not do so in order to save time.

Although many—perhaps the majority—of private garages performing inspections in decentralized programs are competent and conscientious, there is a tendency for the bad inspectors to drive out the good. Inspectors who pass vehicles that should fail are patronized by consumers wishing to avoid expensive repair. Inspectors who do not carry out a full under-hood inspection can complete an inspection in less time, and therefore at lower cost, than their more conscientious counterparts. Studies by the U.S. EPA have found rates of improper inspections exceeding 50 percent in decentralized I/M programs (Tierney 1991; U.S. EPA 1995), even in states such as California that have made strong enforcement and oversight efforts.

An undercover investigation of California's decentralized Smog Check program, a model program in

many respects, showed that fewer than 60 percent of the stations identified any defect in the defective cars presented at the station and only 25 percent found all the defects in the cars. A 1990 U.S. EPA undercover investigation presented cars with missing catalytic converters at 25 inspection stations in Virginia. Only 10 of the 35 inspectors discovered that the converter was missing. These two undercover investigations as well as other evaluations suggest that less than one-fourth of vehicles inspected in decentralized programs received thorough and accurate inspections (Glazer and others 1995).

Comparison of Centralized and Decentralized I/M Programs

Personnel and training requirements differ significantly between centralized and decentralized programs. Depending on the size of the system, the number of licensed inspectors in a decentralized program can range from a few hundred to over 10,000. Centralized programs on the other hand, require from a few dozen to a couple of hundred inspectors. The physical magnitude of training requirements for anti-tampering inspections can be daunting in a decentralized system (Tierney 1991).

Since the number of inspection locations and inspectors is smaller in a centralized program, the skills and competence of the inspectors can be much higher, and

Table 4.2 Estimated Costs of Centralized and Decentralized I/M Programs in Arizona, 1990
(All costs in 1990 U.S. dollars)

Variable	Centralized manual lane		Centralized automatic lane		Decentralized automated garage[a]	
	(1)	*(2)*	*(1)*	*(2)*	*(1)*	*(2)*
Tests per hour	6	12	20	30	2	5
Tests per man-hour	1.5	4	6	15	1	3
Labor cost per test	5.33[b]	2.00	1.33[b]	0.53	12.00[c]	4.00
Equipment cost per test	1.00	0.50	1.00	0.50	5.00	3.00
Facilities cost per test	3.00	2.00	2.00	1.50	10.00	5.00
Data collection cost per test	0.50	1.00	0.10	0.25	0.50	1.00
Administration cost per test	0.35	0.50	0.20	0.30	1.00	2.00
Total cost per test[d]	10.18	6.00	4.63	3.08	28.50	15.00

a. Computerized vehicle inspection station.
b. At $8.00 per man-hour.
c. At $12.00 per man-hour.
d. Does not include overhead or profit.
Source: Rothe 1990

cheating is more easily prevented. Centralized I/M programs tend to have lower rates of improper inspections. It has been estimated (Weaver and Burnette 1994) that providing adequate quality control and oversight for a 400-station decentralized program to handle I/M testing in Bangkok would require a staff of about 100 personnel, and a budget of U.S.$1.2 million per year. Similar oversight of a centralized program with the same testing capacity was estimated to need 20 staff, and an annual budget of U.S.$350,000.

Inspection costs tend to be lower in centralized facilities due to economies of scale. In California's decentralized idle/2,500 rpm program, inspection fees range from U.S.$16 to U.S.$54 per inspection, averaging about U.S.$30. In addition the costs of administration and enforcement are recovered through a U.S.$7.75 charge on each inspection certificate. The nine centralized, contractor-run programs in the United States averaged less than U.S.$8 per inspection in 1993. Centralized programs in Arizona, Connecticut, and Wisconsin use dynamometers and loaded test procedures and charge fees ranging from U.S.$7.50 to U.S.$10, as the costs of administration and enforcement. Costs of IM240 testing in centralized facilities are estimated to be less than U.S.$20 per vehicle.

Comparative cost estimates for centralized and decentralized programs for the state of Arizona are presented in table 4.2. Based on these estimates, as well as on its greater effectiveness, Arizona decided to retain its centralized system. According to a survey of state I/M programs in the United States, decentralized, computer-

ized testing has the highest cost, averaging U.S.$17.70 per vehicle; centralized contractor-run programs average U.S.$8.42 per vehicle, while government run systems claimed the lowest cost at U.S.$7.46 per vehicle (Tierney 1991). Six of the I/M programs in the United States subsidize the I/M test costs out of general tax revenues and do not require a test fee.

The institutional setting also affects the type of test procedures possible. Simple measurements, such as hydrocarbon and carbon monoxide concentrations at idle, can be made with garage-type analyzers. More sophisticated and costly dynamometer testing is required to identify malfunctioning emission controls on new-technology vehicles. Because of the cost of the equipment, centralized facilities are the only practical solution for sophisticated, dynamometer-based testing such as the IM240.

Centralized I/M programs are better suited to handle retesting after repairs, and to determine eligibility for cost waivers. In decentralized programs, where the same garage performs both the repair and the retesting, these activities present a clear conflict of interest. Based on problems where testing and repairs are carried out by the same organization, the U.S. EPA now discounts by 50 percent the estimated benefits of any I/M program that does not separate testing and repair. This discount has had the effect of mandating centralized I/M programs in heavily polluted metropolitan areas. Resistance to mandatory centralized I/M has been very strong in the United States, although these programs generally impose a lower cost on the consumer.

It has been argued that the effectiveness of central I/M programs is offset by increased inconvenience to the vehicle owner, since it is necessary to drive to one of a limited number of facilities and, if the vehicle fails, drive it elsewhere to be repaired and then return to have it reinspected.

However, dissatisfaction with existing centralized programs has been minimal—many consumers seem to prefer centralized inspections. Before Mexico City switched entirely to centralized, contractor-operated I/M, 20 percent of the public chose to have their vehicles inspected at the 16 public stations, despite significantly greater inconvenience (the public stations often had lines and did not accept cash; consumers had first to go to an agency of the Treasury and make the payment, then take the receipt to the inspection facility). This preference was apparently due to the perception that the public stations were more honest and did better inspections, thus protecting the consumer against possible fraud by the private garage and ensuring that the vehicle owner contributed to the fight against air pollution. In New Jersey (United States), where vehicle owners have a choice of free centralized inspections or paying for inspections in private garages, 80 percent of the public chooses centralized inspection. Integrating fee payment, vehicle registration, and vehicle inspection in the same facility could reduce any inconvenience associated with centralized inspections (U.S. EPA 1995).

Centralized inspection programs are sometimes opposed by the private automotive repair industry, which stands to gain customers and revenues from decentralized programs (increasingly complex testing procedures and equipment requirements in the United States have blunted some of this opposition). They may also be opposed by automotive user groups (either because of the convenience issue discussed above or because centralized inspections are harder to circumvent). Fleet operators dislike centralized testing because it often means that they can no longer self-inspect. Because of this opposition, the majority of I/M programs in the United States used decentralized inspection until recently. The Clean Air Act Amendments of 1990 mandated the use of enhanced I/M—utilizing the IM240 test (or an equally effective procedure) in a centralized program — for highly polluted areas. Because of the challenge that effective, centralized test procedures pose to existing vested interests, this requirement has created great political resistance, and many states are openly defying U.S. EPA regulations on this issue. As a result, effective I/M programs and the resulting air quality benefits have been seriously delayed.

Centralized inspection programs are more common in Latin American countries. Chile has a highly effective program of centralized I/M for passenger cars and commercial vehicles, while Mexico has adopted a similar program, initially confined to commercial vehicles. Until recently, passenger cars were allowed to continue using the decentralized facilities in Mexico City as a political expedience. Centralized I/M programs are scheduled to start in several metropolitan regions in Brazil over the next few years.

Inspection Frequency

Until recently, most I/M programs in the United States required annual testing of vehicles, but the U.S. EPA now recommends testing every two years as being almost as effective while having much lower cost. In Japan, vehicles must be submitted for testing when they are three years old and every two years thereafter. European requirements for in-service testing vary by vehicle type; annual testing is mandatory for heavy-duty diesel vehicles, spark-ignition automobiles are tested annually once they are three years old, and testing for light diesel vehicles begins at age four and is required every two years thereafter (CONCAWE 1994). The schedule for compulsory motor vehicle testing in Singapore is shown in table 4.3; it varies by type and age of vehicle.

Vehicle Registration

To simplify administration of I/M and guard against evasion of program requirements, the system should be linked with the vehicle registration system. Ideally, vehicles should be identified by license plate, engine and chassis numbers, and physical description (make, model, year, body type), and this information should be available at the vehicle testing location. Information should also be kept on vehicle test results, retrofit equipment installed, and vehicle disposition. This will require considerable strengthening of vehicle registration programs in many developing countries.

Roadside Inspection Programs

Roadside or on-road vehicle inspection programs provide an important supplement to periodic inspection and maintenance. Periodic inspection is predictably scheduled, giving vehicle owners an opportunity to evade the program. For instance, one common cause of high smoke emissions in diesel engines is tampering with the maximum fuel setting on the fuel injection pump. This provides more fuel to the engine, increasing power output and smoke. Since owners know when the vehicle will be inspected, they can adjust the pump to its proper setting immediately before the inspection, then return it to the higher power setting afterward. Another way of reducing visible smoke is putting a handful of gravel in the exhaust pipe just before the inspection.

Table 4.3 Schedule of Compulsory Motor Vehicle Inspection in Singapore by Vehicle Age

Type of vehicle	Vehicle age		
	Less than three years	*Three to ten years*	*More than ten years*
Motorcycles and scooters	Exempt	Annually	Annually
Cars and station wagons	Exempt	Every two years	Annually
Tuition cars	Annually	Annually	Annually
Private hire cars	Annually	Annually	—
Public service vehicles			
Taxis	Every six months	Every six months	Every six months
Public buses (SBS, TIB, CSS)	Every six months	Every six months	Every six months
Other buses	Annually	Annually	Annually
Trucks and goods vehicles	Annually	Annually	Every six months
All other heavy vehicles	Annually	Annually	Every six months

— Not applicable
Note: Life span of all private hire cars and taxis is seven years.
Source: Registry of Vehicles 1993

Similar tricks are possible, and reportedly common, for meeting carbon monoxide emissions standards with gasoline vehicles, and it is conceivable that a motorcycle white smoke inspection can be passed by similar means.

Since a vehicle owner cannot predict whether he will be targeted by an on-road or roadside inspection, such inspections are difficult to evade. On-road and roadside inspections are especially useful for enforcing vehicle smoke limits, since the visible nature of smoke emissions allows enforcement efforts to target the worst offenders. This reduces the costs and increases the effectiveness of the program and minimizes inconvenience to owners of vehicles that are not polluting excessively.

A roadside smoke inspection program has been implemented in California to control smoke emissions from heavy trucks. Smoke inspection teams are deployed at weight and safety inspection stations and at other roadside locations. A member of the inspection team observes oncoming trucks visually, waving over those producing excessive smoke. These are then subjected to a free-acceleration smoke test using a smoke opacity meter. A truck fails if smoke opacity is in excess of 50 percent for half a second. Vehicles that pass are sent on their way; those that fail receive a citation that can be cleared only after the vehicle is repaired. The minimum penalty is U.S.$300 if proof of correction (repair) is provided in 15 days; otherwise the fine is U.S.$500. A second citation during the same year carries a much larger fine (U.S.$1,800) and requires that the vehicle be presented at a smoke test location for smoke measurement after repairs. Vehicles with outstanding violations may be removed from service by the California Highway Patrol. In southern California, the roadside smoke inspection program is supplemented by a Smoking Vehicle Reporting Program (SVRP) whereby members of the public can report suspected violators to the regulatory agency, which, after verifying the vehicle identity, sends a letter to the vehicle owner warning that the operation of the smoking vehicle is illegal and that appropriate repairs must be made. There is no follow-up enforcement.

Roadside checks for smoke opacity are also appropriate for control of excess white smoke emissions from two-stroke motorcycles. These emissions are believed to result, in many cases, from using too much or the wrong type of lubricating oil. Such practices are more likely to be detected by visual screening on the road than by a periodic I/M test.

Roadside checks for hydrocarbon and carbon monoxide emissions from gasoline vehicles have also been implemented in a number of countries, including Mexico and Thailand. Singapore has one of the most effective on-road smoke enforcement program, using both I/M inspections and mobile inspection teams (box 4.3). Since vehicles with high hydrocarbon and carbon monoxide emissions are not usually as obvious as those producing excessive smoke, these programs must rely on stopping vehicles at random. This reduces the efficiency of the program and increases the inconvenience to drivers of low-emitting vehicles. The development of practical remote-sensing systems for vehicle emissions offers the potential for roadside inspection programs to be targeted more effectively. Remote sensing is discussed later in this chapter.

Emission Standards for Inspection and Maintenance Programs

Implementation of an I/M program requires that standards be established for emissions from in-service vehicles. Ideally, test procedures should be defined based on a strong correlation with actual on-road emissions, and cutpoints for each procedure should be established by taking into account both the technical capabilities of the vehicles and the emissions standards for which they

---

Box 4.3 On-Road Smoke Enforcement in Singapore

Under Section 72 of the Road Traffic Act, if smoke or visible vapor is emitted from any motor vehicle used on a public road and the emission causes or is likely to cause annoyance or danger to the public or injury or damage to any person or property, the owner and the driver of the vehicle is guilty of an offense. The smoke test is a key component of emissions inspection and enforcement. The passing standard for the diesel smoke is 50 Hartridge Smoke Units (HSU). While the routine smoke test during periodic inspections is retained for diagnostic purposes, it is more effective to concentrate on roadside enforcement as an effective deterrent to smoke emissions, particularly from heavy-duty vehicles. The Registry of Vehicles has introduced mobile smoke test units to control smoke emissions from diesel vehicles. These units, deployed at a number of strategic locations, conduct on-the-spot smoke tests on passing diesel-powered vehicles.

The enforcement standard and penalties imposed are as follows:

Smoke level (HSU)	*Penalty*
Less than 50	Pass (no penalty)
51-70	Fines (owner and driver, S$100 each)
71-85	Fines (owner and driver, S$100 each)
	Vehicle prohibited from the road until faults rectified
86 or more	Owner and driver charged in court.
	Vehicle prohibited from road until faults rectified.

The emphasis of enforcement inspection is not on hefty fines and penalties (though these serve a deterrent effect), but on the need for vehicle owners to pay attention to the overall condition of their vehicles. The roadside inspections are supplemented by periodic island-wide visual surveys of vehicles to monitor the effectiveness of roadside smoke testing and enforcement programs.

Source: Registry of Vehicles 1993

---

were designed. Emissions measured by basic I/M test procedures do not correlate well with emissions measured by the more sophisticated tests used to establish vehicle emissions standards, or with actual on-road emissions. These procedures are primarily effective in identifying vehicles with grossly excessive emissions in order to require that these be repaired.

In-use vehicle emission standards are set at several levels to accommodate current and future emission control technologies, as well as the required levels of emission reductions.

Emission standards for I/M programs must consider both the statistical distribution of vehicle emission levels and the maximum failure rate that can be considered politically acceptable. The worst 10 or 20 percent of the vehicles (gross emitters) generally account for a substantial fraction of total emissions. For example, the highest emitting 20 percent of cars in Bangkok accounted for about 50 percent of measured carbon monoxide emissions (figure 4.6). Likewise, 20 percent of the buses accounted for 50 percent of the measured smoke emissions (figure 4.7). By measuring the actual distribution of emissions in the target population, it is possible to set emission standards such that the highest-emitting vehicles fail and must be repaired, without producing

such a high failure rate that the program becomes politically unsustainable. With continued monitoring of vehicle emission levels and failure rates, it should be possible to tighten standards as vehicle conditions improve over time, creating a progressive improvement in vehicle emission levels.

The emission standards recommended by McGregor and others (1994) for Thailand and the failure rates resulting from enforcing these standards on the existing vehicle fleet are shown in table 4.4. Based on the on-road measurements in Bangkok, one would expect failure rates, given these standards, to be extremely high and perhaps politically unacceptable. However, it was also considered that anticipatory maintenance would reduce the failure rate, and that higher failure rates would be politically acceptable for commercial vehicles such as trucks and taxis.

Setting cutpoints (in-use vehicle emission standards) was a major decision in the establishment of the Finnish in-use emissions control program. Gasoline-fueled passenger cars were the only group of motor vehicles subject to in-use control, using the two-speed idle test. Type-specific values from emissions certification data were used in setting standards. Much of the fleet was certified under ECE Regulation 15, which provides carbon mon-

Figure 4.6 Cumulative Distribution of CO Emissions from Passenger Cars in Bangkok

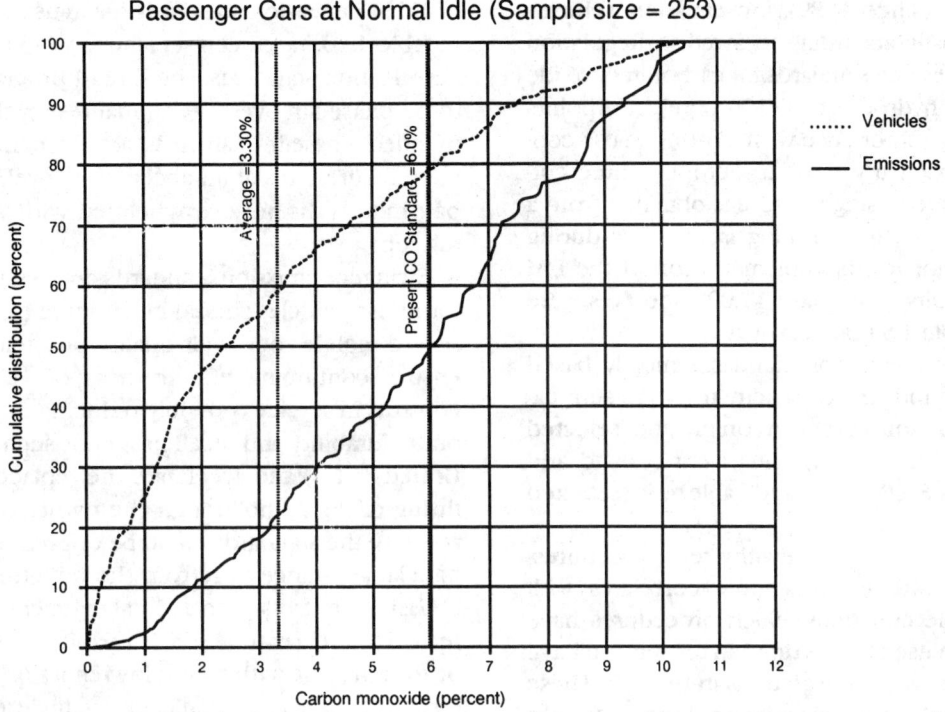

Figure 4.7 Cumulative Distribution of Smoke Opacity for Buses in Bangkok

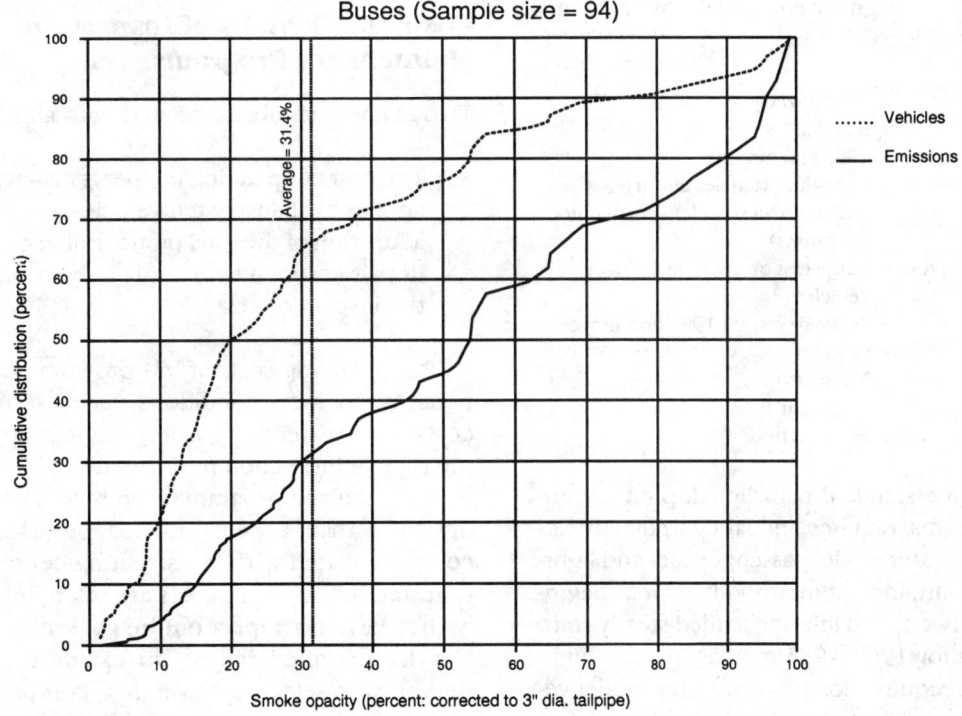

oxide standards at idle (4.5 percent in 15/03 and 3.5 percent in 15/04), but no idle hydrocarbon standard. In later model years (since 1989), low-emission vehicles were certified mostly according to Swedish Regulation A13, which includes idle standards for carbon monoxide (0.5 percent) and hydrocarbons (100 ppm). Cutpoints were also based on experience with existing in-use control programs, particularly those in Europe. Three cutpoints were evaluated using the results obtained from a few prototype test stations making inspections during the 18 months prior to full implementation of the I/M program. The results from these 17,000 short-tests are summarized in table 4.5 (Laurikko 1994).

In-service vehicle emission standards mostly based on two-speed idle and smoke opacity tests, are summarized in table 4.6 (for European Union and selected member states), table 4.7 (Argentina, New Zealand, and East Asia), table 4.8 (Poland), and table 4.9 (selected U.S. jurisdictions).

Unlike the idle and smoke opacity test procedures, the U.S. EPA's IM240 test procedure correlates well with actual vehicle emissions. Such procedures have only recently been used for routine I/M testing and have been adopted by several jurisdictions in the U.S. These tests make it possible to relate emissions from in-service vehicles to new-vehicle emission standards. Failure criteria for these programs are generally set at levels such that any vehicle significantly exceeding the applicable new-vehicle emissions standard would fail. The U.S. EPA's recommended design for enhanced I/M programs (U.S. EPA 1995) is summarized below:

Network Type:	Centralized
Test frequency:	Biennial
Model year coverage:	1968 and newer
Vehicle types:	Light-duty vehicles and trucks
Exhaust tests:	IM240 on all vehicles (at specified cut-points)
Evaporative system tests:	Purge test on 1971 and newer vehicles
	Pressure test on 1981 and newer vehicles
Visual test:	None
Waiver rate:	3 percent
Compliance rate:	96 percent

The performance standard actually adopted for enhanced I/M programs, requires annual centralized testing for all 1968 and later model passenger cars and light-duty trucks in non-attainment metropolitan areas beginning in 1995. The two-speed idle and loaded steady-state tests can be used for 1968-1985 models but the high-tech IM240 test is required for 1986 and later vehicles. Besides the exhaust emission tests, this performance standard requires a visual inspection of the catalyst and fuel inlet checks on 1984 and newer vehicles. Pressure and purge checks are also required to test the efficacy of the evaporative system.

IM240 emission standards for in-use vehicles, shown in table 4.10, have been set at levels 2-3 times above the certification standards. The IM240 program is designed to be phased-in over 2-3 annual test cycles. The recommended phase-in standards are even more lenient to keep failure rates at an acceptable level, so that the repair industry is not overwhelmed with vehicles requiring repair.

Stringent emission standards and testing procedures for in-use vehicles should be adopted in countries with severe mobile source air quality problems. This would ensure continuing effectiveness of vehicle emission controls in service (possibly reducing the need for elaborate durability and recall programs such as those in the United States) and accelerate the replacement or retrofitting of highly polluting. The owner of a vehicle not meeting the standards could be offered technical and financial assistance to retrofit the polluting vehicle with emission controls such as a catalytic converter or an alternative fuel (natural gas, propane) kit, to re-engine it, or to replace it with a new, low-emission vehicle. An alternative approach would use vehicle emission taxes rather than standards. This would offer greater flexibility, while creating an incentive to reduce emissions even below the legislated standards.

Costs and Benefits of Inspection and Maintenance Programs

Two major costs are associated with an I/M program:

- The cost of operating the program – which generally applies to all inspected vehicles, since it tends to be a function of the total number of vehicles examined.
- Repair costs – which apply only to vehicles that fail the inspection test.

The operating costs of I/M programs vary widely, depending on local conditions such as land and labor costs, the existence of safety inspection stations, and the type of inspection performed.

A centralized program is likely to have higher start-up costs than a decentralized program because of costs associated with constructing centralized inspection stations. As these costs are often financed privately, they have no impact on the government budget. In Maryland (United States), for example, a private contractor was selected through a competitive bidding process to construct ten new inspection stations, with an average of five lanes each, to inspect approximately 1.6 million vehicles annually. The contractor retained

Table 4.4 Inspection and Maintenance Standards Recommended for Thailand

Type of vehicle	CO at idle (percent)	CO at high idle (percent)	HC at idle (ppm)	HC at high idle (ppm)	Opacity at peak (percent)	Opacity at high idle (percent)	Failure rate (percent)
Motorcycles and side-car delivery	6	6	14,000	14,000	70	40	56
Tuk-tuks and three-wheeled delivery vehicles	2	n.a.	14,000	n.a.	70	—	19
Passenger cars, see-lors, gasoline pickups	5	5	300	300	—	—	40
Taxis	5	5	300	300	—	—	73
Medium and heavy-duty trucks (diesel)	—	—	—	—	65	—	75
Buses (BMTA)	—	—	—	—	25	—	26
Buses (mini-bus)	—	—	—	—	25	—	46
Pickups (diesel), pickups as taxis & buses	—	—	—	—	65	—	58

— Not applicable
Source: McGreogor and others 1994

Table 4.5 Distribution of Carbon Monoxide and Hydrocarbon Emissions from 17,000 Short Tests on Gasoline Cars in Finland

Carbon monoxide concentration on idle (percent)	Share of total (percent)		Hyrocarbon concentration on idle (ppm)	Share of total (percent)	
	ECE regulation 15-03	ECE regulation 15-04		ECE regulation 15-03	ECE regulation 15-04
0 – 3	57.8	78.7	0 – 400	71.1	90.0
3 – 4	12.8	10.1	400 – 600	16.2	8.2
4 – 5	8.3	4.2	600 – 800	4.6	0.6
5 – 6	6.0	3.0	800 – 1,000	2.4	0.3
More than 6	15.0	4.0	More than 1,000	5.7	1.0
Total	100.0	100.0	Total	100.0	100.0
Failure rate[a]	25 percent	15 percent	Failure rate	5.7 percent	1.9 percent

a. Corresponding to following cutpoints: CO (4.5 percent) and HC (1,000 ppm) for ECE 15-03; and CO (3.5 percent) and HC (6,000 ppm) for ECE 15-04.
Source: Laurikko 1994

U.S.$7.50 of the U.S.$9 inspection fee to recoup the investment in inspection stations and to cover the operating costs, while U.S.$1.50 was returned to the state to cover the administrative costs of the program. Implementation of a centralized program does not always result in high construction costs; New Jersey (United States) incorporated its emissions testing program into an existing centralized safety inspection program, thus avoiding the cost of constructing entirely new facilities.

Table 4.6 In-Service Vehicle Emission Standards in the European Union, 1994

Location/vehicle type	Carbon monoxide (percent)	Smoke
Austria		
Spark-ignition vehicle (idle speed)		
Without catalyst	3.5	
With three-way catalyst	<0.3	
Diesel		Bacharach number ≤4.5 (±1.0 tolerance) for soot measurement
Denmark		
Gasoline-fueled vehicles		
Before 1984	5.5	
1984 and later	4.5	
With three-way catalyst	0.5	
Diesel-fueled vehicles		3.8 Bosch
Finland		
Gasoline-fueled cars, registration years		
Before 1978	Exempted	
1978–85	4.5	-
1986–90	3.5	-
1990 and later[a, b]	0.3	–
Diesel-fueled vehicles, registration years		
Before 1990		7.0 Bosch
1990 and later		3.5 Bosch
1980 and later (effective 1995)		
naturally aspirated		2.5 m^{-1}
turbo-charged		3.0 m^{-1}
Italy		
Gasoline-fueled	4.5	
Diesel-fueled buses		65 percent
Other diesel-fueled vehicles		70 percent
Netherlands		
Gasoline-fueled & LPG vehicles (since 1974)	4.5	
Cars with 3-way catalyst (since 1976)	0.5	
United Kingdom		
Gasoline-fueled vehicles[c]		
First used between Aug 1, 1975 & July 31, 1983	6.0	
First used on or after Aug 1, 1983	4.5	
European Union (recommended application)		
Gasoline-fueled vehicles		
Manufactured up to October 1, 1986	4.5	
Manufactured after October 1, 1986	3.5	
All models fitted with 3-way, closed-loop catalytic converter		
Idle	0.5	
2000 rpm	0.3	
Diesel-fueled vehicles		
Naturally aspirated		2.5 m^{-1}
Turbocharged		3.0 m^{-1}

a. The cutpoint for a three-way catalyst equipped car (1990) is at high idle (2,500 rpm); there is also a limit on excess air/fuel ratio of 1±0.03 at high idle.

b. Finland also has standards for hydrocarbon emissions as follows: before 1978, exempted; 1978-85, 1,000 ppm; 1986-1990, 600 ppm; and 1990 and after, 100 ppm.

c. HC = 0.12 percent after August 1, 1983.

Source: CONCAWE 1994

Table 4.7 In-Service Vehicle Emission Standards in Argentina, New Zealand, and East Asia, 1994

Country/vehicle characteristics	Carbon monox-ide (percent)	Hydrocarbons (ppm)	Smoke
Argentina			
Gasoline-fueled vehicles			
1983–1991	4.5	900	
1992	3.0	600	
1994	2.5	400	
Diesel-fueled vehicles			
1994			5 Bacharrach index and 2.62 m^{-1}
New Zealand			
Vehicle Model			
Pre-1982	4.5	800	
1982–1992	3.5	400	
1992 and later	2.5	400	
China			
Diesel-fueled vehicles (before 1984)			
Free acceleration			5.0 Bosch
Full load			4.5 Bosch
Diesel-fueled vehicles (after 1984)			
Free acceleration			4.0 Bosch
Full load			3.5 Bosch
Japan			
Gasoline-fueled vehicles			
Four-stroke engine	4.5	1,200	
Two-stroke engine	2.5	7,800	
Republic of Korea			
Gasoline and LPG-fueled vehicles (1987)			
Rectification			
Until Sept. 1990	1.3–9.0	1,200–4,800	
From Oct. 1990	1.3–4.4	1,200	
Penalty			
Until Sept. 1990	9.1 and above	—	
From Sept. 1990	4.5 and above	above 4,800	
Gasoline and LPG-fueled vehicles (1988)			
Rectification			
Until Sept. 1990	1.3–9.0	200–880	
From Oct. 1990	1.3–4.4	220	
Penalty			
Until Sept. 1990	9.1 and above	—	
From Sept. 1990	4.5 and above	above 880	
Diesel-fueled vehicles			
Until Sept. 1990			50 percent opacity
From Oct. 1990			40 percent opacity
Singapore			
Gasoline-fueled vehicles	2.5	—	

— Not applicable

LPG = Liquefied Petroleum Gas.

Source: CONCAWE 1994

Table 4.8 In-Service Vehicle Emission Standards in Poland, 1995

Vehicle type	Carbon monoxide (percent)	Hydrocarbons (ppm)	Smoke
Vehicles with spark-ignition engines			
Registered before October 1, 1986			
Motorcycles			
Other Vehicles[a]	5.5		
Registered after October 1, 1986 and before July 1, 1995	4.5		
Motorcycles			
Other Vehicles	4.5		
Registered after July 1, 1995	3.5		
Motorcycles			
Other vehicles[b] (at idle)	4.5		
Other vehicles[b] (at 2000–3000 rpm raised idle)	0.5	100	
Vehicles with diesel engines[c]	0.3	100	
Naturally aspirated			2.5 $^{m\text{-}1}$
Turbocharged			3.0 $^{m\text{-}1}$

a. Includes all heavy-duty vehicles.
b. Excluding heavy-duty vehicles.
c. At free acceleration from low idle speed.
Source: CONCAWE 1994

Table 4.9 In-Service Vehicle Emission Standards for Inspection and Maintenance Programs in Selected U.S. Jurisdictions, 1994

Jurisdiction/vehicle type/model year	Carbon monoxide (percentage)	Hydrocarbons (ppm)	Smoke (percent opacity)
U.S. EPA basic I/M program	1.2	220	
Passenger cars (1981 and later models)			
Arizona			
Passenger cars and light trucks			
1967–71	5.5	500	
1972–74	5.0	400	
1975–78	2.2	250	
1979	2.2	220	
1980 and later models	1.2	220	
Diesel-fueled vehicles			50
Florida			
Gasoline-powered vehicles			
Gross vehicle weight up to 6,000 pounds			
1975–77	5.0	500	
1978–79	4.0	400	
1980	3.0	300	
1981 and later models	1.2	220	
Gross vehicle weight of 6,001 to 10,000 pounds			
1975–77	6.5	750	
1978–79	5.5	600	
1980	4.5	400	
1981–1984	3.0	300	
1985 and later models	1.2	220	
Diesel-fueled vehicles			
Cruise mode			20
Idle mode			5

Note: In the United States, gasoline emission standards apply at both idle and 2,500 rpm. See table 4.10 for IM240 emission standards.
Source: CONCAWE 1992; CONCAWE 1994; Rothe 1990

Table 4.10 U.S. IM240 Emission Standards
(grams per mile)

Vehicle type/ model year	Carbon monoxide	Hydrocarbons	Nitrogen oxides
Light-duty vehicle			
Tier 1 (94/95/96)[a]	15	0.7[b]	1.4
1994–1995[c]	20	0.8	2.0
1986–1993	20	0.8	2.0
Light-duty truck (≤6000lb GVW)			
Tier 1 (94/95/96)[a]	15	0.7[b]	2.0
1994–1995[c]	20	1.2	3.5
1986–1993	20	1.2	3.5
Light-duty truck (>6000lb GVW)			
Tier 1 (94/95/96)[a]	15	0.8[b]	2.5
1994–1995[c]	20	1.2	3.5
1986–1993	20	1.2	3.5

Note: The standards become fully effective in 1997 following a phase-in period during 1995–1997.
a. 1994 and later models.
b. Non-methane hydrocarbons.
c. Not meeting Tier 1 emission standards.
Source: U.S. EPA 1992

British Columbia (Canada) has evaluated the cost of several alternatives for a heavy-duty vehicle I/M program; of the various options available, roadside smoke inspection appeared to be the least capital intensive (table 4.11) and arguably the most cost effective (GVRD 1994).

Decentralized programs are less expensive to set up than centralized programs but are more expensive to operate. Cost elements include licensing for numerous inspection stations, certification of repair facilities, emission test equipment, audit vehicles, and program management (data processing, quality assurance, public information, training, supplies, and so on). Program management costs should generally be borne by the government but it may be advantageous to contract program management to a specialized firm.

Repair expenditures are a legitimate cost of I/M programs. Much of the repair cost information reported in the literature is dated and not too reliable (Faiz and others 1990). Non-uniform reporting methods and the inability to segregate the costs of I/M-related repairs make it difficult to obtain reliable and accurate repair cost data. Repair costs may be distorted by warranty coverage provisions in U.S. emission regulations, which can require the vehicle manufacturer to pay for all repairs needed to pass the I/M test. The warranty coverage, however, remains in effect only under the condition

that the vehicle has not been tampered with or misused and that the vehicle owner follows the maintenance practices prescribed by the manufacturer. U.S. EPA estimates suggest that repair costs associated with the enhanced I/M program (IM240) would range between U.S.$40–$250 depending on the tests that are failed (U.S. EPA 1995).

Experience with Colorado's I/M program shows average repair costs of U.S.$186 for vehicles failing the IM240 test and U.S.$86 for older vehicles tested by the 2-speed idle procedure. For diesel-fueled vehicles, the average repair cost was U.S.$97, and no vehicles required waivers to pass (Colorado 1996).

Emission Improvements and Fuel Economy

The U.S. EPA's MOBILE5a model provides estimates of vehicle emission reductions due to different I/M programs. Estimated emission reductions from I/M programs for U.S. gasoline-fueled automobiles and heavy-duty vehicles (derived from the MOBILE5 model) for various levels of emissions control technology are presented in tables 4.12 and 4.13. Emission reductions attributed by the U.S. EPA to enhanced I/M programs are given in table 4.14. Among European programs, Sweden estimates that its I/M program has reduced carbon monoxide by 20 percent and hydrocarbons by 7 percent based on the ECE test procedure, and Switzerland estimates reductions in hydrocarbons and carbon monoxide of 20 to 30 percent (Berg 1989). Researchers at the Institut National de Recherche sur les Transports et Leur Securite (INRETS) in France tested vehicles "before" (as received from the owner) and "after" they had been tuned to manufacturer's specifications. As shown in table 4.15, carbon monoxide and hydrocarbon emissions and fuel consumption fell significantly following proper tuning, but nitrogen oxide emissions increased especially for gasoline-fueled vehicles (Joumard and others 1990). Where effective nitrogen oxide controls are in place, I/M programs help to reduce nitrogen oxide emissions as well. Current I/M programs yield small reductions in NO_x emissions and these are mostly the result of lower tampering rates. With the advent (in the United States) of advanced I/M programs based on loaded testing, nitrogen oxide emissions are expected to decline substantially.

It is important to note that emission reductions start out slowly and increase gradually over time because I/M programs tend to retard the overall deterioration rate of fleet emissions. Maximum benefits are achieved by adopting the I/M program as early as possible. Fuel savings have been attributed to the improved vehicle maintenance practices associated with an effective I/M program. Fuel savings range from 0 to 7 percent. In cal-

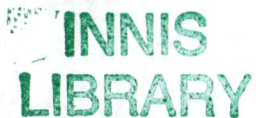

Table 4.11　Alternative Options for a Heavy-Duty Vehicle I/M Program for Lower Fraser Valley, British Columbia, Canada
(Canadian dollars)

Program category	Training/ education[a]	Smoking vehicle road testing			Centralized test stations	
		On-road smoke patrol	Roadside opacity testing	Dynamometer testing facility	LFV[c] registered heavy-duty vehicles	All heavy-duty vehicles operating in LFV
Program alternative	*(1)*	*(2A)*	*(2B)*	*(2C)*	*(3A)*	*(3B)*
Program elements	a) Develop appropriate regulations b) Industry training and public education c) Smoking vehicle reporting program	Alternative 1 plus: d) On-road testing e) Enforcement (tickets) f) Voluntary vehicle repair g) Repair incentives	Alternative 1 plus: d) Random roadside testing e) Enforcement (tickets) f) Voluntary vehicle repair g) Repair incentives	Alternative 1 plus alternatives 2A or 2B plus: h) On road or roadside inspection plus dynamometer testing i) Mandatory repair and dynamometer retesting	Alternative 1 plus: d) Dynamometer testing of LFV registered vehicles only e) Enforcement via registration f) Mandatory repair and dynamometer retesting	Alternative 1 plus d) Dynamometer testing of all HDVs operating in LFV e) Enforcement via registration for local vehicles, tickets for others f) Mandatory repair and dynamometer retesting for locally registered HDVs
Test type	None	Visual (smoke) opacity plus opacity meter	Opacity meter	Dynamometer and analyzers	Dynamometer and analyzers	Dynamometer and analyzers
Regulated pollutants	Smoke, PM_{10}, hydrocarbons	Smoke, PM_{10}, hydrocarbons	Smoke, PM_{10}, hydrocarbons	Smoke, PM_{10}, hydrocarbons	Smoke, PM_{10}, carbon monoxide, nitrogen oxides, hydrocarbons	Smoke, PM_{10}, carbon monoxide, nitrogen oxides, hydrocarbons

Program cost analysis					
Capital costs	$200,000	$1,200,000	$5,400,000	$5,300,000	$11,200,000
Annual operating costs	$1,450,000	$1,200,000	$4,400,000	$3,500,000	$6,300,000
Smoking vehicle reporting program	$120,000	$120,000	$120,000	$120,000	$120,000
Annual program cost[b]	*$1,620,000*	*$1,520,000*	*$5,320,000*	*$4,620,000*	*$8,520,000*
Vehicle repair cost	*$6,200,000*	*$4,200,000*	*$6,200,000*	*$3,500,000*	*$11,500,000*
Vehicle downtime cost	*$4,300,000*	*$2,400,000*	*$4,600,000*	*$3,000,000*	*$10,300,000*
Total annual cost	**$12,120,000**	**$9,120,000**	**$16,120,000**	**$11,120,000**	**$30,320,000**

Pollutants reduced (tonnes/year)	1995	2005	1995	2005	1995	2005	1995	2005	1995	2005
PM_{10}	93	50	62	33	119	63	100	53	133	70
HC	92	77	61	51	117	98	141	116	174	143
NO_x	0	0	0	0	0	0	137	226	178	296

a. Training and education would be a necessary first component of all full-scale programs; with an estimated cost of C$ 400,000.
b. Annual program cost = all operating costs + amortization of total capital costs.
c. LFV = Lower Fraser Valley.
Source: GVRD 1994

culating the cost effectiveness of its programs, the U.S. EPA uses a figure of 3.5 percent for basic I/M programs. For enhanced high-tech I/M programs fuel economy improvements are estimated to range between 6 and 19 percent.

Impact on Tampering and Misfueling

The U.S. EPA has been collecting data on tampering and misfueling to assess the magnitude of the problem. These activities can increase hydrocarbon emissions

Table 4.12 Estimated Emission Factors for U.S. Gasoline-Fueled Automobiles with Different Emission Control Technologies and Inspection and Maintenance Programs
(grams per kilometer)

Emission control (fuel consumption)	Carbon monoxide	Methane	Non-methane volatime organic compounds	Nitrogen oxides
Advanced three-way catalyst control (11.9 kilometers per liter)				
No I/M program	6.20	0.040	0.67	0.52
Basic I/M program[a]	5.36	0.040	0.62	0.51
Enhanced I/M program[b]	4.39	0.030	0.51	0.46
Basic program compared with no program (percent difference)	16	0	8	2
Enhanced program compared with basic program (percent difference)	23	33	22	11
Early three-way catalyst (9.4 kilometers per liter)				
No I/M program	6.86	0.050	0.78	0.66
Basic I/M program[a]	5.95	0.044	0.71	0.64
Basic program compared with no program (percent difference)	15	14	10	3
Oxidation catalyst (6.0 kilometers per liter)				
No I/M program	22.37	0.100	2.48	1.84
Basic I/M program[a]	17.15	0.093	2.14	1.68
Basic program compared with no program (percent difference)	30	7	16	10
Non-catalyst control (6.0 kilometers per liter)				
No I/M program	27.70	0.15	3.08	2.04
Basic I/M program[a]	21.95	0.14	2.86	1.94
Basic program compared with no program (percent difference)	26	7	8	5
Uncontrolled (6.0 kilometers per liter)				
No I/M program	42.67	0.19	5.62	2.7
Basic I/M program[a]	34.47	0.19	5.46	2.7
Basic program compared with no program (percent difference)	24	0	3	0

Note: U.S. EPA MOBILE5a estimates for: ambient temperature 24°C; speed 31kph; and gasoline RVP 62 kPa (9.0 psi).
a. Centralized, biennal testing; 2,500 rpm and idle test procedure; emission cutpoints at 220 ppm for hydrocarbons and 1.2 percent for carbon monoxide. Visual inspection of catalyst and fuel inlet restrictor for catalyst-equipped vehicles. Assumed 96 percent compliance rate with a 3 percent waiver rate.
b. Centralized, annual testing; transient IM240 test procedure; emissions cutpoints at 0.5 gram per kilometer for hydrocarbons, 12.5 grams per kilometer for carbon monoxide, and 1.25 grams per kilometer for nitrogen oxides. Visual inspection of catalyst and fuel inlet restrictor for catalyst-equipped vehicles. Assumed 96 percent compliance rate with a 3 percent waiver rate.

Source: Chan and Reale 1994

Table 4.13   Estimated Emission Factors for U.S. Heavy-Duty Vehicles with Different Emission Control Technologies and Inspection and Maintenance Programs
(grams per kilometer)

Emission control (fuel consumption)	Carbon monoxide	Methane	Non-methane volatile organic compounds	Nitrogen oxides
Three-way catalyst control (2.9 kilometers per liter)				
No I/M program	10.20	0.124	1.41	2.49
Basic I/M program[a]	9.15	0.112	1.34	2.46
Enhanced I/M program[b]	8.69	0.106	1.17	2.42
Basic program compared with no program (percent difference)	11	11	5	1
Enhanced program compared with basic program (percent difference)	5	6	15	2
Non-catalyst control (6.0 kilometers per liter)				
No I/M program	47.61	0.21	5.73	3.46
Basic I/M program	43.04	0.19	5.57	3.46
Basic program compared with no program (percent difference)	11	10	3	0
Uncontrolled (6.0 kilometers per liter)				
No I/M program	169.13	0.44	18.5	5.71
Basic I/M program	161.21	0.42	18.11	5.44
Basic program compared with no program (percent difference)	5	5	2	4

Note: U.S. EPA MOBILE5a estimates for: ambient temperature 24°C; speed 31 kph; and gasoline RVP 62 kPa (9.0 psi).
a. Centralized, biennial testing; 2,500 rpm and idle test procedure; emission cutpoints at 220 ppm for hydrocarbons and 1.2 percent for carbon monoxide. Visual inspection of catalyst and fuel inlet restrictor for catalyst-equipped vehicles. Assumed 96 percent compliance rate with a 3 percent waiver rate.
b. Centralized, annual testing; transient IM240 test procedure; emissions cutpoints at 0.5 grams per kilometer for hydrocarbons, 12.5 grams per kilometer for carbon monoxide, and 1.25 grams per kilometer for nitrogen oxides. Visual inspection of catalyst and fuel inlet restrictor for catalyst-equipped vehicles. Assumed 96 percent compliance rate with a 3 percent waiver rate.
Source: Chan and Reale 1994

tenfold and carbon monoxide emissions twentyfold. Surveys by the U.S. EPA in the 1980s showed that nearly one in five vehicles had at least one emission control disablement and that a significant number of vehicle owners had poisoned their catalytic converters by using leaded fuel. Widespread use of leaded fuel in catalyst-equipped vehicles was also observed in Mexico when the difference in price between the two fuels was large. Fortunately, many countries have adopted tax policies that make the price of unleaded gasoline similar to that of leaded gasoline; this should restrain the tendency to switch fuels that occurred in the United States and Mexico.

Investigations by the U.S. EPA suggest that I/M programs can reduce the rates of tampering and misfueling by half (table 4.16), though they have no significant effect on disablement of the exhaust gas recirculation (EGR) system. Since the I/M programs surveyed in 1982 addressed only hydrocarbon and carbon monoxide

problems, it is not surprising that the programs had no effect on tampering with EGR systems, which are designed to reduce nitrogen oxide emissions. By minimizing tampering and misfueling of newer model cars that use catalysts to control nitrogen oxide emissions, I/M programs should prove effective in reducing nitrogen oxide emissions as well.

Tampering surveys conducted by agencies in the United States suggest that older vehicles have higher rates of tampering than newer vehicles. Furthermore, the under hood visual inspection portion, if carried out diligently, was found to be effective in identifying many elements of a vehicle emission control system that may have been tampered with or modified. It is extremely difficult to ensure adequate visual inspection in decentralized I/M programs, however. Tampering and misfuelling rates tend to be significantly lower in centralized I/M programs

Table 4.14 U.S. EPA's I/M Performance Standards and Estimated Emission Reductions from Enhanced I/M Programs

Test procedure	Estimated reductions in emissions (percent)		
	Carbon monoxide	Hydrocarbon	Nitrogen oxides
Low enhanced I/M performance standard	16.1	9.3	1.5
High enhanced I/M performance standard	35.4	31.9	13.4
Low performance I/M programs			
Biennial idle test			
Without technician training and certification	16.2	9.8	2.4
With technician training and certification	22.3	12.6	6.6
Annual 2-speed idle test and repair			
Without ASM+pressure and purge+technician training & certification	15.7	10.2	2.6
With ASM+pressure and purge+technician training & certification	24.9	21.1	11.6
High enhanced I/M performance standards			
Comprehensive biennial test—IM240 program			
Without technician training and certification	37.8	34.5	14.6
With technician training and certification	43.9	37.3	18.9
Biennial test—ASM			
Without technician training and certification	33.7	31.7	14.3
With technician training and certification	39.8	34.5	18.6

Source: Walsh 1995

Table 4.15 Effect of Engine Tune-Up on Emissions for European Vehicles
(ratio of emissions after tuning to before tuning)

Vehicle type/testing procedure	Carbon monoxide	Hydrocarbons	Nitrogen oxides	Polycyclic aromatic hydrocarbons	Particulate matter	Carbon dioxide	Fuel consumption
Light-duty gasoline-fueled vehicles							
Mean of four hot cycles	0.80	0.89	1.06	—	—	1.01	0.96
ECE 15 hot cycle	0.85	0.85	1.04	0.81	—	0.96	0.93
ECE 15 cold cycle	0.78	1.18	1.02	—	—	1.00	0.98
Light-duty diesel-fueled vehicles							
Mean of four hot cycles	0.97	0.78	1.05	n.a.	0.75	0.94	1.02
ECE 15 hot cycle	0.96	0.81	1.11	1.14	0.64	0.86	1.00
ECE 15 cold cycle	0.84	0.65	1.12	n.a.	0.66	0.89	0.93

— Not applicable
n.a. = Not available or not significantly different.
Source: Joumard and others 1990

Cost-Effectiveness

The U.S. EPA has estimated the cost-effectiveness of hydrocarbon emission reductions from the United States vehicle fleet using enhanced I/M at about U.S.$500 per ton — a very cost-effective option. For less-effective and more expensive decentralized programs, the estimated costs were as high as U.S.$15,000 per ton (U.S. EPA 1995).

The estimated cost per ton of PM-10 reduced in the first year of a centralized I/M program for heavy-duty diesel vehicles varied from C$17,000 for an on-road smoke patrol to C$64,000 for a centralized test station with a dynamometer testing facility (GVRD 1994).

Few cost-effectiveness estimates have been made for I/M programs in developing countries, and none of these have had the benefit of actual test data. A World

Bank study for Mexico estimated the cost per ton of hydrocarbons, nitrogen oxides, and particulate matter reduced by a centralized I/M program in Mexico City. The estimated cost-effectiveness was U.S.$839 per ton for a centralized program addressing high-use commercial vehicles, U.S.$1,720 per ton for a centralized program addressing private passenger cars, and U.S.$2,056 per ton for a decentralized passenger car program. In addition to being more cost-effective, the centralized passenger car program was projected to achieve nearly double the reduction in emissions of the decentralized program.

International Experience with Inspection and Maintenance Programs

Inspection and maintenance programs have been an integral element of air pollution control programs in many industrialized countries, but special care is required in designing the inspection and enforcement mechanisms of I/M programs in developing countries. Otherwise, weak administrative and regulatory arrangements could result in massive evasion of I/M programs or corrupt practices on the part of I/M officials and inspectors. Experience with traffic and safety regulation enforcement in developing countries has shown many instances of these types of problems. If possible, vehicle I/M testing should be carried out in centralized, high-volume, test-only facilities. Automated reading of emission measurements and computerization of the pass-fail decision can help to minimize opportunities for fraud. Close links should also be established between the vehicle registration process and emissions testing in order to avoid the potential for counterfeiting of vehicle inspection certificates and stickers. Finally, inspection facilities should be

operated by a small number of private contractors, with the government's role limited to oversight of the contractor's performance. Since the government could cancel the contract if corruption is discovered, the contractor has a strong incentive to establish pay, management, and supervision practices that minimize this possibility. This is more difficult in situations where the government operates the inspection facility. A brief survey of the international experience with I/M programs is presented in the following sections.

Argentina. At present, about 400,000 buses and trucks operating in the Gran Buenos Aires area and about 250,000 taxis and other commercial vehicles registered in the Capital Federal area are subject to periodic inspections as explained below:

- Buses, both CNG and diesel-fueled, are inspected at 6-month intervals at one of the 12 inspection stations owned and operated by the private firm CENT under the supervision of CONTA (Comision Nacional del Transporte Automotor), an agency of the national Ministry of Transport. Emission checks are performed visually. A new regulation will require all inspection stations to have the basic equipment for carrying out emission checks.

- Taxis and commercial vehicles are inspected by SACTA, a private sector concessionaire under the jurisdiction of the Federal Capital Administration. Inspection charges are U.S.$42 per vehicle and U.S.$21.50 for re-inspection upon rejection. One of the objectives of the technical control is to ensure compliance with exhaust and noise emissions standards. Exhaust emission controls are limited to

Table 4.16 Tampering and Misfueling Rates in the United States
(percent)

Type of tampering	1982		1987–89	
	No I/M	With I/M	Decentralized I/M	Centralized I/M
Overall	16.7	13.9	n.a.	n.a.
Catalytic converter	4.4	1.7	8.0	4.0
Inlet restrictor	5.9	3.1	12.0	6.0
Air pump system	4.6	2.3	15.0	12.0
Exhaust gas recirculation (EGR) system	9.8	10.1	n.a.	n.a.
Evaporative canister	n.a.	n.a.	11.0	8.0
Fuel switching	15.1	6.2	13.0	8.0

n.a. = Not available
Source: Faiz and others 1990; Tierney 1991

carbon monoxide and hydrocarbons for CNG and gasoline-fueled vehicles; diesel-fueled cars are tested for smoke opacity.

The vehicle inspection program suffers from severe non-compliance. While non-compliance with the mandatory annual inspection for taxis is relatively low at less than 5 percent (out of 38,500 vehicles), it is significantly more common among trucks and passenger transport vehicles: only 10 percent of 170,000 registered trucks and about 8 percent of the 21,000 registered passenger transport vehicles show up for annual inspections.

To reduce non-compliance, the municipal Secretariat for Urban Planning and Environment together with the municipal police carry out sporadic road checks. Their results indicate that 25 percent of all vehicles operated in the Federal Capital Area violate emission norms. In 1991, two-thirds of the surveyed cars in circulation were found to be in non-compliance with the CO emission limit of 4.5 percent. In April 1994, 235 out of 898 randomly stopped vehicles, (almost 25 percent) failed to meet emission limits. In addition it was found that 15 percent of the checked buses (80 out of 483 buses), nearly 40 percent of the tested trucks (96 out of 252 trucks), and more than one-third of the inspected cars (59 out of 163 cars) violated the standards. To strengthen enforcement CONTA has introduced a hot line for reporting buses emitting excessive smoke. SACTA also carries out random roadside checks.

Both the provincial and the Federal Capital administrations plan to introduce mandatory inspections for private vehicles. The Province of Buenos Aires has already solicited tenders from private inspection firms. In contrast to Buenos Aires, recent experiences in Mendoza and Cordoba suggest that air pollution can be reduced significantly through improved vehicle maintenance. These cities are reported to have achieved significant improvements in air quality by enforcing compliance with emission standards.

Austria. All cars are subject to an annual safety and carbon monoxide emission inspection.

Brazil. Vehicle inspection and maintenance programs are still at an early stage of development. The State of São Paulo and some other states have had on-road smoke enforcement programs for diesel vehicles, but these programs have had only limited effectiveness. Plans are now underway to implement centralized, contractor-operated I/M inspection in São Paulo, Recife and Belo Horizonte. The centralized I/M program in Sao Paulo is scheduled to start in 1996 and is aimed to identify maladjusted or poorly maintained light duty vehicles with excessive emissions and needing proper mainte-

nance, and to ensure that emission control systems are in place and in good operational conditions. Preliminary estimates show that a centralized I/M program for the São Paulo Metropolitan Region could reduce carbon monoxide, unburned fuel and particulate matter emissions by 10 to 20 percent. This is based on preliminary inspections of vehicles and measurements of emission reductions following repair of failed vehicles. In addition, fuel economy is expected to improve between 5 to 10 percent.

Canada. The Federal Government regulates new vehicle emissions; provinces exercise control over in-use vehicles. Ontario, British Columbia, and several other provinces have evaluated various options for I/M programs for in-use vehicles, in conjunction with safety inspection programs. One of the issues is whether the program should be voluntary or mandatory; some of the factors against a mandatory program in the sparsely populated territories are the high costs of travel for inspection, negative public response to mandatory inspection, and lack of qualified personnel to perform inspections in some communities. Box 4.2 provides an evaluation of British Columbia's AirCare I/M program.

Chile. Technical inspection for all diesel vehicles has been mandatory for several years. Inspection stations and procedures have been recently upgraded, especially for buses. The bus inspection procedure is based on a dynamic test that uses opacimeters. The maximum opacity standard at idle is 2 percent and at full load is 30 percent. Mandatory hydrocarbons and carbon monoxide checks of gasoline-fueled vehicles have been introduced as well. Selective roadside inspections are also carried out by the police to control carbon monoxide emissions from cars and opacity levels of bus emissions (Escudero 1991).

China. At the time of registration each motor vehicle is checked for idling emissions (gasoline-fueled) or free acceleration smoke (diesel-fueled) at local inspection stations. Periodic inspections are required annually for all vehicles and are conducted by the local vehicle management office in centralized inspection lanes utilizing the idle test or the free acceleration test. The quality and effectiveness of these inspections vary significantly from one jurisdiction to another.

Costa Rica. In February 1993 a law specifying vehicle emission standards was enacted by the Congress. Regulations came into force at the end of 1994. Besides the compulsory use of the catalytic converters in new cars (from 1995 only lead-free gasoline is sold) a decentralized I/M program is being implemented with the help of private garages. In addition, traffic police makes roadside random checks with mobile equipment. However the inspectors are not well trained and do not handle the equipment properly, so that even highly polluting cars pass the controls. An important element of the I/M program has been the training of mechanics and certification of garages to perform proper tune-up of gasoline- and diesel-engined vehicles. The public is regularly informed of measures taken through a public information campaign.

Finland. After an 18-month lead-time the in-use control program was initiated on January 1, 1993. Both state-owned inspection stations and private repair garages and dealerships perform the short test (two-speed idle) and issue inspection certificates. Each sector has received an equal share of the workload. If emissions certificates are less than eight months old, cars are exempted from the emissions test when performing the technical inspection. Actual failure rates have been much lower than those calculated from preliminary tests.

About 10 percent of the cars have been ordered to correct carbon monoxide settings. Only about 3 percent of the decisions to fail a vehicle have been made on the basis of excessive hydrocarbon emissions. The failure rate on air-fuel ratio is low, but this indicates only that a small number of catalyst-equipped vehicles have been tested. Regular, non-catalyst cars must be inspected every year, but low-emission vehicles (usually three-way catalyst equipped) are subject to inspection not earlier than the third year of their service life. After that they are exempted for two more years, but after the sixth year they must be inspected annually.

The Finnish program does not permit any waivers, a widely used practice in U.S. programs. However, no case has emerged where a car could not pass the short test after a reasonable tune-up or repair. The decision to exempt cars made before 1978 was probably wise in this respect, since such vehicles are not widely used. The overall response of motorists to this control program has so far been positive.

Until 1995, only gasoline-fueled passenger cars were covered under the control program. Diesel-fueled cars and trucks are only checked if they emitted considerable visible smoke. The program have since been expanded to test diesel-fueled vehicles using the free acceleration method (Laurikko 1994).

Germany. Biennial inspection of automobiles has been required for many years; the program was recently expanded to an annual test. As Germany introduces more rigorous emission standards with increased use of catalyst technology, it is evaluating a variety of techniques to improve its I/M test requirements. If the Ger-

man program is implemented, it will likely serve as a model for other European countries, particularly countries interested in reducing nitrogen oxide and diesel particulate emissions. Details of the German in-service emission test requirements are summarized in CONCAWE 1994.

Hong Kong. In an effort to address the diesel particulate problem, observer teams issue citations to smoking vehicles, requiring such vehicles to report to a central testing station for instrumented (Hartridge) tests. Vehicles that fail the 60 HSU standard must receive necessary repairs or face fines. With the opening of an automated inspection facility, annual inspections will be required for all goods vehicles (only those over 11 years old were tested before), taxis, and buses. Private cars are tested for roadworthiness in private garages if they are more than six years old, but this has little if any impact on emissions. A mandatory I/M program is planned for introduction in 1996.

India. The first stages of a vehicle pollution control program have been completed. Major elements include a 3 percent limit (by volume) on idle carbon monoxide emissions for all four-wheel, gasoline-fueled vehicles, a 4.5 percent limit (by volume) on idle carbon monoxide emissions for all two- and three-wheel gasoline-fueled vehicles, and a 75 HSU limit on smoke density for diesel vehicles at full load (60 to 70 percent of maximum rated engine speed); the limit on the free acceleration test is 65 HSU. These standards went into effect on October 1, 1989. Automobile standards similar to ECE Regulation 15-04 but tailored to Indian driving conditions went into effect in 1991. Diesel vehicle requirements were also tightened at that time. The need for a comprehensive I/M program has increased substantially with the requirement that all new cars in the four major metropolitan areas (Bombay, Calcutta, Delhi and Madras) be fitted with catalysts, from April 1, 1995.

Italy. Since 1989, the Italian Department of Environment has invited all vehicle owners to have their vehicles checked on a voluntary basis. A network of test centers has been setup with the cooperation of motor-manufacturers and the oil industry. After testing, the owner is advised that the vehicle requires adjustments if it exceeds carbon monoxide limits for gasoline-fueled vehicles and smoke opacity limits for diesel-fueled vehicles.

Japan. The history of motor vehicle emission standards, including those for in-use vehicles, is similar to that in the United States. In Japan, inspection programs for carbon monoxide have been in effect since 1970, and hydrocarbon measurements and limits were added in 1975. Vehicles must be submitted for testing once they are three years old and every two years thereafter.

Republic of Korea. An active air pollution control program is in place that includes a system of random roadside inspections. As part of a study on motor vehicle pollution control, a task force has been organized to develop a set of recommendations to improve I/M. While there is an I/M program in place, standards are lax. The government is planning substantial improvements to the I/M program, including the possible addition of the IM240 test.

Mexico. A rudimentary I/M program was initiated in 1988, with pass-fail decisions left to individual mechanics. This approach resulted in a high percentage of improper inspections. To rectify the problem, concessions were granted for 24 high-volume, inspection-only facilities (macrocentros) utilizing computerized emission analyzers. It was mandatory for all high-use commercial vehicles (taxis, minibuses) and government vehicles to be inspected at these facilities, although private automobiles could use them, too. In addition, private garages inspecting private automobiles were required to use computerized analyzers and were subject to surveillance and enforcement activities. Programs to improve mechanic training and repair quality were also developed.

By the end of 1995, however, it became increasingly clear that the private garage system under the dual I/M system, was not working. As one indicator, for example, the failure rate in the private garages averaged about 9 percent whereas in the centralized lanes, the failure rate was about 16 percent. Stations conducting improper or fraudulent inspections were taken to court on several occasions and were subsequently shut down. As a result, the government decided to close all the private garage inspection stations and to switch to a completely centralized system.

The upgraded I/M program has the following key elements:

- Since January 2, 1996 all I/M tests in the Federal District are conducted at one of the existing centralized facilities– 26 facilities with about 142 lanes. It is intended to add 36 new facilities with 3 lanes each by January 1997.
- Standards (cut points) were tightened so that the average failure rate increased to 26 percent.
- The inspection frequency for all but high mileage vehicles was reduced to once per annum to relieve congestion at the lanes; taxicabs and other high use vehicles continue to be tested twice per annum.

- The inspection stations are being upgraded so that all will have a dynamometer and be capable of running the ASM (25/50/15) test by end 1996.
- Private cars receive the two stage idle test whereas the high use vehicles receive a steady state loaded test.
- Independent auditors monitor the tests carried out at each station. These auditors are rotated monthly and their contracts are funded by the companies operating the stations. In addition, each test is recorded on video in real time and the tapes and computer print outs are spot checked to assure that the tests are valid.
- The funding for the centralized facilities will be entirely private with cost recovery from inspection/test fees. Fees will be 70 pesos (U.S.$9.30) for cars and 100 (U.S.$13.25) pesos for high-use vehicles; 10 pesos of the fee will go to the government to pay for the stickers.
- In designing the I/M program, the government has exercised close coordination with the automobile manufacturers.
- In use vehicle standards were tightened by approximately 30 percent. The standards applicable in the DDF prior to January 1995 (old) and January 1, 1995 (new) summarized in table 4.17.

The Philippines. Diesel smoke control has three elements: information, education, and communications, coupled with strong enforcement. Teams operate on key routes in Manila and on the basis of Ringlemann chart readings issue citations and revoke the registration plates of offending vehicles. To reclaim plates, vehicles must be presented to the central testing facility for a more technically-sound Hartridge instrument test. This test must be passed (67 HSU or less) to avoid a fine (200 pesos for the first failure, 500 for the second, and 1,000 for the third). The citation rate was 5,000 vehicles a month in 1990. These limits are quite lenient, so that vehicles passing the test still emit visible smoke (Walsh and Karlsson 1990).

Poland. Motor vehicle emission standards have been gradually tightened over the last five years with a view to harmonize the national requirements with the European Union. A revised regulation specifying emission requirements to be satisfied by new and in-use motor vehicles in Poland came into effect on July 1, 1995. The new regulation requires emission checks for in-use vehicles. The compliance with in-use requirements is checked during mandatory periodic inspections and random road-side checks; the frequency of periodic in-

Table 4.17 In-Use Emission Limits for Light-Duty Vehicles in Mexico

Model year	CO (%)		HC (ppm)	
	Old	New	Old	New
Passenger cars				
1979 and older	6.0	4.0	700	450
1980-1986	4.0	3.5	500	350
1987-1993	3.0	2.5	400	300
1994 and later	2.0	1.0	200	100
Combis and light trucks				
1979 and older	6.0	5.0	700	600
1980-1985	5.0	4.0	600	500
1986-1991	4.0	3.5	500	400
1992-1993	3.0	3.0	400	350
1994 and later	2.0	2.0	200	200

Note: Old prior to January 1995; New from January 1, 1995.
Source: Onursal and Gautam 1996

spections is a function of vehicle category and age, as indicated below:

- For passenger cars and light duty vehicles subject to type approval: three years after the first registration, then after two years, and subsequently once every year.
- For passenger cars and light duty vehicles not subject to type approval: once every year.
- For trucks having a maximum mass exceeding 3,500 kg: once every year, then every six months.

Singapore. Cars are tested for idle carbon monoxide and hydrocarbon emissions, but standards are not yet enforced because the I/M program has not been fully implemented. Standards are lenient—800 ppm for hydrocarbons and 4.5 percent for carbon monoxide. Diesel smoke testing is carried out using the free-acceleration test; mobile vans are employed by the police who stop smoking vehicles and administer a Hartridge smoke test. If the vehicle fails to meet a standard of 50 Hartridge Smoke Units (HSU), the owner must pay a fine and repair the vehicle (Registry of Vehicles 1993).

Sweden. An annual I/M program is limited to measurement of carbon monoxide emissions at idle with lenient standards. However, a more comprehensive but non-sophisticated I/M program has been evaluated for implementation. Catalyst-equipped passenger cars aged two years or older have to be inspected annually by the Swedish Motor Vehicle Inspection Company. Older ve-

hicles (1976-1988 car models) must be submitted for the annual test until they are five years old or have reached 80,000 kilometers, whichever occurs first. It is expected that this limit will be extended to 160,000 kilometers. Failed vehicles have to be rectified and retested within a month. Vehicles not retested within the one month limit are not permitted to ply on the roads.

Switzerland. A mandatory I/M program has been in effect since 1986. This program requires compulsory testing; gasoline-fueled light-duty vehicles are tested annually at authorized garages, while all vehicles with diesel engines must undergo a regular check at an official test station administered by the local canton. Diesel vehicles are tested at three-year intervals except for trucks transporting hazardous materials, which are tested annually.

Taiwan (China). Random roadside inspections have been implemented by local Environmental Protection Administration bureaus. A free motorcycle emissions test and maintenance program has also been promoted. In large cities buses are tested for free. The Environmental Protection Administration (EPA) has set up state-of-the-art vehicle inspection and maintenance stations throughout the island. In addition, inspection licenses are issued to qualified privately-owned garage stations (Shen and Huang 1991).

In order to reduce the pollution from motorcycles, the EPA is actively promoting a motorcycle I/M program. In the first phase, from February through May 1993, the EPA tested 113,147 motorcycles in Taipei City.

Of these, 49 percent were given a blue card indicating that they were clean, 21 percent a yellow card indicating that their emissions were marginal, and 30 percent failed the test. Between December 1993 and May 1994, 142,287 motorcycles were inspected, with 55 percent receiving blue cards, up 6 percent from the earlier program, and 27 percent failing, a drop of 3 percent in the failure rate. The major repair for failing motorcycles was replacement of the air filter.

Thailand. Teams from the Department of Pollution Control and the Bangkok Police carry out roadside checks of carbon monoxide emissions from passenger cars, hydrocarbons from motorcycles, and smoke from diesel vehicles. The motorcycle hydrocarbon standard is 14,000 ppm, which means that most two-stroke motorcycles can pass the test. The smoke opacity limit is Bosch number 5, measured in a free-acceleration test. Commercial vehicles are also subject to annual checks of smoke opacity (for diesels) or carbon monoxide (for spark-ignition vehicles) as part of the licensing process, but these inspections are of limited effectiveness. In Bangkok, the rate of failure for smoke opacity was just 2 percent, whereas on-road emission measurements suggest that more than 50 percent of the trucks have smoke opacity exceeding the established limits. The Land Transport Department is in the early stages of implementing a decentralized periodic I/M test for all types of vehicles in the Bangkok Metropolitan Region. Efforts to improve I/M test procedures and appropriate emissions standards are also under way.

United States. The I/M program is administered by the states and consists of periodic emission testing of in-use vehicles. Inspection and maintenance programs are required as part of the State Implementation Plans (SIP) for those states unable to achieve compliance with carbon monoxide or ozone air quality standards in non-attainment areas.

Inspection and maintenance programs have been established in 140 regions comprising parts or all of 36 states and the District of Columbia. Only four states (Connecticut, Massachusetts, Rhode Island, and New Jersey) and the District of Columbia have state-wide (city-wide in the case of D.C.) programs; the rest concentrate on major population centers. While most U.S. I/M programs have been adopted in response to federal requirements, they differ significantly in terms of implementation. This is because the U.S. EPA, in specifying the program requirements, recognized the wide diversity of existing local conditions. Rather than attempting to dictate the specific details of the program, the U.S. EPA chose to set out performance requirements. These were based on emissions reductions, which were pre-dicted by evaluating the program details using the MOBILE emission factor models.

The type of I/M program, enforcement mechanisms, frequency of inspections, and other features are generally left to the discretion of the responsible state or local administration, subject to U.S. EPA's approval that the requisite performance criteria would be achieved. About 86 million vehicles are subject to I/M programs in the United States (U.S. EPA 1995).

United Kingdom. Emission testing has been introduced as part of the roadworthiness test applied to all four-wheel gasoline-fueled cars three years or older. Standards and testing methods for light-duty diesel vehicles and two-stroke engines are being developed. Inspection tests are conducted with the engine warmed-up and at idle, with the test results compared to the emission standards. Vehicles fail the complete DOT roadworthiness test even if the emissions performance is the only item of failure. A free re-test is allowed provided the vehicle is returned to the same testing station within 10 days.

Emission checks also include roadside testing. Vehicles that fail the roadside check are required to be rectified within ten days or become subject to a fine of up to UK£5000. In extreme cases, the vehicle may be subject to an immediate ban. In a 1994 roadside check in Central London, 20 percent of the gasoline-fueled vehicles and one in three diesel vehicles gave off illegal emissions (Ahuja 1994). Roadside emission checks have been included in the roadworthiness test since November 1991.

Remote Sensing of Vehicle Emissions

Remote sensing of vehicle emissions is a new technology that could improve the effectiveness of vehicle emissions control programs. The technique works by measuring the absorption of a beam of infrared light by the carbon dioxide, carbon monoxide, and hydrocarbons in a vehicle's exhaust plume (figure 4.8). Based on the absorption, a computer is able to calculate the ratios of carbon monoxide and hydrocarbons to carbon dioxide in the exhaust. From this information it is possible to calculate the concentrations of hydrocarbons and carbon monoxide in the exhaust at the instant the vehicle passes the remote sensing device (Stedman and Bishop 1991). This system has been used to characterize the statistical distribution of vehicle emissions in a number of cities, including Chicago and Mexico City (figures 4.9, 4.10, and 4.11). It is expected that a remote sensing system for NO_x emissions would utilize either a beam of ultraviolet light or light from a turnable diode laser projected across the road (U.S. EPA 1995). A prototype system capable of measuring nitrogen oxide concentrations has been developed. A technical assessment of the use of remote sensing for I/M programs is provided in appendix 4.1.

Although remote sensing offers significant potential for vehicle emissions monitoring and control, the technique has limitations. Remote sensor technology measures the instantaneous concentration of hydrocarbon and carbon monoxide in a vehicle's exhaust at the time it passes the sensor. This measurement is a snapshot of emissions and thus not a reliable indicator of overall vehicle emission levels or even "gross emitter" status, since emissions measured by the remote sensing device often do not correlate with the Federal Test Procedure (FTP) measurements or other measures of actual emissions over realistic driving cycles. Even properly functioning vehicles may exhibit high carbon monoxide and hydrocarbon concentrations in the tailpipe under driving conditions such as hard acceleration or cold starting. Many vehicles also exhibit moderately high carbon monoxide concentrations at idle. Some older vehicles experience high hydrocarbon concentrations during deceleration, while newer vehicles cut off fuel during deceleration, making the hydrocarbon to carbon dioxide ratio hard to predict.

Since the air-fuel ratio at normal cruise should be close to stoichiometric (very close for closed-loop vehicles, less so for vehicles without closed-loop control), a high hydrocarbon or carbon monoxide concentration in normal cruise conditions indicates a likely emissions problem. Still, the hydrocarbon or carbon monoxide concentration that would indicate a problem in a vehicle with electronic closed-loop air-fuel ratio control may

be normal for a vehicle without such control. In addition, gasoline-fueled, heavy-duty vehicles may have anomalously high readings because of their heavier engine loading and more lax emissions standards. Also, hydrocarbon concentrations exhibited by two-stroke motorcycles in normal operation would indicate gross malfunction in a four-stroke engine.

Another limitation on current remote sensing devices is that they can measure only one lane, requiring vehicles to pass in single file. Furthermore, no practical method is available for sensing of evaporative and other non-tailpipe emissions. The system does not work on vehicles with elevated exhaust pipe discharge and invalid readings result when the sensor beam is blocked by pedestrians, bicycles, or trailers.

For remote sensing to be reliable and useful, extraneous sources of variation should be eliminated. It is important to distinguish between emission signals that indicate a high emissions malfunction and those that result from normal operation. Aberrant readings caused by acceleration or deceleration can be screened out by adding a leading and trailing photocell to the photocell that triggers data acquisition in the sensor. By comparing the time-distance ratios between the first and second and the second and third detectors, it is possible to determine whether the vehicle is accelerating or decelerating and to reject the data for these conditions. Cases of intrinsically dirty vehicles, such as two-stroke motorcycles and heavy-duty gasoline trucks, can be screened out by capturing the vehicle's image on film and sorting the data into categories based on vehicle type. Similarly, newer vehicles that may have been subject to more stringent emissions standards can be distinguished from older vehicles that

Figure 4.8 Illustration of a Remote Sensing System for CO and HC Emissions

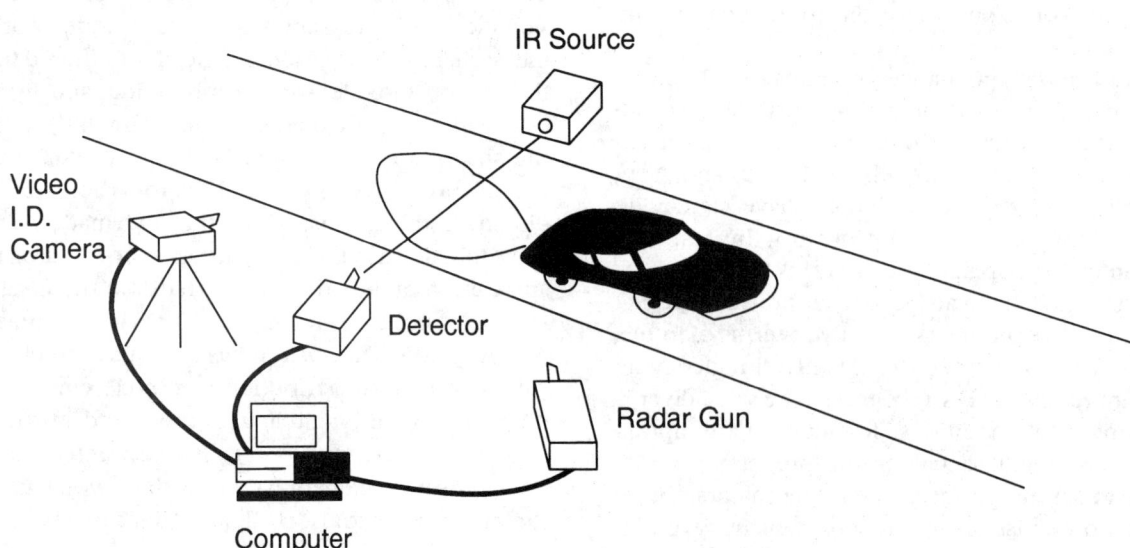

Source: U.S. EPA 1995

Figure 4.9 Distribution of CO Concentrations Determined by Remote Sensing of Vehicle Exhaust in Chicago in 1990 (15,586 records)

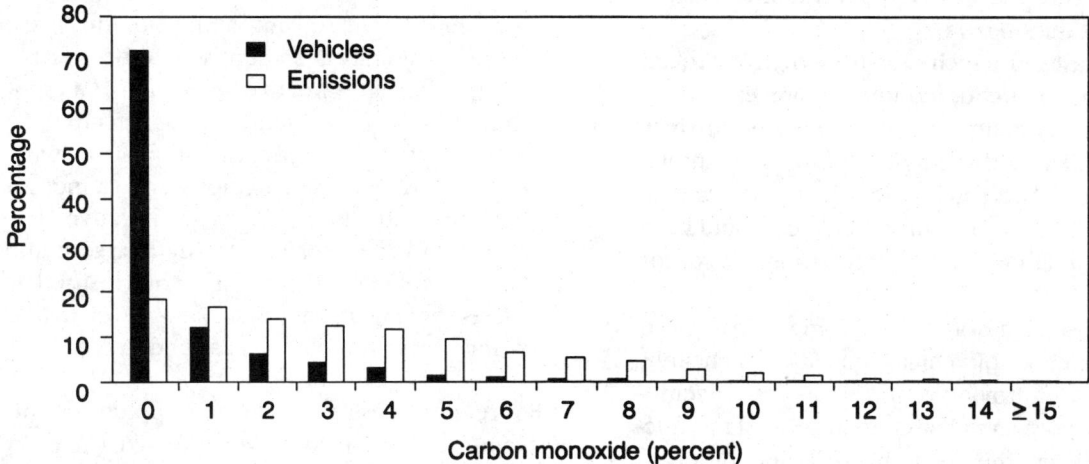

Source: Beaton and others 1992

Figure 4.10 Distribution of CO Concentrations Determined by Remote Sensing of Vehicle Exhaust in Mexico City

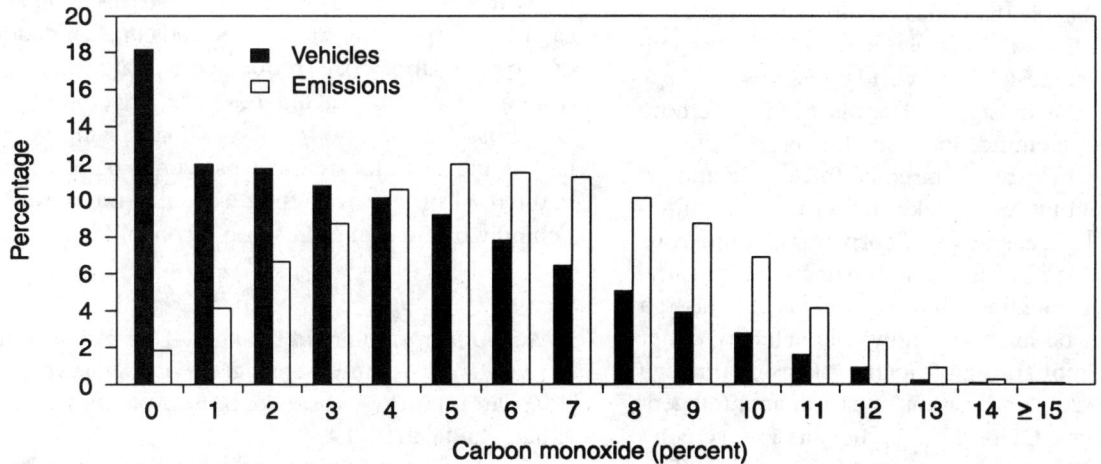

Source: Beaton and others 1992

Figure 4.11 Distribution of HC Concentrations Determined by Remote Sensing of Vehicle Exhaust in Mexico City

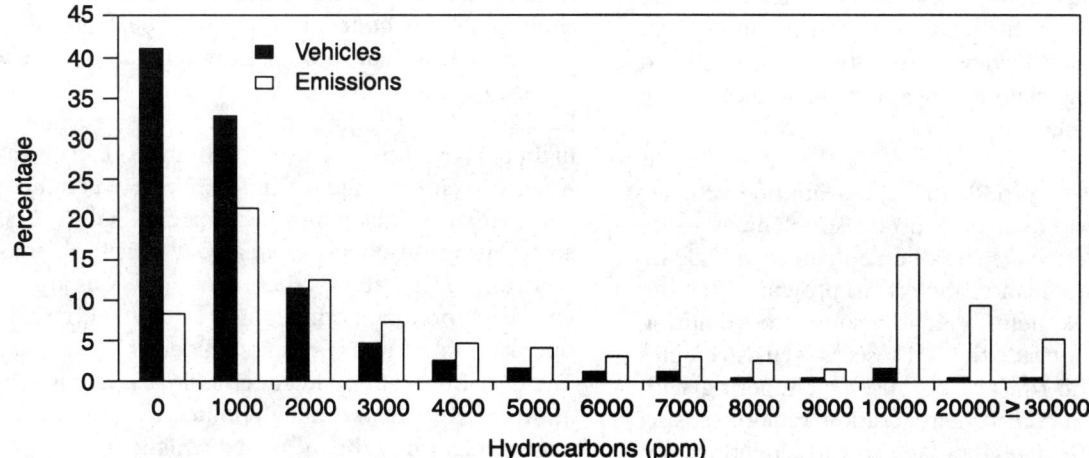

Source: Beaton and others 1992

were not subject to such standards. In cases of doubt, the year of manufacture of the vehicle can be determined by checking the license plate number against the vehicle registration data.

Identifying cases in which unusually high hydrocarbon and carbon monoxide conditions are caused by normal cold-start enrichment is more difficult. It may be possible to identify this condition from the remote-sensing "signature." Alternately, it may be more practical simply to locate the sensors where vehicles would have operated for several minutes before passing the sensor location.

The main advantage of remote sensing is that it can screen a large number of vehicles quickly and cheaply. This is useful for improving vehicle emissions inventories, since it provides an unbiased indication of the fraction of vehicles on the road that fall into different emission classes. Remote sensing can thus complement the results of more detailed surveillance testing using dynamometer measurements. Such testing provides more accurate measurement of total emissions over the driving cycle than does remote sensing but suffers from potential sampling bias and lack of representation of the cycle test of real world emissions.

Potentially the most useful application of remote sensing is direct identification of high-emitting vehicles on the road so they can be targeted for testing and repairs. This could ultimately take the form of a comprehensive surveillance system, incorporating numerous detectors operating continuously and unattended at strategic locations throughout an urban area. High emission readings would be linked to individual vehicles, either through video monitoring of license plates or through electronic vehicle identification as part of an intelligent highway system. Owners of high emission vehicles would be notified to present them for inspection. This arrangement could supplement or, if sufficiently developed, replace routine I/M testing. Targeting inspections at likely high emitters would significantly reduce the social costs of I/M programs. Several recent studies in the U.S. have examined this possibility (Austin and others 1994; California I/M Review Committee 1993; Walsh and others 1993). The main findings of these studies can be summarized as follows:

- Remote sensing to identify gross-emitting vehicles on the road could be a useful supplement to increase the effectiveness of an enhanced vehicle inspection and maintenance (I/M) program. Remote sensing could help to identify some gross-emitting vehicles that evade the I/M tests as well as vehicles that pass the I/M test because of tampering with the emission controls. In addition, remote sensing offers an effective approach to early identification and correction of vehicles that become gross emitters between successive scheduled inspections.

- Neither remote sensing alone nor the use of remote sensing as a "screen" would be a viable substitute for an effective periodic I/M program. Remote sensing presently is less effective than advanced I/M tests in identifying vehicles with high exhaust HC and CO emissions, and cannot identify vehicles with high NO_x or evaporative HC emissions. In order to avoid failing a large number of "clean" vehicles, the failure criteria used for remote sensing must be set at a level that allows many "dirty" vehicles to pass undetected.

For remote sensing to achieve its potential, a number of technical issues need to be addressed. The most critical of these is improving the accuracy of remote sensors in identifying high emitters. Another problem that needs to be solved is sampling across multiple lanes of traffic without creating conspicuous structures or causing a safety hazard. The remote sensing system should also be capable of operating for long periods unattended in order to save labor costs, reduce conspicuousness of the testing system, and avoid roadside driver distraction. One possible solution to the multi-lane sampling problem would be to locate the sensor units behind gratings in the road surface, with the receivers located overhead, behind a traffic sign or in some inconspicuous place.

Evaluation of Remote-Sensing Data

A summary of on-road CO and HC emission statistics derived from remote sensing surveys at several locations around the world is presented in table 4.18. (Zhang and others 1994).

The median values are substantially lower than the mean. This indicates that the distribution of emissions is highly skewed, with a relatively small number of high emission values having a disproportionate impact on the mean. Such a skewed distribution is to be expected, since pollutant concentrations may be quite high but cannot be less than zero. As shown in table 4.18 shows, the mean and median emission values in cities such as Bangkok, Kathmandu, and Mexico City are significantly higher than those measured in most European and American cities, while the difference between the mean and median values is proportionately less. In U.S. and some European countries, nearly all cars have emission controls, so that tailpipe pollutant emissions are usually low. High pollutant emissions are due to either unusual operating conditions or emission control system failure. The distribution of pollutant emissions is highly skewed, as the few vehicles with high pollutant levels account for a large fraction of the measured emissions.

Where average pollutant emissions are high, the effect of the highest-emitting vehicles on the overall average is proportionately less. The two cities in Asia, Bangkok and Kathmandu, stand out with respect to hydrocarbon emissions, due to the high percentage of motorcycles and three-wheelers equipped with two-stroke gasoline and LPG engines. These engines have extremely high hydrocarbon emissions, even when properly tuned and operating normally. The wide range of gross polluter cutpoints depicted in table 4.18 suggests that

emission standards for in-use vehicles should be based on an evaluation of local conditions and not derived from general international practice. Based on the results of remote sensing emission surveys, there may be a case to vary both the stringency of in-use emission standards and the frequency of vehicle inspections by vehicle model, vehicle age and geographical location. This may be particularly relevant to developing countries.

Observations like those in table 4.18 have sometimes been used to support arguments such as that "10 per-

Table 4.18 Remote Sensing CO and HC Emission Measurements for Selected Cities

Location	Date	No. of records	Carbon monoxide				Hydrocarbons			
			Mean (%)	Median (%)	Percentage of gross polluters[a]	Gross polluter cutpoint[b] (%)	Mean (%)	Median (%)	Percentage of gross polluters[a]	Gross polluter cutpoint[b] (%)
Bangkok	Aug-93	5,260	3.04	2.54	21.65	5.24	0.95	0.57	17.93	1.98
Chicago	Jun-92	8,733	1.04	0.25	7.50	4.20	0.09	0.06	15.89	0.13
Denver	Oct-91	35,945	0.74	0.11	6.68	3.42	0.06	0.03	10.92	0.10
Edinburgh	Nov-92	4,524	1.48	0.69	13.40	3.44	0.13	0.08	11.87	0.23
Gothenburg	Sep-91	10,285	0.71	0.14	8.80	2.58	0.06	0.05	19.71	0.09
Hamburg	May-94	11,128	0.57	0.12	6.80	2.43	0.04	0.02	9.88	0.09
Hong Kong	Aug-93	5,891	0.96	0.18	9.17	3.41	0.05	0.04	15.87	0.08
Kathmandu	Aug-93	11,227	3.85	3.69	24.88	5.85	0.76	0.36	11.05	1.06
Leicester	Nov-92	4,992	2.32	1.61	18.15	4.33	0.21	0.13	13.78	0.33
Lisbon	May-94	10,426	1.48	0.38	12.00	4.21	0.06	0.03	10.42	0.12
London	Nov-92	11,666	0.96	0.17	8.38	3.58	0.14	0.07	10.00	1.98
Los Angeles	Jun-91	42,546	0.79	0.15	6.97	3.46	0.07	0.04	10.27	0.15
Lyon	May-94	14,276	0.97	0.22	8.79	3.52	0.07	0.04	11.07	0.14
Melbourne	May-92	15,908	1.42	0.57	12.43	3.52	0.11	0.06	11.05	0.19
Mexico City	Feb-91	31,838	4.30	3.81	24.29	6.58	0.21	0.11	10.20	0.41
Milan	May-94	13,943	1.25	0.39	11.99	3.42	0.06	0.04	10.74	0.12
Rotterdam	May-94	12,882	0.55	0.13	6.99	2.31	0.04	0.02	9.91	0.08
Seoul	Aug-93	3,104	0.82	0.26	9.83	2.62	0.04	0.02	4.80	0.15
Taipei	Aug-93	12,062	1.49	0.88	16.78	3.09	0.06	0.05	20.09	0.09
Thessaloniki	Sep-92	10,536	1.40	0.55	13.32	3.46	0.16	0.08	9.86	0.33
Zurich	Mar-94	11,298	0.83	0.17	6.93	3.66	0.03	0.02	9.34	0.06

a. Percentage of vehicles responsible for half of the emissions.
b. CO and HC emissions values (in percent) corresponding to the gross polluter threshold.
Source: Zhang and others 1994

cent of the vehicles are responsible for 50 percent of the emissions" and that if these "gross polluters" could only be identified and repaired, air quality could be improved substantially at little cost. This argument is only partly valid. Although the distribution of real-world vehicle emissions is certainly skewed, with a majority of emissions produced by a minority of the vehicles, the degree of skewness is exaggerated by remote sensing data. That is because these data confound the variability in emissions between vehicles with the variability in emissions from the same vehicle under different operating conditions[3]. Remote sensing does not distinguish between the variability in instantaneous measurements of emissions and variability in total emissions per kilometer over the vehicle's driving cycle. This can lead to erroneous conclusions regarding the contribution of gross emitters to pollutant emissions.

On-Board Diagnostic Systems

The objective of an on-board diagnostic (OBD) system is to, identify and diagnose emission-related malfunctions in vehicles equipped with sophisticated electronic engine control systems, as such malfunctions can greatly increase emissions. As vehicle engine and emission control systems have grown more complicated, diagnosis and repair of malfunctioning systems has become increasingly difficult. Many emissions-related malfunctions can go undetected in modern vehicles tested with current I/M programs. This is especially true for NO_x related malfunctions, since basic I/M programs do not test for NO_x emissions.

Regulations requiring on-board diagnostic systems have been in effect in California for some time, and were recently strengthened with many additional requirements. These second-generation requirements are known as OBD2. Under the revised requirements, the vehicle's computer system is required to detect failures or loss of efficiency in the major emissions-related components and systems, including the catalytic converter, air-fuel ratio sensor, EGR system, evaporative purge system, and all emissions-related sensors and actuators connected to the computer. The system is also required to detect engine misfire (a common cause of catalytic converter failure). When a failure is detected, the system activates internal malfunction codes and lights a special malfunction indicator light, warning the driver to have

the vehicle repaired. A check of the codes and the malfunction indicator light is included in the California I/M program; vehicles do not pass if the light is illuminated.

The U.S. EPA has also adopted regulations similar to OBD2. On-board diagnostic systems meeting the U.S. EPA regulations were first introduced in light-duty vehicles in 1994 with a phase-in period extended to the 1998 model year. These regulations require that manufacturers protect engine calibrations and operational software from tampering. For light-duty vehicles with diesel engines, the federal OBD rule does not require monitoring of the oxygen sensor or detection of engine misfire but catalyst malfunctions or deterioration must be detected. There are presently no OBD requirements for heavy-duty spark-ignition or diesel engines.

Since the OBD system monitors engine and emission control system functioning over the full range of driving conditions, it covers a much broader range of checks than any I/M test. If successful, these systems could conceivably reduce the I/M process to a simple check of the malfunction indicator light. The status and prospects of technology to meet these requirements are controversial, however. It remains to be seen how effective and tamper-proof these systems will be in large-scale deployment. Effective I/M test procedures will continue to be required for at least the foreseeable future.

An additional benefit of an OBD system is in assisting the service industry to quickly and properly diagnose and repair malfunctions in the vehicle's Emission Control System (ECS). Since the OBD system is an integral part of the vehicle's engine control system, it senses computer electronic signals under a variety of operating speed and load modes and can thus detect and store intermittent faults that are difficult to duplicate in the service garage.

Vehicle Replacement and Retrofit Programs

Although new vehicle emission standards can help cap the growth in vehicle emissions and I/M programs can reduce emissions from in-use vehicles, levels of air pollution are nonetheless likely to remain unacceptably high in many cities in the developing world, because of the large number of uncontrolled vehicles already in operation. Where the severity of the air quality problem so warrants, it may be appropriate to consider vehicle retrofit and replacement programs. Such measures are likely to be most cost-effective when applied to intensively-used vehicles such as taxis, minibuses, buses, and trucks. These vehicles have high emission levels in proportion to their numbers in the vehicle fleet. Retrofitting existing vehicles with new engines or emission control systems or replacing them with new, low-emitting vehicles can often reduce emissions from these ve-

---

3. Consider, for example, a car that normally emits 1 gram per kilometer of CO but produces 200 grams per kilometer under hard acceleration. This is typical of modern emission-controlled vehicles. Now consider a remote sensing sample of 20 such identical vehicles, of which 19 were captured in normal operation, and one under hard acceleration. A naive analysis would suggest that the one vehicle (5 percent of the sample) accounted for more than 80 percent of total CO emissions; this is an absurd result.

hicles by 70 percent or more. Even where active vehicle retrofit and replacement programs are not warranted, care should be taken to ensure that the vehicle tax structure does not encourage the retention of old vehicles in service, and preferably, that it acts to encourage purchase of new, emission-controlled vehicles.

Scrappage and Relocation Programs

Vehicle fleets in many developing countries are characterized by a large number of old, poorly maintained, and high-emitting vehicles. Although the value of these vehicles may not be high, low labor costs for repairs make it feasible to keep them in operation. The value of such marginal vehicles can be affected by tax policies, sometimes in unexpected ways. For example, many developing countries impose high luxury taxes on purchase of new vehicles, which increases the value of vehicles already in the fleet. Similarly, high taxes on ownership of new vehicles, with declining tax rates as the vehicle ages, also tend to increase the value of older vehicles and reduce their scrappage rate. From an emissions perspective, flat or even increasing taxes on vehicle ownership as a function of age would be preferable to a declining tax rate. Even better would be a tax based on vehicle emissions levels.

Other policies can also affect vehicle scrappage rates. Perhaps the most effective way to induce scrappage of older vehicles is to institute a strict inspection and maintenance program, possibly covering both emissions and safety requirements. Flat limits on the age of vehicles permitted to circulate are also possible but not advisable except in unusual cases, as they unfairly discriminate against older vehicles that are properly maintained (Beaton and others 1995). A possible exception might be where new vehicles have significantly stricter emission standards than older vehicles in the fleet. In this case, it may be permissible to allow older vehicles to continue to circulate provided they are retrofitted to meet the same standards as new vehicles.

The Union Oil Company of California (UNOCAL) has demonstrated a successful program in Southern California to retire 1970 and earlier vintage automobiles by buying them from their owners for U.S.$700 each and scrapping them. Nearly 8400 old automobiles were removed from the vehicle fleet in 1990 (U.S. Congress/OTA 1992).

UNOCAL's latest program (SCRAP IV) initiated in January 1995, specifically targets pre-1975 model year vehicles, which can emit 50–100 times more pollutants per mile than new vehicles and account for a disproportionately high volume of all mobile emission sources in the Los Angeles Basin. Vehicles acquired through the program are scrapped and made available for self-service parts dismantling. Vehicles with special collector

value are sold to the public. To qualify for the program, vehicles must be fully functional, not partially dismantled, and driven under their own power. They must have been registered in the local area for at least two years. UNOCAL will use most of the emission credits earned from scrapping older vehicles under SCRAP IV to offset some of the emissions from its Los Angeles marine terminal (Oil and Gas Journal 1995). Removing all high-emitting cars through this approach offers a cost-effective approach to reducing emissions as a substitute for controls on stationary sources or increasingly stringent emissions standards for new cars.

Where the existing vehicle fleet retains significant economic value, relocation of the vehicles to smaller towns and villages outside major urban areas may be a useful approach. This makes it possible to retain much of the economic value of the vehicle for society while still removing it from the major cities, where air pollution is worst. While total emissions remain the same, the lower vehicle ownership and use in the countryside results in lower pollutant concentrations, and the lower human population density means that fewer people are exposed to pollutants. Measures to encourage such a shift include differential vehicle taxes between urban and rural areas, age limits on vehicle registration in urban areas, and the application of differential taxes based on vehicle emissions in urban areas.

Vehicle Replacement

Vehicle replacement is generally the most practical solution where the existing vehicle fleet is old, in poor condition, or difficult to retrofit. Such replacements can not only reduce emissions but can improve traffic safety. In the case of buses and minibuses, replacement can also help to improve the quality and comfort of public transport, possibly helping to encourage a mode shift from private cars. A key concern in evaluating the cost-effectiveness of a vehicle replacement program is the disposition of the vehicles replaced. If the old vehicle is allowed to continue operating as before (with a different owner), emissions will not decrease. Requiring that the old vehicle be scrapped ensures that the full emission reduction is achieved but may greatly increase the cost of the program, depending on the old vehicle's value. A less costly alternative is to require that the old vehicle be sold and re-registered outside the urban area, thus increasing the supply of transport in rural areas and ensuring that the vehicle's emissions do not add to pollution in the city. In the case of taxis, it is probably sufficient to allow the old vehicles to be used as private cars, since these are used much less intensively than taxis.

An example of an emissions-related vehicle replacement program is the Taxi Modernization Program undertaken in Mexico City. Based on an agreement signed in

March 1991 between the Mexican authorities, the taxi associations, and the automotive manufacturers, this program is providing some U.S.$700 million, through the commercial banks, as a line of credit to fund the taxi program. Automobile manufacturers have agreed to make available specified numbers of vehicles per month, at agreed prices. The aim is to replace all pre-1985 taxis with new vehicles meeting Mexican 1991 or 1993 emission standards. Up to 63,000 taxis (93 percent of the fleet) could be replaced under the program. This program includes a combination of regulatory measures (taxicab licenses are denied for pre-1985 vehicles) and economic incentives (special prices have been negotiated with manufacturers and financing will be provided on favorable terms but at a positive real rate of interest).

Hungary has taken a step toward the elimination of two-stroke engines often found in the heavily polluting Trabant and Wartburg automobiles, the East German products common to many countries in the former Soviet bloc. In 1994, businesses that owned two-stroke engined vehicles were required to scrap them. Individual owners were encouraged to replace two-stroke vehicles with four-stroke engines or to install catalytic converters for the two-stroke engines (box 4.4).

Retrofit Programs

Where a vehicle was originally manufactured without emission controls but remains in reasonable condition, retrofitting emission controls may be a cost-effective option to reduce emissions. Depending on the situation, this retrofit may be accomplished by installing additional components, such as air pumps or catalytic converters; replacing the engine with one designed for low emissions; or retrofitting a separate system, such as a liquefied petroleum gas or natural gas conversion kit. Retrofitting is most practical where the control measure in question can be implemented without changing the basic engine design, as in the case of a catalytic converter. A catalytic converter retrofit program was implemented in Germany for several years. This program, driven by tax incentives for vehicle modification, has been considered a modest success. A similar program has been implemented in Sweden, and programs are being considered in Chile and Taiwan (China). In Hungary, the government has initiated a five-year program to persuade owners of older cars to install catalytic converters with financial assistance up to 60 percent of the cost of retrofitting with catalytic converters (Walsh 1995). On the other hand, a program of engine control retrofit in California in the 1960s intended to reduce nitrogen oxide emissions suffered implementation difficulties and was abandoned after a few years.

Retrofit strategies are especially appropriate for heavy-duty vehicles such as trucks, buses, and minibus-

es. This is because these vehicles have high levels of emissions, long lives, and high usage levels and thus produce large amounts of pollution, particularly visible smoke and particulate matter, per vehicle. Retrofitting these vehicles can therefore be reasonably cost-effective. Also, heavy-duty vehicles are normally designed so that extra space is available and major components such as engines are interchangeable, thus simplifying the retrofit process. Passenger cars, in contrast, tend to be designed as an integrated system, making them more difficult to retrofit.

Since strict emissions standards for new heavy-duty engines have been adopted only recently (and not at all in most countries), most engines in service are effectively uncontrolled. To improve air quality in the short term, it may be desirable in some cases to retrofit these engines for lower nitrogen oxide emissions. By retarding diesel fuel injection timing, a simple procedure with most engine designs, nitrogen oxide emissions can be reduced significantly. While retarding injection timing does increase fuel consumption and particulate matter emissions, it is often possible to achieve reductions of 20 to 30 percent in NO_x emissions without marked effects on particulate emissions and fuel consumption. In many cases, further reductions in both nitrogen oxide and particulate emissions can be achieved by upgrading engine technology at the time of major overhauls, either by rebuilding the engine with more advanced components or by replacing it with an engine designed for low emissions. Given that the overhaul is required in any event, the incremental cost of upgrading technology is relatively small, and the emission reductions can be significant.

For medium- and heavy-duty gasoline vehicles, one practical and effective means of reducing emissions is to convert them to burn liquefied petroleum gas (LPG) or compressed natural gas (CNG). This can result in substantial reductions in emissions (especially with the use of high-technology LPG or CNG systems incorporating feedback control and three-way catalysts). These fuels can also offer substantial cost savings. Where they are available from domestic sources, both LPG and CNG are usually less costly than gasoline. Natural gas is usually cheaper than LPG and has superior characteristics as an engine fuel. Although natural gas vehicle conversions are more expensive than those for LPG, these costs can normally be recovered through lower fuel and maintenance costs.

A successful retrofit requires considerable care and engineering development work. Proper design, prototype testing (including emissions testing), and manufacturing are required. Because of the expense involved in development, retrofitting will generally be most cost-effective where a large number of vehicles of similar type and design are available for retrofit. Examples include transit bus fleets, garbage collection fleets, and urban

---

Box 4.4 Replacing Trabants and Wartburgs with Cleaner Automobiles in Hungary

The city government of Budapest provides public transportation passes to motorists who have turned in their highly polluting two-stroke-engine automobiles. A second provision of the same program has allowed motorists to sell their Trabants and Wartburgs to the city for a price higher than the going market rate and to use the proceeds as part of a down payment on a new, Western-made, environmentally friendly car.

Under this program, 1,451 owners of these polluting cars—two-thirds own Trabants and one-third own Wartburgs—have applied to exchange their cars for passes to use in the city's transportation system. For each Trabant, the city will award four year long passes; for each Wartburg, six year long passes will be issued. Pass holders can use them on any of the city's public transport systems. The program is expected to cost the city 90 million forints (U.S.$900,000).

About 120,000 Trabants and Wartburgs were on Budapest's streets in the early 1990s. The two cars, made in the former East Germany, are notorious for spewing pollutants into the environment. Because of their low price, they were cars of choice in Hungary.

At the time of the program's launching in September 1993, the administration displayed five selected car models in the city hall's courtyard. Dealers for 43 car models had submitted vehicles for consideration by the city administration. A committee chose the finalists on the basis of engine characteristics, the existence of a catalytic converter, availability of service, price, and credit conditions. It negotiated with the banks in the city to set up purchase terms. The cars chosen for the program were the SEAT Marbella, Suzuki Swift, Opel Corsa, Renault, and Volkswagen Polo. More than 700 owners of Trabants and Wartburgs sold their cars to the city for coupons worth 20,000 forints (U.S.$200) and 33,000 forints (U.S.$330), respectively. The motorists could add these coupons to cash for a one-third down payment on the purchase price of a new car with the remaining payments spread over five years at annual interest rates of 13 to 15 percent (considered highly favorable in Hungary). The car prices were 60,000 forints to 190,000 forints (U.S.$600-$1,900) lower than their showroom prices.

The Trabants and Wartburgs turned in by the motorists were destroyed. The cost of the program was about 17 million forints (U.S.$170,000). Taking more than 2,000 Trabants and Wartburgs off the streets would eliminate 331,000 kilograms of pollutants per year.

Public transportation will also be a target. Already, the city of Budapest has made plans to buy as many as 400 buses fueled by natural gas. The buses will be purchased over the next two years and will replace existing polluting diesel buses. The use of natural gas instead of diesel fuel will also save money, enabling the city to recover its investment in new buses in about four years.

Source: Walsh 1994

---

delivery fleets. The highest priority for retrofit programs should go to transit buses and other vehicles operating in congested urban areas, particularly those with high-emission, stop-and-go driving cycles. Such programs could be undertaken, at least initially, on a voluntary or quasi-voluntary basis. Government-owned vehicle fleets are especially suitable for such programs.

An example of a large-scale emissions retrofit program is planned in Mexico City. The Environmental Commission for the Mexico City Region has developed a plan to retrofit more than 100,000 gasoline-fueled minibuses and gasoline trucks with LPG and CNG systems. Vehicles to be retrofitted will be those built between 1977 (1982 in the case of minibuses) and 1991 (when catalyst-forcing emissions standards for gasoline vehicles came into effect). Older vehicles will be forcibly retired; younger vehicles are already equipped with catalytic converters, and have much lower emissions. A highly successful program to retrofit taxis with LPG and CNG systems has been implemented in Buenos Aires. Such retrofit programs are also popular in Asian countries and have been used to convert motorized three-wheelers with two stroke engines (variously called helicopters, rickshaws, tempos, tuk-tuks, and, in Bangladesh, Indonesia, Nepal, and the Philippines respectively) to run on LPG.

Intelligent Vehicle-Highway Systems

There has been interest in recent years in the application of advanced information and control technology to improve the efficiency and safety of road transport. Many competing proposals for intelligent vehicle-highway systems (IVHS) have been floated, research and development is underway, and a number of prototype systems are being constructed. To the extent that it increases the carrying capacity of the road system, and therefore increases vehicle kilometers travelled, IVHS is more likely to contribute to increased vehicle emissions. Nevertheless, appropriate application of IVHS technology could contribute to reducing emissions and improving the cost-effectiveness of emissions control in a number of ways, including (Weaver 1991):

*   Implementation of road use/emissions charges.
*   Implementation of transportation control measures.
*   Selective activation of emission controls.
*   Navigation assistance to drivers in avoiding congestion and minimizing VMT.
*   Assistance in minimizing stop-and-go operation.
*   On-board diagnostics and condition monitoring.
*   Vehicle registration enforcement.

Some of these applications (notably driver assistance) are already contemplated as part of the ongoing research and development in IVHS. The proposal to require on-board computer systems to notify air quality authorities when the OBD system indicates a malfunction was mentioned earlier in the discussion of OBD.

To assess the potential benefits of IVHS for reducing vehicle emissions, it is essential to understand the factors that determine vehicle emissions. Potential impacts of IVHS include changes in the number of vehicle-miles travelled, changes in average speed, changes in driving patterns (reduced acceleration and deceleration), and changes in trip patterns (more short trips). The effects of these changes will vary depending on the vehicle type and the rapidly-evolving emission control technology which is evolving rapidly. A California low-emission vehicle, for instance, reacts very differently to changes in operating patterns than a typical European vehicle, or even a U.S. vehicle of three years ago. This is anticipated to be a fertile field for emissions research in the decade ahead.

References

Ahuja A. 1994. "Fume Check Starts in Puff of Smoke." *The Times*, Nov. 3, 1994, London.

Austin, T. and L. Sherwood. 1989. "Development of Improved Loaded-Mode Test Procedures for Inspection and Maintenance Programs." *SAE Paper* 891120, SAE International, Warrendale, Pennsylvania.

Austin, T.C., F.J. DiGenova, and T.R. Carlson. 1994. "Analysis of the Effectiveness and Cost-Effectiveness of Remote Sensing Devices." Report to the U.S. Environmental Protection Agency, Sierra Research, Inc., Sacramento, California.

Beaton, S.P., G.A. Bishop, and D.H. Stedman. 1992. "Emission Characteristics of Mexico City Vehicles," University of Denver, Colorado.

Beaton, S.P., G.A. Bishop, Y. Zhang, L.L. Ashbaugh, D.R. Lawson, and D.H. Stedman. 1995. "On-Road Vehicle Emissions; Regulations, Cost, and Benefits" *Science, 268: 991-93*.

Berg, W. 1989. "Strategies and Experience with Vehicle Emissions Inspection and Maintenance." Daimler-Benz AG, Stuttgart, Germany.

Bosch, Robert AG. 1986. *Automotive Handbook*. 2nd ed. Stuttgart, Germany.

California I/M Review Committee. 1993. "Evaluation of the California Smog Check Program and Recommendations for Program Improvements." Fourth Report of the California I/M Review Committee to the Legislature, February 16, 1993, California.

Chan, L. and M. Reale. 1994. "Emissions Factors Generated from MOBILE5a." Engine, Fuel, and Emissions Engineering, Inc., Sacramento, California.

Colorado. 1996. "Annual Report to the General Assembly of the State of Colorado on Vehicle Emission Inspection and Maintenance Programs." Colorado Department of Public Health and Environment, Denver, Colorado.

CONCAWE (Conservation of Clean Air and Water in Europe). 1992. *Motor Vehicle Emission Regulations and Fuel Specifications–1992 Update*. Report 2/92. Brussels.

_____. 1994. *Motor Vehicle Emission Regulations and Fuel Specifications–1994 Update*. Report 4/94. Brussels.

Escudero, J. 1991. "Notes on Air Pollution Issues in Santiago Metropolitan Region." Letter #168910467. Comision Especial de Descontaminacion de la Region Metropolitana, Santiago, Chile.

Faiz A., K. Sinha, M. Walsh, and A. Varma 1990. "Automotive Air Pollution–Issues and Options for Developing Countries." *Working Paper No. 492*, World Bank, Washington, D.C.

Glazer A., D.B. Kline, C. Lave. 1995. "Clean on Paper, Dirty on the Road: Troubles with California's Smog Check." *Journal of Transport Economics and Policy*, 29(1): 85–92.

Guensler, R. 1994. "Loop Holes for Air Pollution." *ITS Review* No. 1, Volume 18. University of California. Berkeley.

GVRD (Greater Vancouver Regional District, Province of British Columbia). 1994. "Heavy Duty Vehicle Emission Inspection and Maintenance Program Implementation in the Lower Fraser Valley." Environment Canada, Canada

Jacobs, P.E., D.J. Cheznica, and J.D. Kowalski. 1991. "California's Heavy-Duty Vehicle Smoke and Tampering Inspection Program." *SAE Paper* 911669, SAE International, Warrendale, Pennsylvania.

Joumard, R., L. Paturel, R. Vidon, J-P. Guitton, A-I. Saber, and E. Combet, 1990. *Emissions Unitaires de Polluants des Vehicules Legers*. Report 116 (2nd ed.). Institut National de Recherche sur les Transports et Leur Securite (INRETS), Bron, France.

Laurikko, J. 1994. In-use Vehicle Emissions Control in Finland: Introduction and Practical Experience. *The Science of the Total Environment*, 169: 195-204.

Martino, Wakim, and Westand. 1994. "Comparison of the HC and CO Emissions Estimates Using FTP and M240, and Remote Sensing." Paper presented at the 6th CRC On-Road Vehicle Emissions Workshop, March 16–18, San Diego, California.

McGregor, D., C. Weaver, L. Chan, and M. Reale. 1994. "Vehicle I/M Test Procedures and Standards." Report to the World Bank and the Royal Thai Ministry of Science, Technology and the Environment. Engine, Fuel, and Emissions Engineering, Inc., Sacramento, California (Draft).

OECD (Organization for Economic Cooperation and Development). 1989. "Control of Emissions from Vehicle in Use." ENV/AIR/88.11 (second revision), Paris.

Oil and Gas Journal. 1995. "Unocal Extends California Clean Air Campaign" *Comment. Vol. 93. No. 2 p.22.*

Onursal, B and S. Gautam. 1996. Vehicular Air Pollution: Experience from Seven LAC Urban Centers. A World Bank Study (Forthcoming), Washington, D.C.

Potter, C. and C. Savage. 1986. "An Inspection of the Variation of In-Service Vehicle Emissions and Fuel Economy With Age and Mileage Accumulation: Stage 1. Report." *LR562 (AP)*, Warren Spring Laboratory, Department of Trade and Industry.

Registry of Vehicles. 1993. "Measures Introduced to Control Vehicle Emissions in Singapore." Singapore.

Radian. 1994. "Audit results: Air Care I/M Program", Report prepared for British Columbia Ministry of Environment, Lands and Parks and B.C. Ministry of Transportation and Highways, Canada.

Rothe, V. 1990. "Motor Vehicle Inspection Effectiveness." Report Prepared for Arizona Department of Environment Air Quality, Tempe, Arizona.

Shen, S. and Huang, K. 1991. "Taiwan Air Pollution Control Program–Impact of and Control Strategies for Transportation-Induced Air Pollution." Bureau of Air Quality Protection and Noise Control, Environmental Protection Administration, Taiwan (China).

Stedman, D. and G. Bishop. 1991. "Evaluation of a Remote Sensor for Mobile Source CO Emissions." Report No. EPA 600/4-90/032, U.S. EPA Environmental Monitoring Systems Laboratory, Las Vegas, Nevada.

Tierney, G. 1991. "I/M Network Type: Effects on Emission Reductions, Cost, and Convenience." Report No. EPA-AA-TSS-I/M-89-2, U.S. Environmental Protection Agency, Washington, D.C.

University of Central Florida. 1991. "Air Quality in Urban Transportation." Course Notes, International Road Federation, Washington, D.C.

U.S. Congress/Office of Technology Assessment (OTA). 1992. *Retiring Old Cars: Program to Save Gasoline and Reduce Emissions.* OTA-E-536, U.S. Government Printing Office, Washington, D.C.

U.S. EPA. 1992. "Inspection and Maintenance Program Requirements", Final Rule. Federal Register, Part VII, Volume 57, No. 215, United States Environmental Protection Agency, Washington, D.C.

_____. 1993. "High-Tech I/M Test Procedures, Emission Standards, Quality Control Requirements and Equipment Specifications – Revised Technical Guidance."

Report No. EPA-AA-EPSD-IM-93-1, United States Environmental Protection Agency, Washington, D.C.

_____. 1995. "EPA I/M Briefing Book: Everything You Ever Wanted to Know About Inspection and Maintenance", Report No. EPA-AA-EPSD-IM-94-1226, United States Environmental Protection Agency, Office of Air and Radiation Washington, D.C.

Walsh, M. and J. Karlsson. 1990. "Motor Vehicle Pollution Control in Asia: The Lessons of Europe." *SAE Technical Paper* 900613. Presented at the SAE International Congress and Exposition, Detroit, MI., February 26-March 2.

Walsh, M.P., R.F. Klausmeier, and J.H. Seinfeld. 1993. "Report of the Peer Review Panel, Fourth Annual Report on the California Inspection and Maintenance Program", Memorandum to J.H. Presley, Chairman, California Senate Special Committee on Vehicle Emissions and Vehicle Inspection and Maintenance, Jananne Sharpless, Chairwoman, California Air Resources Board.

Walsh, M. 1994. "Technical Notes." 3105 N. Dinwiddie St, Arlington, VA 22207, mimeo.

_____. 1995. "Technical Notes." 3105 N. Dinwiddie Street, Arlington, VA 22207, mimeo.

Weaver, C.S. and A. Burnette. 1994. "Organizational Requirements, Planning and Budget for Quality Assurance and Quality Control for the Privatized Vehicle Inspection and Maintenance Program", Report to the Royal Thai Department of Land Transport and the World Bank, Radian Corporation, Texas.

Weaver, C. 1991. "Smart Vehicle Contributions to Air Quality Attainment." Report to the U.S. EPA under contract 68-W8-0113. Engine, Fuel, and Emissions Engineering, Inc., Sacramento, California.

Weaver, C. and R. Klausmeier. 1988. Heavy-Duty Diesel Vehicle Inspection and Maintenance Study. (4 volumes). Report to the California Air Resources Board. Radian Corporation, Sacramento, California.

Weaver, C.S. and L.M. Chan. 1994. "Analysis of CARB ASM and IM240 Data", Memorandum, Engine, Fuel, and Emissions Engineering, Inc., Sacramento, California.

_____. 1995. "Company Archives" Multiple Sources, Engine, Fuel and Emissions Engineering, Inc., Sacramento, California.

Zhang, Y., D.H. Stedman, G.A. Bishop, P.L. Guenthes, and S.P. Beaton. 1994. "Worldwide On-Road Vehicle Exhaust Emissions Study by Remote Sensing." Department of Chemistry, University of Denver, Colorado.

Appendix 4.1

Remote Sensing of Vehicle Emissions: Operating Principles, Capabilities, and Limitations

Operating Principles

The technology for remote sensing of vehicle exhausts is based on non-dispersive infra-red (NDIR) spectroscopy. It measures the carbon monoxide to carbon dioxide ratio (CO/CO_2) and the hydrocarbon to carbon dioxide ration (HC/CO_2) in the exhaust of any on-road vehicle. The remote sensing system consists of an infra-red (IR) source that emits a horizontal beam of radiation across a single traffic lane, approximately 25 cm above the road surface, into a receiver on the opposite side. The beam is split in the receiver into four channels having wavelength specific detectors for CO, CO_2, HC, and a reference signal. The IR absorption caused by CO, HC, and CO_2 in the exhaust plume is determined using separate band pass optical filters centered at 4.6, µm, 3.4 µm, and 4.3 µm, respectively, and compared to a reference absorption at 3.9 µm (that does not absorb vehicle exhaust gases), to eliminate any effects related to source fluctuations or dust and smoke behind the vehicle. The data from all four channels are fed to a computer which converts the radiation absorbed by CO, CO_2, and HC into CO/CO_2 and HC/CO_2 ratios. A video camera system is interfaced for the purposes of vehicle identification. On-site calibration is performed daily with a certified gas cylinder containing known concentration ratios of CO, CO_2, propane (for HC), and nitrogen (Zhang and others 1994). The details of the instrumentation have been described elsewhere (Guenther and others 1991). A simple way to conceptualize the remote sensing system is to imagine a typical garage-type NDIR instrument in which the separation of the IR source and detector is increased from 10 cm to several (6-12) meters. Instead of exhaust gas passing through a flow cell, a car now drives between the source and the detector.

As a vehicle passes through the infra-red beam, the remote sensing device measures the ratio of CO (and exhaust HC) to CO_2 in front of the vehicle and in the exhaust plume behind. The "before" measurement is used as a base and the vehicle's CO emission rate is calculated by comparing the "behind" measurement with the expected CO to CO_2 ratio for ideal combustion. Exhaust HC is calculated in a similar manner by comparing the total carbon content of exhaust HC, CO, and CO_2 to the carbon content of the fuel burned by the vehicle (U.S. EPA 1995).

As the effective plume path length and the amount of plume observed depend on air turbulence and wind, the sensor can only directly measure ratios of CO or HC to CO_2. These ratios, termed Q for CO/CO_2 and Q' for HC/CO_2, are useful parameters to describe the combustion system. Using fundamental principles of combustion chemistry, it is possible to determine many parameters of the engine and the emissions control system including the instantaneous air-fuel ratio, the grams of CO or HC emitted per unit of gasoline burned and the volume percent CO and HC in the exhaust gas. Most vehicle equipped with advanced emission control systems will have near-zero Q and Q' since they emit little CO or HC. To obtain a Q greater than near-zero, the engine must operate on the fuel-rich side of stoichiometry and the emission control system, if present, must not be fully operational. A high Q' can be associated with either fuel-rich or fuel-clean air-fuel ratios combined with a missing or malfunctioning emission control system. A lean air fuel ratio, while impairing driveability, does not produce CO in the engine. If the air-fuel ratio is lean enough to induce misfire then a large amount of unburned fuel in the form of hydrocarbons (HC) is present in the exhaust manifold. If the catalyst is absent or non-functional, then high HC will be observed in the exhaust without the presence of high CO. To the extent that the exhaust system of the misfiring vehicle contains some residual catalytic activity, the HC may be partially or completely converted to a $CO-CO_2$ mixture.

The remote sensor is effective across traffic lanes up to 15 meters in width. It can be operated across double lanes of traffic with additional video hardware;

however, the normal operating mode is for single lane traffic. It operates most effectively on dry pavement, as rain, snow, and very wet pavement scatter the IR beam. These interferences can increase the frequency of invalid readings, ultimately to the point that all data are rejected as being contaminated by too much "noise."

Capabilities and Limitations of Remote Sensing

Tailpipe CO and HC Emission

Remote sensing can provide reliable measurements of on-road CO and HC emissions. In a seminal experiment in Los Angeles (Lawson and others 1990) to test the capability of remote sensing to measure tailpipe CO emissions, the following relationship was obtained by regressing percent tailpipe CO against percent remote sensor CO:

Tailpipe %CO = 1.03 [Remote sensor % CO] + 0.08   (A4.1.1)

with a correlation coefficient of 0.97 for a sample size of 34. This experiment confirmed the ability of remote sensing to measure instantaneous CO tailpipe values at different vehicle speeds. It was observed that remote sensing CO measurements were generally higher than no-load idle CO measurements obtained during roadside smog checks, as shown by the following relationship, with a correlation coefficient of 0.67:

Remote sensor %CO = 0.73 [Roadside idle %CO] + 2.51   (A4.1.2)

In a follow-on study (University of Denver 1994), the roadside measurements were compared in a blind test to those measured by a vehicle equipped with a tailpipe probe, trunk mounted CO and HC monitors, and computer control of the vehicle's air-fuel ratio. Compared to the vehicle with known emissions, the remote sensing measurements were found to be accurate within $\pm$ 5 percent for CO and $\pm$ 15 percent for HC. In an experiment involving remote sensing of 58,063 individual vehicles in Los Angeles in 1991, a sample of 307 vehicles identified as high emitters was subjected to a more detailed inspection—41 percent of these vehicles had emission control equipment that had definitely been tampered with, and an additional 25 percent had defective equipment, although not necessarily the result of deliberate tampering. Furthermore, 85 percent of these high emitters failed the tailpipe test and overall 92 percent failed the roadside inspection. Of the 25 vehicles which were identified by remote sensing as high emitters but passed the roadside inspection, four subsequently failed the IM240 test (indicating they were correctly identified as high emitters) and 10 were in cold start mode. Of the

74 vehicles which failed the remote sensing test and were subsequently given an IM240 test, 23 emitted more than 100 grams of CO per mile; 6 emitted more than 20 grams of HC per mile; and 3 emitted more than 10 grams of NO_x per mile.

These pilot studies and several other experiments conducted worldwide (Zhang and others 1994) have established remote sensing as a reasonably reliable and low-cost technology—about U.S.$0.50 a test—for rapid assessment of CO and HC exhaust emissions and for the detection of gross emitters for more detailed evaluation and testing.

Instantaneous Versus Drive Cycle Emissions

Remote sensing has the advantage of being able to test large numbers of vehicles quickly and relatively cheaply. One of the greatest strengths of remote sensing technology may also be its intrinsic weakness, i.e. the ability to take an instantaneous reading or snapshot of what is coming out of a vehicle's exhaust pipe. This can be a liability because a given vehicle has widely variable instantaneous emission rates depending on how it is being operated at any given moment, whether the engine is cold or warm, and whether the catalyst is warmed up, among other things.

HC and CO emissions in the exhaust manifold are a function of the air-fuel ratio at which the engine is operating. These "engine-out" emissions are altered by tailpipe emission controls.

CO emissions are caused by the lack of sufficient air for complete combustion. If the air-fuel mix is uniform, CO is formed uniformly throughout the volume of the combustion chamber.

For HC the situation is more complex. In the main part of the combustion chamber, away from the walls, essentially all the HC is burned; however, the flame front initiated by the spark plug cannot propagate within about one millimeter of the relatively cold cylinder walls. This phenomenon causes a "quench layer", a thin layer of unburned fuel, next to the walls and in the cylinder orifices. Upon opening the exhaust valve, the rising piston scrapes this quench layer containing unburned HC and excess HC is emitted. Unburned HC emissions increase significantly if the vehicle is operating with a rich mixture. There is a second peak in HC emissions which occurs in the fuel-lean zone. This phenomenon is known as "lean burn misfire" or "lean miss"; it is the cause of hesitation experienced at idle before a cold vehicle has fully warmed up. When this misfiring occurs a whole cylinder full of unburned air-fuel mix is discharged into the exhaust manifold. Misfiring also occurs if a spark plug lead is missing or if the ignition system in one cylinder is otherwise fatally compromised. Severe fuel economy losses occur when significant misfiring is taking place.

The fact that there are two regions of high HC and only one of high CO indicates that one would not expect a high correlation between HC and CO exhaust emissions measured by remote sensing. High HC would be expected for some very low CO vehicles as well as for high CO vehicles. One would not expect to see many very low HC readings in the presence of high CO. This conclusion is confounded however, by the presence of catalytic converters in the exhaust system. If a vehicle running with a rich mixture has a functioning air injection system and catalyst then both the HC and CO will be removed. If the catalyst is functioning, but there is no air injection, then some or all of the HC will be converted to CO. In this case, the CO will remain in the exhaust since there is inadequate oxygen for its oxidation. Similarly, it is possible for a catalyst-equipped vehicle operating in the lean burn misfire region to emit CO into the air, even though it was not emitting CO into its own exhaust manifold.

To deal with this wide range of emissions, the US Federal Test Procedure (FTP) attempts to simulate emissions from a vehicle under both cold and warm conditions, with engine operation ranging from idling to accelerating and steady-state driving at various speeds. The emission measurements are averaged to obtain a composite emissions rate. Even this test, which typically costs more than U.S.$1000 to run and takes almost a full day, has been criticized as inadequately simulating the full range of important driving conditions, such as hard acceleration, high speeds, heavy loads, short trips that begin and end before the engine has warmed fully. Normal, clean vehicles operating over the FTP cycle will occasionally have high CO, HC and NO_x emissions, albeit not as frequently or as high as dirty vehicles; conversely, very dirty vehicles driven over the FTP cycle will occasionally have low emissions, although not as frequently as clean vehicles.

A short test procedure, no matter how accurate its readings, will only be able to discriminate between high and low emitters if it tests vehicles under conditions sufficiently broad to identify overall emissions. For a long time, it was believed that a simple idle test was a good indicator of emissions because in typical stop and go city driving, average speeds were quite low and for a large proportion of the time the engine was idling or the idle jets on the carburetor were dominating the air-fuel mixture—an important determinant of engine-out emissions from precontrolled cars. It is now known that while the idle test does a fairly good job of identifying vehicles with excessive HC and CO emissions as determined by the FTP, it is possible to lower the idle emissions of a vehicle without significantly impacting FTP emissions. Also the idle test cannot measure NO_x emissions or evaporative emissions.

In a study by Martino and others (1994), it was noted that the correlation between the FTP and remote sensing is widely variable (highly dependent upon the driving mode at the time that remote sensing was carried out), and frequently quite low. For example, at light and heavy accelerations, the correlations for CO were 0.05 and 0.08 respectively; for HC under the same conditions, the correlations were 0.23 and 0.19, respectively. For idle conditions, the CO correlation between remote sensing and the FTP was quite good at 0.86 and it was even better for a 30 miles per hour cruise, 0.93. For HC, however, under the same conditions the correlations were poorer, 0.21 and 0.68, respectively.

This correlation data should be interpreted cautiously, however, because they are based on a relatively small sample of vehicles. While recognizing this limitation, the results are not very encouraging in terms of the ability of remote sensing to accurately estimate a vehicle's rate of emissions unless the driving conditions can be carefully controlled.

Recognizing the good correlation between the IM240 and the FTP, other researchers have been attempting to predict IM240 emissions using one or more remote sensing measurements; these results have also been mixed (General Motors and others 1994). The criteria used in this study was the ability of remote sensing to predict whether a vehicle was a high emitter or not[1]. For CO, a high emitter was defined as any vehicle emitting more than 20 grams per mile (gpm) on the IM240; 3 percent CO was the corresponding remote sensing cut point. For HC, a high emitter was defined as any vehicle emitting above 1 gpm on the IM240; the corresponding remote sensing cut point was 0.1 percent. It was found that the remote sensors seemed to be quite unreliable, if only one remote sensing reading was used for an individual vehicle. For example, for CO, 9 out of the 24 "clean" vehicles were inappropriately failed, an error of commission rate of 38 percent while 2 out of 20 "dirty" vehicles were passed inappropriately, an error of omission rate of 10 percent. An examination of the data showed that if the remote sensing cut points were raised to reduce the errors of commission, the proportion of dirty vehicles missed by the test would increase significantly. For hydrocarbons, the errors of omission were quite high (54 percent), while the errors of commission were very low, 4 percent. If the number of remote sensing tests on each vehicle is increased to three or more, the performance improves. For CO, the errors of commission dropped to 3 percent but the errors of omission increased to 33 percent. For HC, the errors of commission remained low at 3 percent but the errors of omission increased to 75 percent.

---

1.   If a truly clean vehicle is predicted to be a high emitter, this would be an error of commission. If a truly dirty vehicles is predicted to be clean, this would be an error of omission.

Tailpipe NO_x Emissions

Remote sensing currently is not effective in identifying vehicles emitting excessive quantities of oxides of nitrogen (NO_x). However, appropriate devices have been developed and are in the demonstration stage; it is expected that NO_x identification will be possible in the future. The remote sensing system for NO_x will use either a beam of ultraviolet light, or light from a turnable diode laser (U.S. EPA 1995). A NO_x unit has been constructed and tested in Denver (Colorado), Dearborn (Michigan), and El Paso (Texas). The CO to CO_2 ratio determined by IR remote sensing systems would still be needed to calculate NO_x emissions.

Evaporative Hydrocarbon Emissions

Remote sensing is unable to identify vehicles with excessive quantities of evaporative emissions. This technique cannot be used for under the hood inspection to test for evaporative and other non-tailpipe emissions. Fuel evaporation is a significant source of HC emissions that can exceed tailpipe emissions from controlled cars in hot climates.

Conclusions

It is clear that remote sensing offers the advantage of testing emissions from a large numbers of vehicles quickly and relatively cheaply. However, the critical issue is how well it can be used to test a broad representative cross-section of the overall vehicle population for all pollutants on a routine basis without excessively disrupting the free movement of goods and people. There are several logistical and technical concerns in this respect. In its current use, remote sensing requires that vehicles pass the sensor in single file. It is not practical to constrict heavily traveled roadways to a single lane during rush hours. Concepts of how remote sensors could be used in multiple-lane traffic situations have yet to be tested. Placement on single-lane on and off ramps is a possibility, but this raises concerns over the proportion of the vehicle population that could be covered by such a scheme. Locating remote sensors in risky areas is also a concern from a security standpoint. The cost of creating a remote sensing network would not be small. Many permanent stations would be needed for adequate coverage and they would have to operate over extended time periods.

Remote sensing test results appear to be poorly correlated with average emissions during stop-and-go driving as measured by the IM240 test. The proportion of vehicles that are falsely failed by remote sensing tend to be high and could lead to unnecessary repairs and cause inconvenience to car owners. Unless fairly stringent cutpoints are used, only a small portion of gross emitters are likely to be detected by remote sensing. However, use of a "high emitter profile" could rectify this problem as well as reduce false failures. Furthermore,

remote sensing technology is incapable of detecting even the most common forms of tampering.

In addition to concerns over identifying excess emitters, there are concerns about repairing "dirty" vehicles detected by remote sensing tests. Repairs could be ineffective unless a good confirmatory test such as the IM240 is performed after the repair.

Despite its limitations, remote sensing can play a very useful role in identifying gross polluters for further testing and in better targeting I/M programs, by reducing the need for testing the entire vehicle fleet at regular intervals. Remote sensing is a powerful data gathering tool and offers unique opportunities for future emission reductions. Its role as an adjunct to periodic I/M program could be highly cost effective. The relatively low-cost of acquiring and operating remote sensing technology makes it particularly attractive for use in developing countries.

References

Guenther, P.L., D.H. Stedman, J. Hannigan, J. Bean, and R. Quine. 1991. "Remote Sensing of Automobile Exhaust." *American Petroleum Institute Publication No. 4538,* Washington, D.C.

General Motors, Ford, Chrysler, and Michigan Department of State. 1994. "Real-World Emissions Variability as Measured by Remote Sensors." Paper presented at the 4th CRC On-road Vehicle Emissions Workshop, March 16-18, 1994, San Diego, California.

Lawson, D.R., P. Groblicki, D.H. Stedman, G.A. Bishop, and P. Guenther. 1990. "Emissions from In-Use Motor Vehicles in Los Angeles: A Pilot Study of Remote Sensing and Inspection and Maintenance Program." *Journal of the Air and Waster Management Association Vol. 40. No.8*

Martino, Wakim, and Welstand. 1994. "Comparison of HC and CO Emissions Estimates Using FTP, IM240, and Remote Sensing." Paper presented at the 4th CRC On-Road Vehicle Emissions Workshop, March 16-18 1994, San Diego, California

University of Denver. 1994. "On-Road Remote Sensing of CO and HC Emissions in California". Final Report, Contract No. A032-093, Research Division, California Air Resources Board, Los Angeles.

U.S. EPA. 1995. "EPA I/M Briefing Book: Everything You Ever Wanted to Know About Inspection and Maintenance", Report No. EPA-AA-EPSD-IM-96-1226, United States Environmental Protection Agency, Office of Air & Radiation, Washington, D.C.

Zhang, Y., D.H. Stedman, G.A. Bishop, P.L. Guenther, and S.P. Beaton. 1994. "Worldwide On Road Vehicle Exhaust Emissions Study by Remote Sensing." Department of Chemistry, University of Denver, Colorado.

5

Fuel Options For Controlling Emissions

Fuel composition and characteristics play an important role in engine design and emissions performance. Changes in the composition and properties of gasoline and diesel fuel can affect vehicle emissions significantly, although the relationships among fuel properties, engine technologies and exhaust emissions are complex (ACEA/EUROPIA 1994; 1995a). Changes in one fuel characteristic may lower emissions of one pollutant but may increase those of another (for example, decreasing aromatics content in gasolines lowers CO and HC emissions but increases NO_x emissions). In some instances, engines in different vehicles classes respond very differently to changes in fuel properties (e.g. increasing the cetane number in diesel fuels lowers NO_x emissions for both light and heavy duty DI engines, but not for light duty IDI engines). In response to environmental concerns, conventional automotive fuels have undergone substantial modifications in recent decades and are expected to be improved further. Fuel reformulation is an important element of air quality programs in many jurisdictions. The potential for reduction in pollutant emissions through cost-effective fuel modifications is typically in the order of 10 to 30 percent—less than what can be achieved through vehicle emission controls, but still significant. A major advantage of fuel modifications for emissions control is that they often can take effect quickly and begin reducing pollutant emissions immediately, whereas vehicle emission controls generally must be phased in with the turnover in the vehicle fleet. Another advantage of fuel modifications is that they can be targeted geographically or seasonally by requiring the more expensive "clean" fuels only in highly polluted areas or during seasons with a high incidence of elevated pollution episodes. In addition, fuel modifications are usually easier to enforce, since fuel refining and distribution systems are highly centralized.

Major efforts have been made worldwide to gradually remove lead from gasoline, both to reduce lead emissions and to make possible the use of vehicle emission control technologies such as catalytic converters. Possible further changes to reduce emissions from gasoline include reduced volatility, increased oxygen content, reduced aromatics, and more widespread use of detergent additives. Conventional diesel fuel also can be improved by reducing the sulfur and aromatic content and by using detergent additives.

The substitution of cleaner-burning alternative fuels for conventional gasoline and diesel fuel has attracted great attention during the past two decades. Motivations for this substitution include conservation of oil products and energy security, as well as the reduction of gaseous and particulate emissions and visible smoke. Care is needed in evaluating the air quality claims for alternative fuels, however. While many alternative fuels do reduce emissions considerably, this is not always the case. In some cases CO, HC, NO_x, and aldehyde emissions may increase with the use of alternative fuels.

Alternative fuels commonly considered for vehicular use are natural gas (in compressed or liquefied form), liquefied petroleum gas (LPG), methanol (made from natural gas, coal, or biomass), ethanol (made from grain or sugar), vegetable oils, hydrogen, synthetic liquid fuels derived from the hydrogenation of coal, and various blends such as gasohol. Although the potential benefits of many of these fuels have been greatly overstated, others (notably natural gas and LPG) do have significant emission advantages and may have significant economic advantages as well. Hydrogen, although potentially a very clean fuel, is difficult to store and requires large amounts of energy to produce. Electricity (although not strictly a fuel) can reduce pollutant emissions dramatically if used for vehicle propulsion, although this reduction often comes at a high cost, both financially and in flexibility of operation. Although the pollutant emissions produced in generating electricity should not be ignored, these are generally less than those produced by internal combustion engines in road vehicles.

Environmental assessment of alternative fuels should not be based solely on vehicle end-use emission characteristics but should account for pollutant emissions associated with the production, storage, and distribution of these fuels. This is especially true for assessment of emissions contributing to global warming.

Gasoline

Gasoline is a mixture of 200 to 300 hydrocarbons that evaporates between ambient temperature and 200°C (Celsius). The mixture is produced by the distillation and chemical processing of crude oil; in addition, various chemicals and blending agents are added to improve its properties as a motor fuel (box 5.1). Gasoline characteristics having an important effect on engines and emissions include its knock resistance rating or octane number, volatility characteristics, and chemical composition. The use of tetra-ethyl lead as a gasoline additive is especially important because of its implications for catalytic converters, gasoline composition, antiknock performance, valve wear, engine life, and lead emissions. Worldwide, motor vehicles consume over 2 billion liters of gasoline every day, about two-thirds of which is unleaded.

Lead and Octane Number

A fuel's octane number is a measure of its resistance to detonation and "knocking" in a spark-ignition engine. Knock reduces engine power output, and severe or prolonged knock damages the engine. The tendency to knock increases with increasing engine compression ratio. Higher octane fuels are more resistant to knocking and thus can be used in engines with higher compression ratios. This is desirable because higher compression ratios improve thermodynamic efficiency and power output.

Typically, raising the compression ratio of an engine from 7.5 to 9 increases the octane requirement by about 10 units (numbers) and results in a 10 percent increase in engine efficiency (Armstrong and Wilbraham 1995). The engine octane requirement is also influ-

---

Box 5.1 Gasoline Blending Components

Gasoline is produced by blending the output from several refinery processes, each containing different hydrocarbons. The properties of gasoline can be modified by changing the proportions of the blending components, which include:

- Butane—a gas often sold as liquefied petroleum gas (LPG), which has a high octane number and can be used as motor fuel. Butane is dissolved in other gasoline components and provides volatility for cold starts.
- Straight-run gasoline—from the primary distillation of crude oil; has a low octane number.
- Isomerate—produced by converting straight-chain paraffins to branched-chain paraffins by isomerising straight-run gasoline fractions. Provides clean-burning, higher-octane quality.
- Reformate—from the catalytic reforming of n-alkanes into iso-alkanes, and cyclo-alkanes into aromatics. The process, known as platforming when the catalyst contains platinum, produces high-octane gasoline, but the yields fall as the octane number of the product increases. Catalytic reforming is the most important process for producing gasoline; a typical summer-grade premium gasoline contains about 70 percent reformate.
- Catcracked gasoline—from catalytic cracking of residual oil containing large molecules with boiling points above 370°C to form unsaturated olefins and branched molecules with high octane rating.
- Hydrocracked gasoline—from the hydrocracker which breaks up large molecules but also adds hydrogen. Contains mainly low-octane alkanes and is unsuitable for gasoline but is a good feedstock for the catalytic reformer.
- Polymerate—produced by reacting light alkenes (olefins such as propene and butene) to give heavier olefins, which have a good octane quality and do not increase the vapor pressure of gasoline.
- Alkylate—from processes where light iso-alkanes from primary distillation, such as isobutane, are coupled with light alkenes (olefins such as butene) from the catcracker to make high-octane components.
- Oxygenates—certain alcohols and ethers containing carbon, hydrogen, and generally one oxygen atom. Can be used as octane boosters or gasoline extenders.

Conventional thermal cracking, which requires heating the feed to 500°C at pressure of up to 70 bars, is now virtually obsolete for producing motor gasoline. Breaking and coking are two forms of thermal cracking used to convert marginal fuel oils and other residual products into more profitable products such as gasoline, naphtha, and gas oil.

Source: Owen and Coley 1990; Shell 1989

enced by environmental and engine operating characteristics, such as:

- An increase in humidity from 20 to 90 percent (at 20°C) decreases octane demand by about 5 units (numbers).
- An increase in altitude from sea level to 1,000 m reduces octane demand by about 10 units.
- An increase in engine spark advance by 4 degrees can increase octane requirement by about 5 units.
- An increase in engine cooling jacket temperature from 60°C to 90°C can increase octane requirement by 6-7 units.
- Engine deposit accumulation at 10,000 miles can increase the octane requirement by as much as 10 units (Eager 1995).

Engines designed for use with high-octane fuels can produce more power and have lower fuel consumption than engines designed for lower-octane fuels. For a given engine design, however, there is no advantage in using a higher octane fuel than the engine requires. Higher-octane "premium" gasoline—unless it is required to avoid knocking—is not necessarily of higher quality and provides no benefit in most cases. Certain high-performance cars are an exception to this rule; they are designed for high-octane gasoline but are equipped with knock sensors that permit operation on regular gasoline by retarding the ignition timing if knock is detected. Such vehicles exhibit better performance and fuel economy if operated on the high-octane gasoline for which they are designed.

Octane number is measured by the research and motor octane tests. The results of these tests are expressed as either the research octane number (RON) or the motor octane number (MON) of the fuel. Both tests involve comparing the antiknock performance of the fuel to that of a mixture of two primary reference fuels, iso-octane (octane rating of 100) and n-heptane (octane rating of 0), with the octane number defined as the percentage of iso-octane in the reference octane-heptane mixture that gives the same antiknock performance as the fuel being tested. For fuels with octane numbers above 100, mixtures of iso-octane and tetra-ethyl lead are used to extend the octane scale to 130. The research and motor tests differ in detail: the research test reflects primarily low-speed, mild driving, while the motor test reflects high-speed, intensive driving. Most fuels have a higher RON than MON, and the difference between these two values is known as the sensitivity. In the United States and parts of Latin America gasoline antiknock ratings are combined in an Anti Knock Index (AKI), which is expressed as the average of RON and MON, denoted by $(R + M)/2$. Elsewhere it is common to use RON as the antiknock rating.

In order to ensure that vehicle fuels have octane quality compatible with end-user expectations, the oil and automotive industries have cooperated to define standards for automotive gasoline. The octane value selected as a standard represents a compromise between the greater engine power and efficiency achievable with a higher compression ratio (calling for higher-octane fuel) and the greater costs and energy consumption involved in producing higher octane. In the United States, this compromise resulted in a standard value of 87 for $(R+M)/2$ for unleaded regular gasoline, equivalent to about 91 RON. In Europe, where high fuel taxes make engine efficiency important, the standard is 95 RON.

Tetra-ethyl lead (TEL) and tetra-methyl lead (TML) as fuel additives have been used to reduce the knocking tendencies of gasoline since 1922. A typical lead content of 0.4 gram per liter (g/l) boosts a gasoline's RON by about six units. In the days before advanced refining technology, the antiknock properties imparted to gasoline by tetra-ethyl lead made possible the development of efficient, high-compression gasoline engines. With the development of advanced refining technologies, lead compounds are no longer needed to achieve high octane levels, although lead is still by far the cheapest means of producing high-octane gasoline. Based on typical experience in the United States, Australia, Europe, and elsewhere, the added refining cost of producing gasoline of adequate octane quality without lead is about 0.5 to 3.0 U.S. cents per liter, or equivalent to 2-10 percent of the gasoline price before tax. The cost in developing countries with simpler hydroskimming refineries could be somewhat higher because of the larger capital investment required (Jamal 1995).

The past fifteen years have seen a substantial reduction in the use of lead in gasoline as the adverse health impacts of lead have become better known. Lead deposits poison catalytic converters and exhaust oxygen sensors—a single tankful of leaded fuel is enough to reduce converter efficiency significantly. Lead in the environment is poisonous to humans, and lead antiknocks in gasoline are by far the largest source of lead aerosol—among the most toxic air pollutants in many developing countries. An impressive array of evidence has linked even subacute blood lead levels with a variety of ills, including reduced mental capacity in children and high blood pressure in adults (Schwartz and others 1985). The U.S. Surgeon General has lowered the allowable level of lead in blood, above which medical action is recommended, from 25 mg/deciliter to 10 mg/deciliter—a level comparable to the average blood lead level among urban dwellers in many developing countries.

The demonstrated link between lead use in gasoline and average blood lead levels in humans caused the U.S. Environmental Protection Agency (EPA) in 1985 to limit the lead content of gasoline in the United States to 0.1 gram per gallon, or about 1/30th the typical lead concentration before control. In 1990 the U.S. Congress prohibited the addition of lead to gasoline effective January 1, 1995. Canada banned the use of leaded gasoline in road vehicles in 1990, Colo1mbia and Guatemala in 1991, and Austria (the first European country) in 1993. The Republic of Korea and Sweden eliminated lead in gasoline in 1994. Slovakia and Thailand banned leaded gasoline in 1995. Antigua and Suriname reportedly have banned the use of leaded gasoline as well. Brazil and Japan eliminated leaded gasoline in the 1980s. Nearly all European countries and many of the more industrialized countries of Asia have limited the lead content of gasoline to 0.15 g/l (which yields a 2-3 unit boost in a gasoline's RON). Many European countries have either banned the use of leaded regular gasoline (while continuing to allow more-expensive leaded premium gasoline for high-performance vehicles) or driven it out of the market through tax incentives. Unleaded gasoline is gradually starting to penetrate developing country markets. Unleaded gasoline is marketed in nearly all the countries in Latin America and the Caribbean, and in Eastern Europe. Regular unleaded gasoline is also available in most countries in East Asia and has been introduced in the large metropolitan cities in India. Worldwide, the unleaded share of the gasoline market increased from about 50 percent in 1991 to over 67 percent in 1996. A survey of the international use of lead in gasoline is presented in appendix 5.1. The typical lead content of gasoline sold in Europe, East Asia and the Pacific varies from 0.15 to 0.45 g/l; in Latin America and the Caribbean from 0.1 to 0.84 g/l; in South Asia, Middle East, and North Africa from 0.2 to 0.82 g/l; and in Sub-Saharan Africa from 0.4 to 0.84 g/l (NRDC and CAPE 21 1994). As of 1994, unleaded gasoline was not marketed in any Sub-Saharan African country.

In addition to its effects on fuel octane level, lead in gasoline has many other effects in the engine (Weaver 1986). It serves as a lubricant between exhaust valves and their seats made from soft metals in some older model cars, thus helping to prevent excessive wear, known as valve seat recession. This protective function is the reason the U.S. EPA limited lead content to 0.1 gram per gallon rather than banning it outright and that European specifications require a minimum of 0.15 g/l (box 5.2).

The danger of valve seat recession due to unleaded gasoline use has been overstated in many cases by lead additive companies seeking to protect their markets. Valve seat recession is only likely to occur when engines with soft cast-iron valve seats are driven for long periods at very high speed, as on an Autobahn or a U.S. Interstate Highway. Few developing countries have highway networks that would permit such driving. In the United States, the experience of numerous fleets that converted to unleaded gasoline in the early 1970s shows that the incidence of valve-seat recession was very low, occurring in only a small proportion of vehicles (Weaver 1986). Where valve seat recession does occur, it can be remedied, and recurrence prevented by machining out the valve seats and replacing them with hardened inserts. The cost of this operation is typically a few hundred dollars. Protection may also be provided to vulnerable vehicles by using potassium or sodium-based valve seat protection additives in gasoline. Such additives have been used in several countries where the phase-out period for leaded gasoline has been short (Armstrong and Wilbraham 1995).

Although they protect soft valve seats from recession, lead deposits have a number of negative effects.

Box 5.2 Low-Lead Gasoline as a Transitional Measure

Vehicles equipped with catalytic converters require unleaded gasoline to prevent the catalyst from being poisoned by lead deposits. Vehicles without catalytic converters do not require unleaded gasoline but generally can use it without harm. Even engines without hardened valve seats can be run on unleaded fuel provided that leaded fuel is used every fourth or fifth tank fill; in addition, some adjustment of engine timing may be needed. Reducing or eliminating gasoline lead is desirable for public health reasons, but in the short term (three to five years) reductions in lead use may be constrained by refining capacity. Some of the octane shortfall may be recovered by importing high-octane blending components such as methyl tertiary-butyl ether (MTBE), ethanol, or high-octane hydrocarbon blendstocks, or by importing unleaded gasoline.

The octane boost due to lead does not increase linearly with lead concentration. The first 0.1 g/liter of lead additive provides the largest octane boost, with subsequent increases in lead concentration giving progressively smaller returns. Less lead is required to produce two units of low-lead gasoline as compared to one unit of high-lead and one unit of unleaded gasoline of the same octane value. Thus if octane capacity is limited, the quickest and most economical way to reduce lead emissions generally is to reduce the lead content of all leaded gasoline grades as much as possible, rather than to encourage the owners of non-catalyst cars to use unleaded fuel. This also helps to reserve supplies of unleaded gasoline (which may be difficult to produce and distribute in sufficient quantities) for catalyst-equipped vehicles. Decreasing the lead content in leaded gasoline also reduces the difference in refining cost between leaded and unleaded gasoline. If this is reflected in retail prices, it will reduce the temptation for owners of catalyst-equipped vehicles to misfuel with leaded gasoline.

Source: Weaver 1995

These include corroding exhaust valve materials, fouling spark plugs, and increasing hydrocarbon emissions. To prevent the buildup of excessive deposits, tetra-ethyl lead is used in combination with lead "scavengers" (ethylene dichloride and ethylene dibromide). These scavengers form corrosive acids upon combustion and are major contributors to engine and exhaust system rusting, corrosive wear, and reduced oil life. On balance, the negative maintenance effects of lead and its scavengers outweigh the positive effects, since the use of unleaded gasoline typically reduces maintenance costs. Switching from leaded to unleaded gasoline may increase engine life by as much as 150 percent. Thus there is little technical argument for retaining lead in gasoline if the refining capacity exists to provide the required octane in some other manner.

The cost savings associated with maintenance reductions from unleaded gasoline could be significant. A Canadian analysis estimated savings—from spark plug changes every other year with unleaded gasoline instead of every year with leaded gasoline, oil and filter change once a year instead of twice a year, muffler replacements once instead of twice in five years, and no replacement of exhaust pipes compared with every five years with leaded gasoline—at Canadian 2.4 cents per liter, for 16,000 km yearly driving. Oil industry estimates of maintenance savings ranged from Canadian 0.3 to 2.1 cents per liter, all expressed in terms of 1980 Canadian dollars (OECD 1988).

Refiners have used a number of techniques to replace the octane value formerly contributed by lead additives (see box 5.1). Catalytic cracking and reforming are used to increase the concentrations of high-octane hydrocarbons such as benzene, toluene, xylene, and other aromatic species, and olefins. Alkylation and isomerization are used to convert straight-chain paraffins (which have relatively low octane) to higher-octane branched paraffins. In the past increased quantities of light hydrocarbons such as butane also were blended. Use of oxygenated blending agents such as ethanol, methanol (with cosolvent alcohols), and especially methyl tertiary-butyl ether (MTBE) has increased greatly. In addition, the antiknock additive methylcyclopentadienyl manganese tricarbonyl (MMT) is permitted in leaded gasoline in the United States and in both leaded and unleaded fuel in Canada. In Sweden lead has been replaced by sodium- and potassium-based additives.

Some of these solutions have created environmental problems of their own. For example, the increased use of benzene and other aromatics has led to concern over human exposure to benzene and to controls on gasoline aromatic content. Xylenes, other alkyl-aromatic hydrocarbons, and olefins are much more reactive in producing ozone than most other hydrocarbons. Use of oxygenates may result in increased emissions of aldehydes, which are toxic and may raise exhaust reactivity. Increased use of light hydrocarbons in gasoline produces a higher Reid vapor pressure (RVP)—a measure of vapor pressure at a fixed temperature ($100°F/37.8°C$) and vapor to liquid ratio (4:1)—and increased evaporative emissions. Average RVP levels in U.S. gasoline increased significantly when lead was phased out until regulations were adopted to limit gasoline RVP levels. The potential for expanded use of MMT as an antiknock has raised concerns in the United States about the potential toxicity of manganese in the environment. Authorities in Canada have concluded, however, that the resulting manganese concentrations are not a health threat.

Fuel Volatility

Fuel volatility, as measured by RVP, has a marked effect on evaporative emissions from gasoline vehicles both with and without evaporative emission controls. In tests performed on European vehicles without evaporative emission controls, it was found that increasing the fuel RVP from 62 to 82 kPa (9 to 11.9 psi) roughly doubled evaporative emissions (McArragher and others 1988). The effect is even greater in controlled vehicles. In going from 62 to 82 kPa RVP fuel, average diurnal emissions in vehicles with evaporative controls increased by more than five times, and average hot-soak emissions by 25 to 100 percent (U.S. EPA 1987). The large increase in diurnal emissions from controlled vehicles is due to saturation of the charcoal canister, which allows subsequent vapors to escape to the air. Vehicle refueling emissions are also strongly affected by fuel volatility. In a comparative test on the same vehicle, fuel with RVP of 79 kPa (11.5 psi) produced 30 percent more refueling emissions than gasoline with RVP of 64 kPa (9.3 psi)—1.89 g/liter versus 1.45 g/liter dispensed (Braddock 1988).

Based on these investigations, the U.S. EPA has established nationwide summertime RVP limits for gasoline. These limits are 7.8 psi (54 kPa) in warm climates and 9.0 psi (62 kPa) in cooler regions. Still lower RVP levels were required as of January 1, 1995, in "reformulated" gasoline sold in areas with serious air pollution problems—7.2 psi (49.6 kPa) for warm areas and 8.1 psi (55.8 kPa) in cooler areas. Sweden and Finland also have introduced environmental gasolines with maximum caps on RVP; 70-80 kPa in summer and 90-100 kPa in winter (CONCAWE 1994).

An important advantage of gasoline volatility controls is that they can reduce emissions from in-use vehicles and from the gasoline distribution system. Unlike new vehicle emissions standards, it is not necessary to wait for the fleet to turn over before they take effect. The emission benefits and cost-effectiveness of lower volatility are greatest where few of the vehicles in use are equipped with evaporative controls, as in most developing countries. In uncontrolled cars without a carbon canister, evaporative emissions increase exponentially with increasing fuel RVP and ambient temperatures. Even where evaporative controls are common, as in the United States, volatility control may still be beneficial to prevent in-use volatility levels from exceeding the control levels.

According to Phase 1 results of the U.S. Air Quality Improvement Research Program (AQIRP)[1], a reduction in fuel volatility (RVP) of 1 psi (6.9 kPa) from 9 psi to 8 psi lowered total evaporative emissions by 34 percent with no significant effect on fuel economy. Lowering RVP by 1 psi also reduced exhaust hydrocarbon emissions by 4 percent and CO by 9 percent. NO_x was unaffected (AQIRP 1993; AQIRP 1991a). The lowering of vapor pressure had no significant effect on emissions of exhaust toxics (benzene, 1, 3 butadiene, formaldehyde, acetaldehyde, and polycyclic organic matter).

In its analysis of the RVP regulation, the U.S. EPA (1987) estimated that the long-term refining costs of meeting a 62 kPa (9 psi) RVP limit throughout the United States would be about 0.38 U.S. cents per liter, assuming crude oil cost U.S.$20 a barrel. These costs were largely offset by credits for improved fuel economy and reduced fuel loss through evaporation, so that the net cost to the consumer was estimated at 0.12 U.S. cents per liter.

Gasoline volatility reductions are limited by the need to maintain adequate fuel volatility for good vaporization under cold conditions; otherwise, engines are difficult to start. Volatility reductions below 58 kPa (8.4 psi) have been shown to impair cold starting and driveability and increase exhaust VOC emissions, especially at lower temperatures. For this reason volatility limits are normally restricted to the warm months, when evaporative emissions are most significant. The range of ambient temperatures encountered also must be considered in setting gasoline volatility limits. Gasoline volatility controls could prove particularly beneficial in hot tropical countries with high temperatures but low ranges of diurnal and seasonal variations in temperature.

Olefins

Olefins, or alkenes, are a class of hydrocarbons that have one or more double bonds in their carbon structure. Examples include ethylene, propylene, butene, and butadiene. Olefins in gasoline are usually created during the refining process by cracking naphthas or

---

1.   The AQIRP was established in 1989 by fourteen oil companies and three domestic automakers to develop data to help U.S. legislators and regulators meet national clean air goals. The U.S. $40 million AQIRP program includes three components: extensive vehicle emission measurements using statistically designed experiments, air quality modeling studies to predict the effect of measured emissions on ozone formation, and economic analysis of the fuel-vehicle systems. The vehicles tested in the program are primarily late-model U.S. cars equipped with fuel injection systems, catalytic converters, and carbon canisters.

other petroleum fractions at high temperatures. Olefins are also created by partial combustion of paraffinic hydrocarbons in the engine. Compared with paraffins, volatile light olefins have very high ozone reactivity. Because of their higher carbon content, they also have a slightly higher flame temperature than paraffins, and thus NO_x emissions may be increased somewhat.

The AQIRP examined the impacts of reducing olefins in gasoline from 20 to 5 percent by volume (Hochhauser and others 1991; AQIRP 1993). The results show that while there tends to be a reduction in NO_x emissions from both current and older catalyst-equipped vehicles (about 6 percent), volatile organic compound (VOC) emissions tended to rise by 6 percent in both vehicle classes. This was ascribed to the fact that a reduction in olefin content implies an increase in paraffins. The olefins react much more readily in a catalytic converter than do paraffins. Increasing the paraffin content of the fuel therefore tends to reduce the overall VOC efficiency of the catalytic converter. The result of this change is higher paraffinic VOC emissions (which have substantially reduced reactivity compared with olefinic VOC emissions) and an associated reduction in vehicle exhaust reactivity. Olefin reduction had no significant effect on total exhaust toxic level but lowered 1,3 butadiene by 30 percent.

For non-catalyst cars, the effect of reduced olefins on emissions is small and not fully documented (ACEA/EUROPIA 1994).

Aromatic Hydrocarbons

Aromatic hydrocarbons are hydrocarbons that contain one or more benzene rings in their molecular structure. In order to meet octane specifications, unleaded gasoline normally contains 30 to 50 percent aromatic hydrocarbons. Aromatics, because of their high carbon content, have slightly higher flame temperatures than paraffins and are therefore thought to contribute to higher engine-out NO_x emissions. Aromatics in the engine exhaust also raise the reactivity of the exhaust VOC because of the high reactivity of the alkyl aromatic species such as xylenes and alkyl benzenes. Reducing the content of aromatic hydrocarbons in gasoline has been shown to reduce NO_x emissions, exhaust reactivity, and benzene emissions. Of the fuel properties tested by AQIRP, reduced aromatic content had the largest effect on total toxics due to the lowering of exhaust benzene emissions. Reduction of aromatics from 45 percent to 20 percent lowered the total air toxic emissions from catalyst-equipped cars by 23 to 38 percent. Benzene comprised 74 percent of the toxic emissions from current U.S. model cars with fuel-injected engines and new emission-control technology and 56 percent from older catalyst-equipped cars with carbureted engines (AQIRP 1990; 1991b; 1993).

Statistically controlled test carried out under the European Program on Emissions, Fuels and Engine Technologies (EPEFE)[2] show that reducing the aromatic content in gasolines reduced HC and CO emissions but increased NO_x emissions over the full European driving cycle. The response of NO_x emissions to increasing aromatic content, however, varied with the driving cycle—over the ECE (urban) cycle, NO_x emissions increased with increasing aromatics but over EUDC (extra-urban), there was a decrease in NO_x emissions with increasing aromatics. A reduction in the aromatic content of gasoline lowered benzene emissions but increased aldehyde emissions slightly. The EPEFE investigation also concluded that the catalyst light-off time was substantially reduced by lowering the aromatic content of gasoline. (ACEA/EUROPIA 1994; ACEA/EUROPIA 1995a).

Vehicle exhaust accounts for 85-90 percent of benzene emissions; the remainder is from evaporative and distribution losses. Benzene from fuel and dealkylation of higher aromatics in the combustion process contribute about equally to benzene emissions. A third small source comprises the partial combustion products of other hydrocarbons. Limiting the benzene content of fuel remains the most effective and widely used approach to controlling benzene emissions. Lowering the benzene fraction in gasoline from 3 to 2 percent reduced benzene emissions by about 17 percent (ACEA/EUROPIA 1994).

In the United States the benzene content in reformulated gasoline has been limited to 1 percent by volume maximum, effective January 1, 1995. Total aromatics are indirectly controlled by the requirement that emissions of air toxics be reduced by 15 percent. In California the reformulated gasoline required from March 1996 will have a 1 percent volume benzene limit, while total aromatics will be limited to 25 percent. In the European Union a 5 percent limit on benzene content has been in effect since 1989. In some countries, notably Austria, Finland, Germany, Italy, and Sweden, lower benzene (1 to 3 percent) and aromatic limits (Italy has limited total aromatics to 33 percent) have been introduced or are under consideration (CONCAWE 1995). The benzene content in gasoline is limited to 5 percent in Australia and New Zealand also. The Republic of Korea limits the benzene content to 6 percent and total aromatics to 55 percent by volume. Thailand has limited benzene to 3.5 percent volume since 1993; aromatics were reduced to 50 percent maximum in 1994 and are expected to be reduced to 35 percent by 2000 (CONCAWE 1994).

Unleaded gasoline typically has a significantly higher content of benzene (3 to 5 percent) and aromatics (35 to 56 percent) than leaded gasoline. The cost of reducing the benzene and aromatic content of gasoline depends on the refinery configuration. Simple refineries produce gasoline by reforming and isomerization, and their output is determined by the type of crude oil used. Complex refineries increase the total yield of gasoline by cracking residues otherwise only useful as fuel oil.

According to oil industry estimates, the incremental cost (in 1989 U.S. dollars) of producing gasoline with a 3 percent benzene content and 30 percent aromatics, was about 0.1 U.S. cent per liter for European refineries; for gasoline with 1 percent benzene content and 30 percent aromatics, the incremental cost was between 0.6 and 1.35 U.S. cents per liter (CONCAWE 1989a).

Distillation Properties[3]

Gasoline contains a wide range of hydrocarbons containing from four to eleven carbon atoms and therefore exhibits a wide boiling range. The "front-end" volatility of a gasoline is the temperature at which 10 percent of the fuel evaporates (T10) and is important because it is related to cold-start and vapor lock (a hot weather problem). The temperature at which 50 percent is evaporated (T50) or "mid-range" volatility has been related to short-trip fuel economy, warm-up, and cool weather driveability. "Back-end" volatility, expressed as the temperature at which 90 percent of the fuel is evaporated (T90), has been related to engine deposits and engine oil dilution (Aguila and others 1991). The allowable

---

2. EPEFE is a major collaborative research program by the European automotive and oil industries with the active support of the European Commission to establish correlations among fuel quality, engine design, emissions, and air quality. The main objective of EPEFE is to identify cost-effective measures to meet rational air quality objectives based on scientifically sound information. A further purpose is to provide information and data to support European emissions legislation for the year 2000 and beyond. The research program aims to quantify the reduction in traffic emissions that can be achieved by combining advanced fuels with the vehicle/engine technologies under development in Europe for the year 2000. The set of test vehicles and engines used in the study reflect the wide range in vehicle sizes and engine displacements available in Europe. Some of the vehicles were equipped with prototype emission technologies currently under development (e.g. close-coupled catalysts for gasoline vehicles, oxidation catalysts for light-duty diesel vehicles, and high-pressure ignition systems for heavy-duty diesel engines). All light duty vehicles exceeded the requirements of the 1996 European emission standards while heavy-duty diesel engines exceeded the Euro 2 emission standards that came into force in EU in October 1995. In total, EPEFE examined 12 test gasolines with 16 gasoline-powered vehicles, and 11 test diesel fuels with 19 light-duty vehicles and 5 heavy-duty engines (ACEA/EUROPIA 1995a).

3. The distillation curve of a fuel can be analyzed in terms of temperature (T) or evaporative (E) values. For example, T50 is the temperature at which 50 percent of the fuel volume is evaporated, while E100 is the portion of the fuel evaporated at 100°C, in the standard distillation test. These two measures are representative of mid-range volatility and are inversely related. A gasoline with a high E100 value will have a low T50 value. Both measures are widely used in international gasoline specifications (Armstrong and Wilbraham 1995). Front-end volatility is characterized by E70 or T10 values while back-end volatility is represented by E150 or T90 values.

range of gasoline distillation properties is generally defined by national fuel specifications, many of which are based on American Society for Testing and Materials (ASTM) Standard D 439. Efforts to reduce the allowable range of distillation properties are sometimes advocated to reduce emissions. Such modifications can be extremely expensive, however, because they shrink the available pool of gasoline blending components. In one AQIRP investigation, lowering T90 from 360°F to 280°F reduced HC emissions but increased CO emissions in older catalyst-equipped vehicles. In general, lowering T90 always reduced hydrocarbon emissions, but the reduction was much larger when the aromatic content was high (AQIRP 1993; 1991c).

Another AQIRP study on the effects of gasoline T50 and T90 on exhaust emissions of fuel-injected catalyst-equipped U.S. cars and advanced technology prototypes showed that non-methane hydrocarbon emissions decreased significantly when T50 was lowered from 215°F to 185°F or T90 reduced from 325°F to 280°F. CO emissions also decreased when T50 was reduced but increased when T90 was lowered; a significant interaction was found between T50 and T90 with the effect of one parameter dependent on the level of the other. NO_x emissions tended to increase somewhat when T50 was reduced in the current catalyst-equipped cars. Total toxic emissions (benzene, 1,3 butadiene, formaldehyde, and acetaldehyde) decreased significantly with a lowering of T90 (AQIRP 1995a). EPEFE experiments have shown that increasing the content of mid-range distillates (E100) reduced total HC and benzene emissions but increased NO_x emissions. Carbon monoxide emissions were lowest at E100 of 50 percent (v/v), for a given aromatic content. (ACEA/EUROPIA 1995a).

Oxygenates

Blending small percentages of oxygenated compounds such as ethanol, methanol, tertiary butyl alcohol (TBA), and methyl tertiary-butyl ether (MTBE) with gasoline reduces the volumetric energy content of the fuel while

---

Box 5.3 Use of Oxygenates in Motor Gasolines

Oxygenated supplements, which cover a range of lower alcohols and ethers, can substitute for lead additives in gasoline. Oxygenates are produced from a variety of feedstocks; ethanol is derived mostly from renewable agricultural feedstocks, and methanol is derived primarily from natural gas. Ether production facilities are integrated within a refinery. Large-scale units use butane isomerization/dehydrogeneration technology, where both the butane and methanol feedstocks are derived from lower-value gas sources. The behavior of oxygenates in terms of blending and vehicle performance is different from hydrocarbons-only gasolines. The commonly used oxygenates in gasoline blends are methanol, ethanol, isopropanol, tertiary butyl alcohol (TBA), methyl tertiary butyl ether (MTBE), ethyl tertiary butyl ether (ETBE), di-isoprophyl ether (DIPE), and iso-butyl alcohol. Tertiary amyl methyl ether (TAME) also has been used in small amounts. The volume of oxygenates in gasoline blends varies from 3 to 22 percent, although commonly-used blends have 10-15 percent oxygenates. The oxygen content of the blend is generally about 1 to 2 percent by weight. Oxygenates serve three basic objectives: extending the gasoline stock or serving as a fuel substitute (for example, ethanol in Brazil), boosting octane value, and providing an effective means of reducing harmful emissions (for example, MTBE in unleaded gasoline in Mexico City). Some of the important vehicle performance characteristics of oxygenated fuels include:

Antiknock performance. Oxygenated fuels perform better than hydrocarbon-only fuels at low olefin levels. They give better antiknock performance in unleaded gasolines; this is particularly the case with increasing MTBE content. MTBE has proven to be an effective substitute for lead in gasoline because it has a high octane rating (RON 119, MON 101) and is less water-sensitive than alcohol. It is not clear, however, whether oxygenates give better antiknock performance under lean-burn conditions.

Driveability. Vehicles fitted with fuel injection systems have better tolerance in cold weather conditions for low-volatility fuels containing oxygenates. At high altitudes and in hot weather, oxygenated fuels (except for certain methanol blends) give similar or better handling performance compared with wholly hydrocarbon fuels.

Fuel economy. In the case of commercial oxygenated gasoline (which must comply with existing gasoline specifications), fuel economy is essentially unchanged with increasing amounts of oxygen. There could be marginal benefits in terms of reduced energy consumption with increased fuel oxygenate content.

Exhaust emissions. CO and HC levels are progressively reduced as oxygen content is increased, while NO_x emissions may increase, particularly when oxygenates are blended with fuels with low aromatic content. Ethanol blends increase aldehyde emissions with increasing concentration; with other oxygenates there is only a marginal increase, which may be corrected with exhaust oxidation or three-way catalysts. Oxygenates are particularly useful in lowering emissions from older vehicles, especially toxic emissions such as benzene. For the same fuel oxygenate content, MTBE and ETBE exhibit similar exhaust emission characteristics.

Source: Lang and Palmer 1989; ARCO 1993; AQIRP 1993; Hutcheson 1995

improving the antiknock performance and thus makes possible a potential reduction in lead and harmful aromatic compounds. Assuming no change in the settings of the fuel metering system, lowering the volumetric energy content will result in a leaner air-fuel mixture, thus helping to reduce exhaust CO and HC emissions (box 5.3). This approach has attracted considerable attention in the United States and has been adopted by a number of jurisdictions to reduce wintertime CO emissions. (CO emissions are highest at low temperatures, with low traffic speeds, and at high altitude).

Exhaust HC and CO emissions are reduced by the use of oxygenates, but NO_x emissions may be increased slightly by the leaner operation. AQIRP experiments tested the effects of adding 10 percent ethanol (3.5 wt. percent oxygen) and 15 percent MTBE (2.7 wt. percent oxygen) to industry-average gasoline. The ethanol addition lowered non-methane hydrocarbon (NMHC) and CO emissions in late-model catalyst-equipped, gasoline-fueled vehicles by 5.9 and 13.4 percent, respectively, and increased NO_x emissions by 5.1 percent. The MTBE addition lowered NMHC and CO emissions by 7.0 and 9.3 percent, respectively, and increased NO_x emissions by 3.6 percent (Hochhauser and others 1991; AQIRP 1991a, 1991b, 1992a). For a cross-section of European cars without catalysts, adding 15 percent MTBE to gasoline reduced CO emissions by 4 – 43 percent and HC emissions by up to 14 percent. For catalyst-equipped cars, the average reduction was 10 percent in CO emissions and 6 percent in HC emissions whereas NO_x emissions increased by 5 percent (ACEA/EUROPIA 1994). In tests performed in Mexico City, the addition of 5 percent MTBE to leaded gasoline produced a 14.7 percent reduction in CO and an 11.6 percent reduction in HC emissions from non-catalyst gasoline vehicles. The octane boost due to the MTBE also made possible a 0.066 g/l reduction in lead.

Colorado (United States) initiated a program to mandate the addition of oxygenates to gasoline during winter months, when high ambient CO levels tend to occur. The mandatory oxygen requirement for January-March 1988 was 1.5 percent by weight, equivalent to about 8 percent MTBE. In the following years the minimum oxygen content was 2 percent by weight, equivalent to 11 percent MTBE. These requirements were estimated to reduce CO exhaust emissions by 24 to 34 percent in vehicles already fitted with three-way catalyst systems. This program led the U.S. Congress to mandate the use of oxygenated fuels (minimum 2.7 percent oxygen by weight) in areas with serious winter CO problems. Other countries that require mandatory use of oxygenates in gasoline are Brazil (22 percent ethanol), South Africa (8 to 12 percent alcohol), Thailand (5.5 to 10 percent MTBE), and Korea (0.5 percent minimum oxygen). Reformulated gasolines used in Sweden

and Finland have specific oxygen requirements (2 to 2.7 percent).

Although exhaust HC emissions tend to be lower with oxygenate blended fuels, the use of alcohols as blending agents may increase evaporative emissions considerably. Because of their non-ideal behavior in solution, blends of ethanol and methanol with gasoline have higher vapor pressure than either component alone. Evaporative emissions from alcohol-gasoline blends also tend to be 50 to 65 percent higher, even when the gasoline composition has been adjusted to keep RVP the same. This is due to the effect of alcohol blending on the shape of the distillation curve. The alcohols also reduce the effectiveness of the charcoal canister. The alcohols are more strongly retained by the evaporative canister and are difficult to strip during purging, thus reducing the canister's adsorptive capacity.

Due to the increased evaporative emissions, alcohol blends tend to produce higher total HC emissions than straight gasoline unless ambient temperatures are so low that evaporative emissions are negligible. Similar adverse effects have not been reported for MTBE and other ethers. Corrosion, phase separation on contact with water, and materials compatibility problems sometimes experienced with alcohol fuels—are much less serious for the ethers. For this reason MTBE and other ethers are strongly preferred as oxygenated blending agents by many fuel marketers, as well as for air quality purposes. The costs of using ethers are also relatively moderate (about 1-3 U.S. cents per liter in 1993 prices), so this can be a cost-effective approach as well.

Sulfur

Sulfur in gasoline is undesirable for several reasons. The most important of these is that it can bind to the precious metals in the catalytic converter under rich conditions, temporarily poisoning it. Although this poisoning is reversible, the efficiency of the catalyst is reduced while operating on high-sulfur fuel. A 1981 study by General Motors showed emission reductions of 16.2 percent for HC, 13.0 percent for CO, and 13.9 percent for NO_x with old catalysts by switching from fuel containing 0.09 percent sulfur to fuel with 0.01 percent sulfur (Furey and Monroe 1981). Even larger reductions were seen in vehicles with relatively new catalysts. In the EPEFE study, a reduction of sulfur content from 328 to 18 ppm resulted in emission reductions of 9.2 percent for HC, 9.4 percent for CO, and 11.2 percent for NO_x for European catalyst-equipped light-duty vehicles. (ACEA/EUROPIA 1995a).

Similar results have been reported from modern, fuel-injected vehicles with three-way catalysts, tested as part of the AQIRP (AQIRP 1993). This study showed that reducing fuel sulfur content can contribute directly to reductions in mass emissions (HC, CO, and NO_x), toxic emissions (benzene, 1,3-butadiene, formaldehyde, and acetaldehyde), and potential ozone formation. The sulfur reduction study used test fuels with nominal fuel sulfur levels of 50, 150, 250, 350, and 450 ppm in ten late-model U.S. vehicles. Reductions in HC, NMHC, CO, and NO_x were 18, 17, 19, and 8 percent, respectively, when the fuel sulfur level was dropped from 450 ppm to 50 ppm. Reducing the fuel sulfur level also reduced benzene emissions by 21 percent and acetaldehyde emissions by 35 percent. Formaldehyde emissions were increased by 45 percent, while changes in 1,3-butadiene emissions were insignificant. Reactivity per gram of non-methane organic gas (NMOG) increased slightly, but this was counterbalanced by the decrease in NMOG mass, so that total ozone-reactivity per mile was reduced by 9 percent (AQIRP 1992b). The EPEFE results corroborate the AQIRP findings except that emissions of 1,3 butadine, formaldehyde, and acetaldehyde were unaffected by changes in the sulfur content of gasolines used in EPEFE test (ACEA/EUROPIA 1995a).

In addition to its effects on catalyst efficiency, sulfur in gasoline contributes directly to sulfur dioxide, sulfate, and hydrogen sulfide emissions. Under lean conditions fuel sulfur forms particulate sulfates and sulfuric acid in catalytic converters. Under rich conditions hydrogen sulfide is formed by the reduction of sulfur dioxide and sulfates stored on the catalyst substrate. The offensive odor of hydrogen sulfide in the exhaust contributes to a popular perception that catalysts do not work and may lead to increased tampering with emission controls. In 1989, 41 percent of the gasoline produced in the United States had a sulfur content above 300 ppm and about 9 percent above 700 ppm. The cost of removing sulfur, a nonlinear function of the sulfur content, has been estimated by AQIRP as follows:

- 0.4 U.S. cent a gallon (0.1 U.S. cent per liter) to reduce the sulfur content of gasoline from 340 ppm to 160 ppm,
- 2.1 to 2.9 U.S. cents a gallon (0.5 to 0.8 U.S. cent per liter) to reduce sulfur from 340 ppm to 50 ppm,
- 4.6 to 5.1 U.S. cents a gallon (1.2 to 1.3 U.S. cents per liter) to reduce sulfur from 340 ppm to 20 ppm.

The cost of reduction to the 20 ppm level is almost twice that of the 50 ppm level because essentially all of the gasoline blending components must be hydrotreated to remove sulfur (AQIRP 1992c).

Fuel Additives to Control Deposits

Deposits on inlet valves and fouling of fuel injectors can be reduced significantly through the use of gasoline additives, with a beneficial impact on emissions. Emissions from light-duty vehicles were significantly lowered by the use of gasoline with deposit-control ad-

ditives – CO emissions by 10-15 percent, HC by 3-15 percent, and NO_x by 6-15 percent (ACEA/EUROPIA 1995b).

Reformulated Gasoline

The potential for reformulating gasoline to reduce pollutant emissions has attracted considerable attention worldwide, and is the subject of major cooperative research programs between the oil and auto industries in the United States and Europe. The most significant potential emission reductions for gasoline reformulation thus far have been through reducing volatility (to reduce evaporative emissions), reducing sulfur (to improve catalyst efficiency), and adding oxygenated blendstocks (with a corresponding reduction in toxic, high-octane aromatic hydrocarbons, which otherwise would be required). There is also evidence that changes in mid-range distillation characteristics may reduce emissions in fuel-injected catalyst-equipped vehicles, but the effects are relatively small and have not been demonstrated in vehicles with carburetors. These same changes led to large increases in hot-soak evaporative emissions from older vehicles, so that the net effect of reducing distillation temperatures in these vehicles was to increase HC emissions.

Research by AQIRP on the effects of gasoline reformulation on exhaust emissions of catalyst-equipped older, current, and advanced (U.S. tier 1) cars has shown that non-methane hydrocarbon (NMHC) emissions were 12-27 percent lower with reformulated gasoline than with industry-average gasoline; furthermore CO emissions were reduced by 21 to 28 percent and NO_x emission by 7 to 16 percent.[4] This study also demonstrated that the differences in exhaust emissions between reformulated gasoline with MTBE (11 percent) and a similar gasoline without oxygenates were generally not significant. Formaldehyde emissions however, were 13 percent lower with the oxygenate-free reformulated gasoline (AQIRP 1991c; 1995b).

Gasoline reformulation, however, is a somewhat vague concept that could include any or all of the modifications discussed above. Such reformulations are likely to have the greatest effect—as well as the greatest cost-effectiveness—when applied to the fuel normally used by existing vehicles without emission controls. In most instances this fuel is leaded gasoline. Modern emission-controlled vehicles require unleaded gasoline,

---

4.   Properties of California Phase 2 reformulated gasoline used in AQIRP study: aromatics (25.4 vol%), olefins (4.1 vol%) benzene (0.93 wt%), sulfur (31 ppm), MTBE (11.2 vol%), RVP (6.8 psi), T10 (142°F), T50 (202°F), T90 (239°F), API Gravity (59.9).
Properties of industry - average gasoline used in the AQIRP study: aromatics (32 vol%), olefins (9.2 vol%) benzene (1.53 wt%), sulfur (339 ppm), RVP (8.7 psi), T10 (114°F), T50 (218°F), T90 (330°F), API Gravity (57.4).

are less sensitive to changes in gasoline parameters, and have relatively low emissions to start with, so that even a large percentage reduction in emissions may be insignificant in absolute terms. This suggests that a "two gasoline" policy might be the most advantageous, with the fuel (whether leaded or a special unleaded blend) used by the older, uncontrolled vehicles being specially formulated to reduce their emissions, and the unleaded gasoline intended for emission-controlled vehicles subject to less stringent restrictions.

The AQIRP has estimated the costs of modifying certain gasoline properties, such as aromatics, oxygenates, olefins, T90, and sulfur (AQIRP 1992c). The incremental costs of controlling individual gasoline parameters, as well as the costs of controlling multiple parameters, are summarized in table 5.1. Some combinations can be controlled at little or no cost beyond the cost of controlling a single parameter, while the cost of controlling other combinations is higher. This is due to the synergy that occurs among gasoline properties when one is controlled and others are allowed to change. It should be noted that single-property cost comparisons do not represent optimal or even likely solutions but may facilitate the ranking of options when fuel reformulation proposals are considered. For most fuels studied, the potential fuel economy penalty is 2 – 5 U.S. cents per gallon, resulting from the lower energy content of these fuels (AQIRP 1993).

In the United States strict requirements have been established for reformulated gasolines to be sold in areas with serious air pollution. The U.S. EPA has modeled the effect of changes in gasoline composition on emissions of hydrocarbons and toxic air contaminants. Phase I re-

formulated gasoline is required to reduce these emissions by 15 percent. Phase II gasoline, to be required in 2000, must achieve a 25 percent reduction. California has adopted its own gasoline reformulation requirements; Phase II California gasoline is expected to yield a reduction of about 30 percent compared with the fuels sold in 1990. The costs of these changes are high, however: federal Phase II gasoline is expected to cost about 3 U.S. cents per liter extra, and California Phase II fuel may cost as much as 5 U.S. cents more per liter to refine.

In Europe, the overall effect of reformulated gasolines, achieved by adding 10 to 15 percent oxygenates (by volume) and reducing the concentration of aromatics, has been a car-weighted average reduction of 11 percent in both HC and NO_x emissions and a 23 percent reduction in CO emissions. These reductions pertain primarily to cars without catalytic converters or carbon canisters. The additional cost of producing these cleaner gasolines is estimated at about 0.5 to 1.0 percent of the retail price (ARCO 1993).

The major benefit of gasoline reformulation is to reduce emissions (especially evaporative emissions) from vehicles with little or no emission control. Reformulation of unleaded gasoline intended for modern emission-controlled vehicles is unlikely to be cost-effective, compared with most other potential emission control measures, and should therefore be considered only where more cost-effective measures are insufficient. For instance, the cost of reducing emissions with California reformulated gasoline is estimated at U.S.$125,000 per ton of NMOG emissions eliminated for vehicles meeting current Tier 1 standards, and

Table 5.1 Incremental Costs of Controlling Gasoline Parameters
(1989 U.S. dollars)

Property controlled	Control level	Manufacturing cost (cents/gallon)
Aromatics	20%	2.3–4.7
MTBE	15%	2.2–3.1
Olefins	5%	2.2–3.1
T90	280°F (155°C)	5.0–8.8
Sulfur	50 ppm	2.1–4.6
T90 and aromatics	280°F (155°C) and 20%	5.0–9.4
Olefins and aromatics	5% and 20%	3.8–7.9
Aromatics, MTBE, olefins, T90	20%/15%/5%/280°F (155°C)	7.0–11.6
Aromatics, MTBE, olefins, T90	20%/15%/9%/310°F (173°C)	2.3–4.7

Note: 1 U.S. gallon = 3.79 liters
Source: AQIRP 1993

U.S.$420,000 per ton for future low-emission vehicles. The estimated cost of reducing NMOG emissions with federal reformulated gasoline range from U.S.$50,000 per ton for Tier 1 vehicles to U.S.$168,000 per ton for low-emission vehicles (LEVs). By comparison, the cost of modest reformulation of leaded gasoline intended for uncontrolled vehicles in Mexico was estimated at less than U.S.$5,000 per ton (Weaver and Turner 1994).

Diesel

Automotive gas oil or high-speed diesel consists of a complex mixture of hydrocarbons evaporating typically between 180°C and 400°C. Diesel fuels are made mainly from straight-run refinery components—hydrocarbons derived directly from the distillation of crude oil. Two main hydrocarbon fractions are used to make diesel fuels—the middle distillates or gas oils, and the residual oils. To these are added small quantities of components from other refining processes such as catcracking and hydrocracking. High-speed diesel engines used in road vehicles run on distillate fuel from gas oil, while low-speed diesel engines used in ships and electric generation use heavy residual fuel oil. Diesel fuels are usually blends because the pattern of fuel demand does not match the output of a simple distillation refinery, and more complex refining patterns have to be used. Also, there is competition between products because the

fractions yielding diesel fuels are also used to make domestic and industrial heating oils and aviation fuel, as shown in figure 5.1 (Shell 1988). Diesel fuel properties are highly dependent on the type of crude oil from which the diesel fuel is refined (table 5.2).

The relationship between fuel characteristics, engine performance, and exhaust emissions is complex, and there often is a tradeoff between measures to control one pollutant and its effect on others. Diesel engines are generally quite robust and tolerant of a wide range in fuel characteristics and quality. Thus fuel properties tend to have a minor influence on emissions compared with the influence of engine design and operating conditions. Nevertheless, the quality and composition of diesel fuel can have important effects on pollutant emissions. The effects of fuel on diesel emissions have received a great deal of study in the past decade, and a large amount of new information has become available. These data indicate that the fuel variables having the most important effects on emissions are sulfur content, cetane number, and the fraction of aromatic hydrocarbons contained in the fuel. Cetane number and aromatic hydrocarbon content are themselves closely related—fuels with high cetane tend to have low aromatic hydrocarbon content, and vice versa. Other fuel properties such as density, back-end volatility and viscosity, also affect emissions, but generally to a much lesser extent. In addition, the use of fuel additives may have a significant impact on emissions.

Figure 5.1 Range of Petroleum Products Obtained from Distillation of Crude Oil

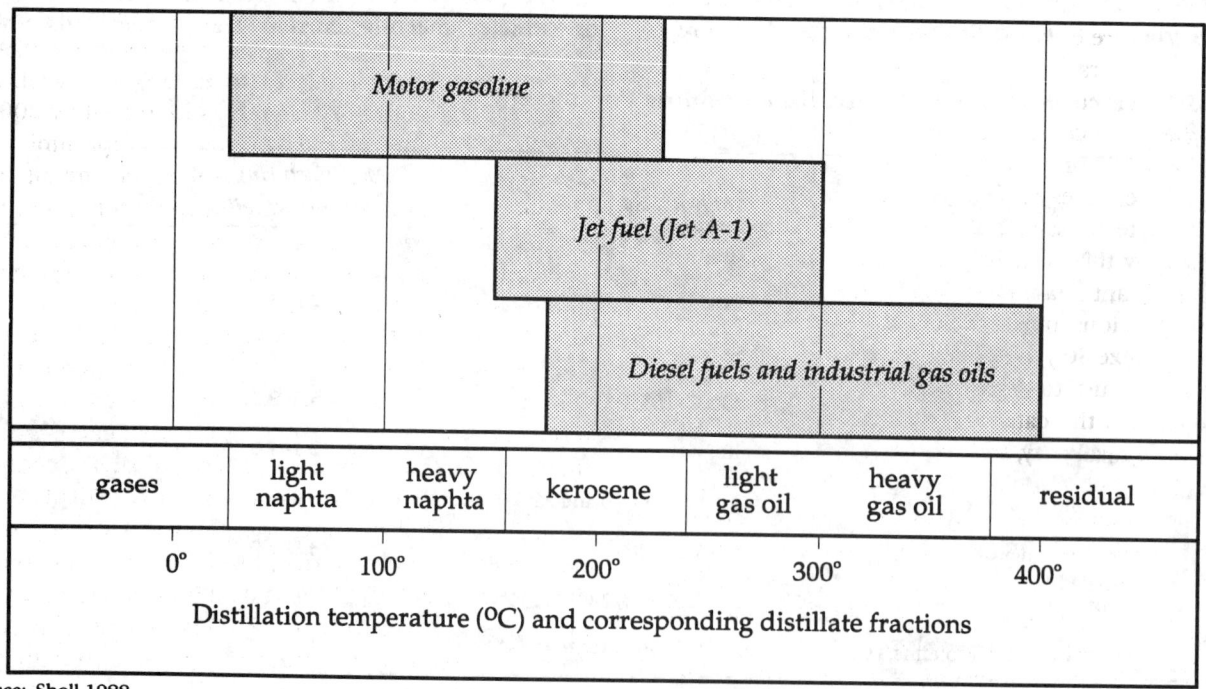

Source: Shell 1988

Table 5.2 Influence of Crude Oil Type on Diesel Fuel Characteristics

Crude oil source	Hydrocarbon type	Cetane number	Sulfur content (percent)	Calorific value
North Sea	Paraffinic	High	0.10–3.0	Low
Middle East	Paraffinic	High	0.86–1.40	Low
Nigeria	Naphenic	Low	0.12–0.34	Medium
Venezuela/Mexico	Naphenic/aromatic	Very low	0.30–1.3	High
Indonesia/Australia	Paraffinic	High	0.10–0.30	Low
Russia	n.a.	n.a.	0.69	n.a.

n.a. = Not available
Source: Owen and Coley 1990

Sulfur Content

Diesel fuel for highway use normally contains between 0.1 and 0.5 percent sulfur by weight, although some, mostly developing countries (for example, Brazil, India, Pakistan) allow 1.0 percent or even higher amounts of sulfur in diesel oil. Sulfur in diesel fuel contributes to environmental deterioration both directly and indirectly. Most of the sulfur in the fuel burns to SO_2, which is emitted to the atmosphere in diesel exhaust. Because of these emissions, diesel vehicles are significant contributors to ambient SO_2 levels in some areas.

Most of the fuel sulfur that is not emitted as SO_2 is converted to various metal sulfates and to sulfuric acid during or immediately after the combustion process. Both materials are emitted in particulate form. The typical rate of conversion in a heavy-duty diesel engine is about 1 to 3 percent of the fuel sulfur; in a light-duty engine it is 3 to 5 percent. Even at this rate sulfate particles typically account for 0.05 to 0.10 g/kWh of particulates in heavy-duty engines, equivalent to about 10 percent of total particulate emissions. The effect of sulfate particles is increased by their hygroscopic nature—they tend to absorb significant quantities of water from the air.

Certain precious-metal exhaust catalysts such as platinum can oxidize SO_2 to SO_3, which combines with water in the exhaust to form sulfuric acid. The rate of conversion with the catalyst is dependent on the temperature, space velocity, and oxygen content of the exhaust and on the activity of the catalyst. Generally, catalyst formulations that are most effective at oxidizing hydrocarbons and CO are also most effective at oxidizing SO_2. The presence of a significant amount of sulfur (greater than 0.05 percent) in diesel fuel thus limits the potential use of catalytic converters or catalytic trap-oxidizers for oxidization controlling PM, CO, and HC emissions from diesel vehicles.

Aside from its particulate-forming tendencies, SO_2 is recognized as a hazardous pollutant in its own right. The health and welfare effects of SO_2 emissions from diesel vehicles are probably much greater than those of an equivalent quantity emitted from a utility stack or industrial boiler, since diesel exhaust is emitted close to the ground in the vicinity of roads, buildings, and concentrations of people.

In order to reduce SO_2 and particulate emissions and to ensure compatibility with advanced diesel emission control systems such as trap oxidizers and oxidation catalysts, diesel engine manufacturers, the U.S. oil industry, and the U.S. EPA agreed to limit diesel fuel sulfur content to 0.05 percent by weight starting in October 1993. Measures to progressively reduce diesel fuel sulfur content to 0.05 percent by October 1996 have been adopted in the European Union as well. A further reduction to 0.03 percent is expected by 2000 (Walsh 1996). Switzerland adopted a sulfur limit of 0.05 percent in 1994, and Austria reduced the allowable sulfur limit from 0.15 percent to 0.05 percent in 1995. Japanese industry has agreed to reduce sulfur content of diesel fuels from 0.2 percent to 0.05 percent by 1997. Thailand has adopted a phased reduction of sulfur in diesel from the current level of 0.5 percent to 0.25 percent in 1996 and 0.05 percent in 2000. Finland and Sweden have introduced virtually sulfur-free diesel fuels (0.001-0.005 percent sulfur), backed by tax incentives to encourage their production and use, particularly in urban areas (CONCAWE 1994). There has been little progress in reducing the sulfur content of diesel fuel in most developing countries, although India, Pakistan, Philippines, and Sri Lanka planned to reduce the sulfur content of diesel from 1.0 percent to 0.5 percent by 1996, initially in the metropolitan areas.

Sulfur is removed from fuel through a hydro-desulfu-rization (HDS) process. Low-pressure HDS plants can remove 65 to 75 percent of sulfur, reduce aromatic levels by 5 to 10 percent, and increase cetane number by 1 to 2. Newer HDS plants operating at medium to high pressures can remove more than 95 percent of sulfur and 20 to 30 percent of aromatics.

The costs of reducing the sulfur content of diesel to 0.05 percent are considered moderate—less than one U.S. cent per liter—and the estimated cost-effectiveness is attractive compared with other diesel control measures. A 1989 study by CONCAWE concluded that the costs of reducing diesel-fuel sulfur content to 0.05 percent in Europe would be between 0.9 and 1.4 U.S. cents per liter, equivalent to U.S.$6,000–$9,000 per ton of sulfur removed (CONCAWE 1989b). The fixed costs of retooling refineries to produce low-sulfur diesel, however, can be quite large, requiring substantial up-front investment. For example, hardware modifications to an Asian refinery to reduce the sulfur content of diesel from 0.5 to 0.2 percent were estimated to raise the production cost of diesel by 1.6 to 1.9 U.S. cents per liter (Barakat 1995).

Cetane Number

The cetane number is a measure of a fuel's ignition quality and indicates the readiness of a diesel fuel to ignite spontaneously under the temperature and pressure conditions in the engine's combustion chamber. The higher the cetane number, the shorter the delay between injection and ignition and the better the ignition quality. The cetane number of a fuel is determined by comparing its ignition quality under standard operating conditions with a blend of two reference fuels—the straight chain—paraffin n-cetane (with a value of 100 by definition), which ignites very quickly, and a branched paraffin, heptamethyl nonane (with an assigned value of 15), which has a long ignition delay. Cetane number is determined by the composition of the diesel fuel. Research has shown that higher-cetane fuel is associated with improved combustion, improved cold starting, reduced white smoke, less noise, and reduced HC, CO, and particulate emissions particularly during the vehicle warm-up phase (IEA 1993). The time to start, the time to achieve a stable idle, and time to reach 50 percent of the initial smoke level all increase progressively with decreasing cetane number. Black smoke and gaseous emissions increase with lower cetane fuels, especially at cetane levels bellow 45. Variations in cetane number have a less pronounced effect on passenger car emissions (IDI engines) compared to heavy-duty truck emissions (DI engines) (Owen and Coley 1990)

European diesel fuel has cetane numbers ranging from 43 to 57, with an average of 50. U.S. diesel fuel tends to have lower cetane, with a minimum of 40 and

an average around 43. The difference is due to the fact that European diesel contains primarily straight-run hydrocarbons, while U.S. diesel contains a higher percentage of cracked products. Since October 1993, diesel fuel used in road vehicles in the United States has been subject to a minimum cetane index[5] requirement of 45. This has helped to limit the aromatic content and maintain the cetane value of U.S. diesel fuel. The cetane rating of diesel fuel used in developing countries varies considerably, but is commonly closer to the European than the U.S. practice. For example, diesel fuel produced in Pakistan (by the Pakistan Refinery in Karachi) has a typical cetane number of 52, comparable in quality to European diesel fuels (Piracha 1993), whereas Brazil's diesel fuel tends to be of lower quality because of the higher content of heavy distillates. As in Europe, diesel fuel in most developing countries contains mostly straight-run hydrocarbons.

There is a tradeoff between cetane number and emission benefits. Given the nonlinearity of the relationship, emission improvements beyond a certain cetane level may not materialize. There is little evidence that emissions would improve significantly with fuel cetane numbers above current European levels. In the most advanced diesel formulations (Finland, Sweden) the cetane number is in the 47 to 50 range. High cetane fuels tend to be paraffinic and exhibit relatively high cloud points,[6] which could be problematic in terms of cold weather operability. Also, cetane-rich paraffinic crudes tend to be high in sulfur content (Hutcheson and van Passen 1990). It is estimated that a two-point reduction in the cetane number (from a median European value of 50 to 48) could increase NO_x emissions by 2 percent and PM emissions by up to 6 percent. Such a two-point lowering of the cetane number could result from the increased use of cracked components in diesel fuel in Europe (Holman 1990).

Aromatic Hydrocarbons

Aromatic hydrocarbons are hydrocarbon compounds containing one or more "benzene-like" ring structures. They are distinguished from paraffins and napthenes, the other major hydrocarbon constituents of diesel fuel, which lack such structures. Compared with these other components, aromatic hydrocarbons are denser, have poorer self-ignition qualities, and produce more soot in

---

5.   The term cetane index is an approximation of cetane number based on an empirical relationship with density and volatility parameters. The cetane index is not a good estimate of the cetane number of diesel fuels containing ignition-improver organic nitrate additives such as 2-ethyl hexyl nitrate (2-EHN).

6.   The cloud point is the temperature at which wax starts to separate out of solution as the fuel cools. Pour point is the temperature at which the amount of wax out of solution is sufficient to gel the fuel. These measure have been used to define the low-temperature properties of diesel fuels.

burning. Ordinarily, straight-run diesel fuel produced by simple distillation of crude oil is fairly low in aromatic hydrocarbons. Catalytic cracking of residual oil to increase gasoline and diesel production results in increased aromatic content. A typical straight-run diesel might contain 20 to 25 percent aromatics by volume, while a diesel blended from catalytically cracked stocks could have 40 to 50 percent aromatics.

Aromatic hydrocarbons have poor self-ignition qualities, so that diesel fuels containing a high fraction of aromatics tend to have low cetane numbers. Typical cetane values for straight-run diesel are in the 50 to 55 range; those for highly aromatic diesel fuels are typically 40 to 45 and may be even lower. This produces more difficulty in cold starting and increased combustion noise and HC and NO_x emissions due to the increased ignition delay.

Increased aromatic content is correlated with higher carbonaceous particulate emissions. Aromatic hydrocarbons have a greater tendency to form soot in burning, and the poorer combustion quality also appears to increase particulate soluble organic fraction (SOF) emissions. Much of the increase in PM and other emissions with increasing aromatic content appears to be due to the accompanying deterioration in cetane quality. Much (but not all) of this deterioration can be recovered through the use of cetane-enhancing additives. Since the cost of these additives is much less than the cost of the extra processing needed to reduce aromatic content, this may be a cost-effective solution. Performance with additives, however, does not always match results obtained with natural cetane values (Owen and Coley 1990).

The content of polyaromatic hydrocarbons may also be an important factor affecting soot formation. Increased aromatic content also is correlated with increased SOF mutagenicity, possibly due to increased polynuclear aromatic hydrocarbon (PNA) and nitro-PNA emissions. There is also evidence that highly aromatic fuels have a greater tendency to form deposits on fuel injectors and other critical components. Such deposits can interfere with proper fuel-air mixing, greatly increasing PM and HC emissions.

To reduce diesel emissions, the California Air Resources Board recently adopted regulations limiting the aromatic hydrocarbon content of diesel fuel in California to 10 percent by volume (the previous value was 30 percent). This regulation has been fairly expensive—costs have exceeded 3 U.S. cents per liter of fuel—and the cost-effectiveness of reducing aromatic hydrocarbons is relatively poor. The new low-aromatic California fuel also has been blamed for a rash of fuel pump seal leaks that occurred in California shortly after it was introduced—apparently as a result of shrinkage of the elastomers making up the seals when exposed to the lower-aromatic fuel. A number of European countries are introducing compositional specifications for diesel fuels to enhance emissions performance. Finland's reformulated diesel limits the aromatics content to 20 percent by volume. Swedish diesel fuels for use in urban areas are limited to an aromatic content of 5 to 20 percent (CONCAWE 1995). The total aromatic content of diesel fuel tends to have a strong correlation with cetane quality and fuel density and by itself may not have a significant effect on PM emissions (Armstrong and Wilbraham 1995).

Other Fuel Properties

Diesel fuel consists of a mixture of hydrocarbons with different molecular weights and boiling points. As a result, as some of the fuel boils away on heating, the boiling point of the remainder increases. This fact is used to characterize the range of hydrocarbons in the fuel in the form of a "distillation curve" specifying the temperature at which 10 percent, 20 percent, and so on of the hydrocarbons have boiled away. A low 10 percent boiling point (T10) is associated with a significant content of relatively volatile hydrocarbons. Fuels with this characteristic tend to exhibit somewhat higher HC emissions than others. Formerly, a high 90 percent boiling point (T90) was associated with higher particulate emissions. More recent studies (Wall and Hoekman 1984) have shown that this effect is spurious—the apparent statistical linkage was due to the higher sulfur content of these high-boiling fuels. A high 90 percent boiling point may have an effect on cold starting, however, because the heavier hydrocarbons readily form wax crystals at low temperatures, and these crystals can block fuel filters, cutting off fuel flow to the engine.

Other fuel properties also may have an effect on emissions. Fuel density, for instance, may affect the mass of fuel injected into the combustion chamber and thus the air-fuel ratio. This is because fuel injection pumps meter fuel by volume, not by mass, and the denser fuel has a greater mass for the same volume. Too high a density will result in over-fueling (richer air-fuel ratio), black smoke and excessive particulate emissions. Too low a density leads to power loss as measured by fuel consumption (Hutcheson and Van Passen 1990). In heavy-duty engines, fuel density can affect injection timing and hence engine performance and emissions.

Fuel viscosity can affect fuel injection characteristics (timing and amount of fuel injected) and the mixing rate as a function of the spray shape and droplet size of the injected fuel. An increase in viscosity increases smoke and CO emissions but tends to have little effect on HC and NO_x emissions (Owen and Coley 1990). The corrosiveness, cleanliness, and lubricating properties of the fuel can all affect the service life of the fuel injection equipment—possibly contributing to excessive in-use emissions if the equipment is worn out prematurely. For the most part, however, these properties are adequately controlled by existing diesel fuel specifications. A generalized assessment of the influence of individual diesel fuel properties on emissions is shown in table 5.3.

Because diesel engines can tolerate a wide range of middle distillate compounds, fuel makers (and more commonly truckers in developing countries because of fuel-price differentials) are sometimes tempted to adulterate diesel with other fuels. These adulterants may include kerosene (especially where kerosene is subsidized for domestic use) and heavy fuel oil. Addition of heavy fuel oil to automotive diesel greatly increases the tendency to form deposits in the engine, with a corresponding increase in emissions. Excessive kerosene in diesel may damage the fuel injection pump due to inadequate lubrication, as well as produce power loss and possible safety hazards due to its lower flash point. Kerosene is sometime added to diesel oil to improve cold-weather performance.

Fuel Additives

Changing the basic formulation of diesel fuel to reduce emissions is expensive and time-consuming because it often requires major new refinery investment. For this reason the potential for achieving significant emission reductions through diesel fuel additives has drawn much attention. A major impetus for this work was provided by the California Air Resources Board, which in 1990 adopted regulations requiring diesel fuel to have a maximum 10 percent aromatics content or an additive providing equivalent emissions benefits by 1994.

Many diesel fuel additives have been and are being promoted as effective in reducing smoke and particulate emissions. Until recently, however, emission tests of most of these additives showed them to have only minor effects on particulate emissions, or to have other drawbacks that argued against their use. Historically, the types of diesel fuel additives that have been shown to affect emissions have included barium and other metallic smoke suppressants, cetane improvers, and deposit control additives.

Smoke suppressant additives are organic compounds of calcium, barium, or (sometimes) magnesium. Added to diesel fuel, these compounds inhibit soot formation during the combustion process and thus greatly reduce emissions of visible smoke. Their ef-

Table 5.3 Influence of Diesel Fuel Properties on Exhaust Emissions

Fuel Property	Smoke		Gases			Particulate matter	Noise
	White	Black	HC	CO	NO_x		
Density							
Up	Increase	Increase	-	-	-	Increase	-
Down	Decrease	Decrease	-	-	-	Decrease	-
Distillation							
IBP up	Increase	Decrease	-	-	Decrease	-	-
10% (T10); Up	Increase	Increase	Decrease	-	Increase	-	-
50% (T50); Up	Increase	Increase	Decrease	-	Decrease	Increase	-
90% (T90); Up	-	Increase	-	-	Increase	Increase	-
FBP; Up	-	Increase	-	-	Increase	Increase	-
Viscosity							
Up		Increase		Increase		Increase	Increase
Down		Decrease		Decrease		Decrease	Decrease
Cetane number[b]							
Up	Decrease	-	Decrease	Decrease	Decrease	Decrease	Decrease
Down	Increase	-	Increase	Increase	Increase	Increase	Increase

Note: IBP = initial boiling point; FBP = final boiling point.
a. Cold-engine.
b. Aromatic content is inversely correlated with the cetane number and density.
Source: Owen and Coley 1990

fects on SOF are not fully documented, but one study (Draper, Phillips, and Zeller 1988) has shown a significant increase in polycyclic aromatic hydrocarbon (PAH) content and mutagenicity of SOF with a barium additive. Particulate sulfate emissions are greatly increased with these additives because they all form stable solid metal sulfates that are emitted in the exhaust. The overall effect of reducing soot and increasing metal sulfate emissions may be either an increase or decrease in the total particulate mass, depending on the soot emissions level at the beginning and the amount of additive used. The principal smoke-suppressant additive, barium, is a highly toxic heavy metal, raising additional concerns about its possible widespread use. Copper based additives in diesel fuel can be effective in reducing particulate emissions but may result in significant increase in dioxin emissions. Trace amounts of chlorides (0.9 ppm) in diesel fuel from the dioxins (2,3,7,8 tetracholorodibenzo-p-dioxin) when catalyzed by the copper additives.

Cetane enhancers are used to improve the self-ignition properties of diesel fuel. These compounds (generally organic nitrates) are added to low-grade diesel fuels to reduce the adverse impact of poor cetane quality on cold starting, combustion noise, and emissions. Cetane improvers have been shown to reduce particulate emissions when added to substandard fuels to bring their cetane levels up to the normal diesel range. Within the normal range, however, they tend to have less effect.

Deposit control additives can prevent or reverse the increase in smoke and particulate emissions (and fuel consumption) that results from the formation of deposits on fuel injectors and also can reduce the need for maintenance for the same reason. These additives (often packaged in combination with a cetane-enhancer) help prevent and remove coke deposits on fuel injector tips and other vulnerable locations—thus maintaining new-engine injection and mixing characteristics. A study for the California Air Resources Board estimated that the increase in PM emissions due to fuel injector problems from trucks in use was more than 50 percent of new-vehicle emissions levels (Weaver and Klausmeier 1988). A large share of this excess is unquestionably due to fuel injector deposits. Although their use is quite likely justified for their long-term benefits alone, deposit control additives typically have little or no direct effect on particulate emissions.

Information is also available on a new class of diesel fuel additives which reduce smoke and particulate emissions without using barium or other metals. These all-organic smoke suppressant additives may offer significant scope for reducing particulate emissions from existing vehicles in a simple and cost-effective manner. Andrews and Charalambous (1991) have documented their tests on such an additive, which was shown to act as a cetane improver and as a detergent—helping to keep injectors clean. Most importantly, it produced a 40 to 60 percent direct and immediate reduction in diesel particulate emissions over the entire load range when mixed at 0.2 percent (v/v) concentration in the fuel. Both the solid carbon (soot) and the soluble organic fraction of the particulate matter were reduced by the additive. Although the precise actions of the additive are not known, it is suspected that it may function by inhibiting soot formation during the combustion process.

Similar results were reported for an (apparently) different additive by Smith and others (1991). The team studied the effects of a proprietary ashless fuel additive developed by Exxon. Measurements were taken in a number of heavy-duty truck engines and light-duty passenger car engines, including both naturally aspirated and turbocharged models. Significant particulate reductions were found, ranging from 15 to 40 percent depending on the engine and test cycle. Again, these reductions were immediate (and thus not the result of cleaning up injector deposits, although the additive is claimed to do that as well).

Some refiners have received permission to meet the California diesel fuel requirements by marketing diesel fuel blends having higher aromatic levels than the 10 percent allowed in the regulation but lower sulfur than the regulatory limit of 0.05 percent. These fuels also incorporate proprietary fuel additives to attain the same emissions benefits as the 10 percent aromatic fuel.

Effect of Diesel Fuel Properties on Emissions: Summary of EPEFE Results

The European Programme on Emissions, Fuels, and Engine Technologies (EPEFE) has examined the effect of variations in European diesel fuel properties on emissions of light duty diesel vehicles and heavy-duty diesel engines (Camarsa and Hublin 1995, ACEA/EUROPIA 1995a). The test vehicles and engines used in the EPEFE study conformed to EU 1996 emission limits (based on the complete European driving cycle) for light-duty diesel vehicles, and Euro 2 limits (based on the 13 -mode test) for heavy-duty diesel engines. Properties of the diesel test fuels are summarized in table 5.4.

With respect to emissions from light-duty diesel vehicles, the results of the EPEFE study are summarized in

Table 5.4   Properties of Diesel Test Fuels Used in EPEFE Study

Diesel fuel type	Density (g/l)	Polyaromatics	Cetane number	T90 (°C)	Sulfur (ppm)
Low polyaromatics and low density	826–829	0.9–1.1	49.5–58.0	326–347	404–469
Low polyaromatics and high density	857	1.1	50.0	348	415
High polyaromatics and high density	855	7.3–8.0	50.2–59.1	344–371	420–442
High polyaromatics and low density	829	7.1–7.7	50.2–50.6	346–349	402–416

Source: Camarsa and Hublin 1995

Table 5.5   Change in Light-Duty Diesel Vehicle Emissions with Variations in Diesel Fuel Properties
(percent)

Diesel fuel property[a]	CO	HC	NO$_x$	PM	CO$_2$
Density 855 to 828 g/l	-17.1[b]	-18.9[c]	+1.4	-19.4	-0.9
Polyaromatics 8 to 1 percent	+4.0[c]	+5.5	-3.4	-5.2	-1.08
Cetane 50 to 58	-25.3	-26.3	-0.18 (NS)	+5.2	-0.37 (NS)
T95 370 to 325°C	-1.8	+3.4	+4.6	-6.9	+1.59
Sulfur 2000 to 500 ppm	—	—	—	-2.4	—

— Not applicable
NS = Non-significant.
a. Baseline properties: density 855 g/l; polyaromatic content 8 percent; cetane number 50; T95 370°C; sulfur 2000 ppm.
b. Negative values indicate a decrease in emission.
c. Positive values indicate an increase in emission.

table 5.5. The main conclusions pertaining to light-duty diesel fuel effects are:

- *Particular matter.* Reducing fuel density is the most significant factor in lowering PM emissions. Reductions in polyaromatic content, T95, and sulfur also lowered PM emissions but to a lesser degree, whereas an increase in cetane number increased PM emissions.
- *Carbon monoxide.* Increasing cetane number or decreasing density is a major factor in reducing CO emissions; decreasing polyaromatic content increased CO emissions, but this effect was less pronounced compared to cetane number and density.
- *Nitrogen oxides.* Decreasing polyaromatic content reduced NO$_x$ emissions, whereas decreasing T95 and density increased NO$_x$ emissions; cetane number had no significant effect on NO$_x$ emissions.
- *Volatile organic compounds.* Increasing cetane number or decreasing density reduced HC emissions significantly, including toxic emissions of benzene, 1,3 butadane, formaldehyde, and acetaldehyde; decreasing polyaromatic content increased HC emissions including benzene emissions, while formaldehyde and acetaldehyde emissions were reduced; decreasing T95 also lowered formaldehyde and acetaldehyde emissions.

Reducing diesel fuel density and increasing cetane number are the two most important parameters in reducing light-duty diesel vehicle emissions (Camarsa and Hublin 1995).

With respect to emissions from heavy-duty diesel engines, results of the EPEFE study are summarized in table 5.6. The main conclusions with respect to heavy-duty diesel fuel effects are:

- *Particulate matter.* Reducing the content of polyaromatics or sulfur reduced PM emissions; lighter-density diesel fuel had lower PM emissions but this effect was not statistically significant.
- *Nitrogen oxides.* All fuel parameters in the EPEFE study affected emissions; reducing density, polyaromatics or T95 decreased NO$_x$ emissions, so did an increase in cetane number but the effect was not very strong; these effects, however, were not as large as those measured for CO and HC emissions.
- *Carbon monoxide.* Increasing cetane number was the only factor that reduced CO emissions; reducing T95 and density increased CO emissions.
- *Hydrocarbons.* Reducing T95 or density increased HC emissions, while a reduction in polyaromatic content or an increase in cetane number reduced HC emissions.

Table 5.6   Change in Heavy-Duty Diesel Vehicle Emissions with Variations in Diesel Fuel Properties
(percent)

Diesel fuel property[a]	CO	HC	NO_x	PM	CO_2
Density 855 to 828 g/l	+5.0[b]	+14.25[c]	-3.57	-1.59	+0.07
Polyaromatics 8 to 1 percent	0.08 (NS)	-4.02	-1.66	-3.58	-0.60
Cetane 50 to 58	-10.26	-6.25	-0.57	0 (NS)	-0.41
T95 370 to 325°C	+6.54	+13.22	-1.75	0 (NS)	+0.42
Sulfur 2000 to 500 ppm	—	—	—	-13.0	—

— Not applicable
NS = Non-significant.
a. Baseline properties: density 855 g/l; polyaromatic content 8 percent cetane number 50; T95 370°C; sulfur 2000 ppm.
b. Negative values indicate a decrease in emissions.
c. Positive values indicate an increase in emissions.
Source: Camarsa and Hubbin 1995; ACEA/EUROPIA 1995a

Increasing cetane number and decreasing polyaromatics are the two most significant parameters in reducing heavy-duty diesel engine emissions (Camarsa and Hublin 1995). The absence of any effect on PM emission from changes in cetane number is different from the results of a number of U.S. studies. This difference most likely is due to the higher cetane number of the EPEFE test fuels compared to the diesel fuels in the United States. Increasing cetane number from 50 to 58 seems to have little effect on PM emissions, but increasing it from 40 to higher levels such as 45 or 50 has a significant effect.

Individual vehicles and engines showed a wide range of response to the fuel properties investigated in the EPEFE study. In general DI and IDI light-duty vehicles showed the same trend concerning the effect of fuel properties on regulated emissions except for the NO_x response to cetane number. The lowest sensitivity to fuel property changes was associated with light-duty IDI engines with mechanical injection controls; light-duty diesel vehicles equipped with IDI engines achieved consistently low absolute emission rates in grams per kilometer.

The EPEFE test results for light-duty diesel vehicles show that the effect of fuel density on engine emissions, to some extent, was caused by the physical interaction of fuel density with the fuel management system. Vehicle sensitivity to variations in fuel density can be influenced by the choice of a specific engine tuning/calibration set; sensitivities were lowest when the engines were set to high density tuning and highest at low density tuning. Except for NO_x lowest emissions were generally observed using low density fuels with low density engine tuning. When engines were appropriately tuned to use fuel of a given density, emissions were generally lower than when density and tuning were not matched. At both high and low density tunings, mass emissions of PM, HC, and CO were higher with high density fuels although high density fuels gave lower NO_x emissions.

In case of heavy-duty diesel engines, reducing fuel density lowered engine power and increased fuel consumption. Adjustments of the injection system to the same mass fuel delivery and dynamic injection timing eliminated the difference in emission levels between the low and high density fuels tested in the EFEFE study. The effect of fuel density on engine performance and emissions was caused by the physical interaction with the fuel injection system, which is purely hydraulic in nature. Fuel density did not appear to have any effect on the combustion process (ACEA/EUROPIA 1995a).

Alternative Fuels

The possibility of substituting cleaner-burning alternative fuels for gasoline and diesel has drawn increasing attention over the past decade. A vast array of scientific and popular literature has been devoted to the subject of alternative fuels (OECD 1995; Hutcheson 1995; OTA 1995; EIA 1994; Maggio and others 1991; OTA 1990; Transnet 1990; Sperling 1989; World Bank 1981). Alternative fuels have the potential to conserve oil products and preserve energy sources, as well as reduce or eliminate pollutant emissions. Some alternative fuels have the potential for large, cost-effective reductions in emissions of regulated pollutants (CO, HC, NO_x) but may cause a sharp increase in emissions of toxic pollutants (table 5.7). Care is needed, however, in evaluating the air quality claims for alternative fuels—in many cases the same or even greater emission reductions could be achieved using a conventional fuel with an advanced emissions control system. Which approach is the more cost-effective will depend on the relative costs of conventional and alternative fuels.

In many parts of the world at present, natural gas and LPG are competitive (in resource terms) with gasoline or diesel fuel and may therefore be attractive from an economic as well as an environmental perspective. Ethanol, methanol, and hydrogen are generally more expensive than gasoline. Table 5.8 compares the costs of the main alternative fuels in the United States with that of gasoline in 1992.

The basic physical properties of the main alternative fuels are compared with gasoline and diesel fuel in table 5.9. As this table shows, all of the main alternative fuels have lower energy density, and therefore require a greater volume of on-board storage than gasoline or diesel fuel to achieve a given operating range.

While the use of alternative fuels can make low emissions easier to achieve, a vehicle using "clean" alternative fuels will not necessarily have low pollutant emissions. Much depends on the level of emission control technology employed. While strict "technology-forcing" emission standards have produced advances in gasoline and diesel engine emission control technology, alternative fuels have not been subjected to similar regulatory oversight and control until recently. This is especially true of retrofit equipment for converting existing vehicles to run

Table 5.7 Toxic Emissions from Gasoline and Alternative Fuels in Light-Duty Vehicles with Spark-Ignition Engines
(mg/km)

Compound	Gasoline	RFG	M85	M100	E85	CNG	LPG
Benzene	7.95	4.88	4.38	0.32	1.21	0.242	0.242
Toluene	33.66	3.45	8.66	2.11	0.75	0.695	0.695
m&p Xylenes	4.57	4.77	1.54	0.30	1.30	0.705	0.033
0-Xylenes	1.95	1.58	0.46	0.16	0.39	0.399	0.101
1,3-Butadiene	0.19-0.50	0.24	0.44	2.05[a]	0.12	0.093-0.404	—
Formaldehyde	4.78	0.60	13.87	21.76	3.15	2.712	4.870
Acetaldehyde	0.94	0.50	10.02	0.27	13.32	0.529	0.641
Acrolein	1.12	—	4.44	0.09	—	0.330	0.118

— Not available

Notes: RFG = reformulated gasoline; M85 = 85 percent methanol blend; M100 = pure methanol; E85 = 85 percent ethanol blend; CNG = compressed natural gas; LPG = liquefied petroleum gas.

a. Other sources suggest that emissions of 1,3 butadiene from M100 could be virtually nil.

Source: U.S. EPA 1993; OECD 1995

Table 5.8 Wholesale and Retail Prices of Conventional and Alternative Fuels in the United States, 1992

Source	Gasoline[a]	Methanol[b]	Ethanol[c]	LPG[c]	CNG[d]	LNG[e]	Hydrogen[f]
Wholesale (U.S.$/gal)	0.51-0.68	0.32-0.42	1.29-1.45	0.25-0.45	0.25-0.50	0.40-0.55	0.25[g]
Wholesale (U.S.$/therm[h])	0.41-0.54	0.56-0.74	1.70-1.91	0.29-0.53	0.26-0.52	0.53-0.72	0.85
Retail (U.S.$/gal)	0.97-1.32	0.80-0.92	n.a.	0.95-1.10	0.40-0.90	n.a.	9.60-16.00
Retail (U.S.$/therm)	0.78-1.06	1.41-1.62	n.a.[i]	1.12-1.29	0.41-0.93	n.a.	33.10-55.17

Note: 1 US gallon = 3.79 liters

n.a. = Not available

a. Wholesale and retail prices - *Oil & Gas Journal*, December 21, 1992, page 114.

b. Wholesale prices - *Oxy-Fuel News*, October 5, 1992, page 9. Retail prices - California 1992.

c. Ethanol and LPG wholesale prices - *Oxy-Fuel News*, October 5, 1992, pp. 8-9. Retail prices - California 1992.

d. Wholesale and retail prices - industry estimates.

e. Wholesale prices - industry estimates.

f. Wholesale and retail prices are based on quotes from industrial gas suppliers.

g. Natural gas and hydrogen are priced in dollars per 100 ft^3.

h. 1 therm = 100,000 Btu.

i. Not available at retail outlets.

Source: Seisler and others 1993

Table 5.9 Properties of Conventional and Alternative Fuels

Property	Gasoline	Diesel	Methanol	Ethanol	Propane (LPG)	Methane (CNG)	RME
H/C ratio	1.9	1.88	4.0	3.0	2.7	4.0	n.a.
Energy content (LHV) (MJ/kg)	44.0	42.5	20.0	26.9	46.4	50.0	36.8
Liquid density (kg/l)	0.72-0.78	0.84-0.88	0.792	0.785	0.51	0.422	0.86-0.90
Liquid energy density (MJ/l)	33.00	36.55	15.84	21.12	23.66[a]	21.13[b]	32.4-33.1
Boiling point (°C)	37-205	140-360	65	79	-42.15	-161.6	n.a.
Research Octane Numbers	92-98	~25	106	107	112	120	n.a.
Motor Octane Numbers	80-90	-	92	89	97	120	n.a.
Cetane Numbers	0-5	45-55	5	5	~2	0	45-59
Stoichiometric air-fuel ratio	14.7	14.6	6.5	9.0	15.7	17.2	13.0
Reid Vapor Pressure (psi)	8-15	0.2	4.6	2.3	208	2,400	0.5

n.a. = Not available

Notes: LHV = lower heating value; LPG = liquefied petroleum gas; RME = rapeseed methyl ether; CNG = compressed natural gas.

a. Energy density of propane at standard temperature and pressure: 0.093MJ/l.

b. Energy density of methane at standard temperature and pressure: 0.036MJ/l; at 200 bar pressure: 70.4MJ/l.

Source: EIA 1994; OECD 1995; Hutcheson 1995

Table 5.10 Inspection and Maintenance (Air Care) Failure Rates for In-Use Gasoline, Propane, and Natural Gas Light-Duty Vehicles in British Columbia, Canada, April 1993
(percent)

Model year	Gasoline	Propane	Natural gas
1974 or older	31	19	11
1975-81	37	42	24
1982-87	25	47	23
1988-93	6	44	34

Source: B.C. Ministry of Energy, Mines and Petroleum Resources 1994

on alternative fuels. Regulations requiring emissions certification for such equipment are presently effective only in California and Mexico; a similar requirement will apply in the rest of the United States beginning in 1997. Because of the absence of regulation, many vehicle conversion kits are sold with relatively poor air-fuel ratio control, resulting in unnecessarily high emissions. Installed in a modern, emission controlled gasoline vehicle, such kits can even increase emissions compared to those that the vehicle produced before being converted (table 5.10). Regulatory pressure, however, has resulted in the introduction of a number of retrofit kits incorporating electronic controls and well-designed emission control systems. Vehicles equipped with these kits have demonstrated the ability to comply with California LEV emission standards. Of course, the lowest emissions and the most efficient operation are obtained where an engine is designed by the manufacturer for alternative fuel use.

Natural Gas

Natural gas, which is 85 to 99 percent methane, has many desirable qualities as a fuel for spark-ignition engines. Clean-burning, cheap, and abundant in many parts of the world, it already plays a significant role as a motor fuel in Argentina (box 5.4), Canada, Italy, New Zealand, Russia, and the United States. Recent advances in the technology for natural gas vehicles and engines, new technologies and international standardization for storage cylinders, and the production of original equipment manufacture (OEM) natural gas vehicles in a number of countries have combined to boost the visibility and market potential of natural gas as a vehicle fuel.

There are over one million natural gas vehicles in operation worldwide; Argentina and Italy account for more than 50 percent of the global fleet. The penetration of natural gas vehicles in heavy-duty fleets is still minuscule. A CNG fleet of 2,000 buses and trucks in Sichuan Province, China is a notable example (IANGV/IGU 1994).

Most of the natural gas vehicles (NGVs) in operation worldwide are retrofits, converted from gasoline vehicles. The physical properties of natural gas make such a conversion relatively easy. Conversion costs typically

---

Box 5.4 CNG in Argentina: An Alternative Fuel for the Buenos Aires Metropolitan Region

In 1985 a program of tax exemptions was introduced in Argentina to promote the replacement of petroleum fuels by compressed natural gas (CNG). The program was quickly adopted by mid-sized trucks and taxis. By the end of 1994, 210,000 vehicles in the Buenos Aires Metropolitan Region had been converted to CNG usage. Of the nearly 40,000 officially-registered taxis, about 65 percent use CNG, with the remainder running on diesel; of the 15,000 registered buses, only about 300 run on CNG. The CNG program has substituted for about 12 percent of diesel fuel use in the Buenos Aires Metropolitan Region. This translates into a 6 percent reduction in particulate emissions.

CNG-fueled cars are preferred by taxi owners because the nominal cost of CNG (U.S.$0.26 per cu.m) is lower than diesel (U.S.$0.27 per liter) and is substantially below the price of gasoline (U.S.$0.81 per liter). CNG is supplied to gas stations at a price of U.S.$0.13 per cu. m., compared with U.S.$0.21 per liter for diesel. In addition, a new CNG taxi is 30 to 35 percent less expensive than a new diesel-fueled taxi.

Despite the fuel price advantage, the potential for switching buses to CNG seems limited, mostly due to the inconvenience associated with refueling. Refueling times for CNG are long. Refueling a bus with CNG during the day takes about 15 minutes with a fast-fueling pump and 20 minutes or more with a normal pump. During the evening hours, when pressure is higher, tanks are filled within 8 minutes.

A CNG station, including compressors, requires an investment of about U.S.$1.5 million. In addition, CNG buses tend to be more expensive than diesel-fueled buses: the price of a new diesel bus is about U.S.$85,000, compared with U.S.$90,000 for a new CNG bus. CNG vehicles are somewhat less fuel efficient than diesel and gasoline-fueled vehicles, due to the extra weight of the gas cylinders.

CNG-fueled vehicles emit no lead and produce fewer NO_x, CO, and HC emissions than gasoline-fueled vehicles. There is concern about possibly higher NO_x emissions from burning CNG compared with conventional fuels under real-life operating conditions. On balance, however, there appears to be a greater health risk attributable to lead and particulates than NO_x, CO, or HC in Buenos Aires. Thus until further evidence becomes available, it is appropriate to regard CNG as an environmentally cleaner fuel and to signal this virtue through appropriate pollution-based fuel tax differentials on both gasoline and diesel.

Source: World Bank 1995

---

range from U.S.$1,500 to U.S.$4,000 a vehicle, and are due mostly to the cost of the on-board fuel storage system. With the savings on fuel, many high-use vehicles can recover this cost in a few years, sometimes in less than 2-3 years, depending on relative fuel prices.

In recent years several thousand factory-built, light-duty NGVs have been produced in the United States, mostly by Chrysler. Ford also began limited mass production of an optimized natural gas passenger car in mid-1995. The Chrysler and Ford vehicles incorporate fuel metering and emission control systems similar to those in fuel-injected gasoline vehicles. These vehicles are by far the cleanest non-electric motor vehicles ever made—easily meeting California's stringent ultra-low emission vehicle (ULEV) standards. The cost of these vehicles in their present, limited-volume production is U.S.$4,000 to U.S.$6,000 more than gasoline-fueled vehicles, or about 20 percent of the selling price. With full mass production this additional cost is expected to drop to U.S.$1,500 – U.S.$2,500 per vehicle. CNG technology is particularly in demand for city buses, taxis, urban delivery vans and trucks and waste collection vehicles, especially in areas where emissions are tightly controlled (IEA 1993).

Over the last ten years, the market for natural gas vehicles has grown from about 10 countries to over 40, although most of the activity is concentrated in about 20 countries. Besides the United States, OEM natural gas vehicles are manufactured in fourteen countries, notably Argentina, France, Italy, Japan, and Russia. Worldwide OEM production includes 15 manufacturers in the area of light-duty engines for cars and taxis, and 24 in the area of heavy-duty engine for buses and trucks. The main manufacturers and suppliers of conversion kits are Argentina, Canada, China, Russia, France, Italy, Netherlands, New Zealand, and USA.

Engine technology and performance. Natural gas engine technology has been reviewed extensively elsewhere (IANGV/IGU 1994; Weaver and Turner 1995). Natural gas engines can be grouped into three main types on the basis of the combustion system used: stoichiometric, lean-burn, and dual-fuel diesel. Most natural gas vehicles now in operation have stoichiometric engines, which have been converted from engines originally designed for gasoline. Such engines may be either bi-fuel (able to operate on natural gas or gasoline) or dedicated to natural gas. A dedicated engine can be optimized by increasing the compression ratio and making other changes, but this is not usually done in retrofit situations because of the cost. Nearly all light-duty natural gas vehicles use stoichiometric engines, with or without three-way catalysts, as do a minority of heavy-duty natural gas vehicles.

Lean-burn engines use an air-fuel mixture with more air than is required to burn all of the fuel. The extra air dilutes the mixture and reduces the flame temperature, thus reducing engine-out NO_x emissions and exhaust temperatures. Because of reduced heat losses and various thermodynamic advantages, lean-burn engines are generally 10 to 20 percent more efficient than stoichiometric engines. Without turbocharging, however, the power output of a lean-burn engine is less than that of a stoichiometric engine. With turbocharging the situation is reversed. Because lean mixtures knock less readily, lean-burn engines can be designed for higher levels of turbocharger boost than stoichiometric engines and thus can achieve higher power output. The lower temperatures experienced in these engines also contribute to longer engine life and reliability. For these reasons, most heavy-duty natural gas engines are of the lean-burn design. These include a rapidly growing number of heavy-duty, lean-burn engines developed and marketed specifically for vehicular use.

Dual-fuel diesel engines are a special type of lean-burn engine in which the air-gas mixture in the cylinder is ignited not by a spark plug but by the injection of a small amount of diesel fuel, which self-ignites. Most diesel engines can readily be converted to dual-fuel operation, retaining the option to run on 100 percent diesel fuel if gas is not available. Because of the flexibility this allows, the dual-fuel approach has been popular for heavy-duty retrofit applications. Until recently dual-fuel engine systems tended to have very high HC and CO emissions due to the production of mixtures too lean to burn at light loads. Developments such as timed gaseous fuel injection systems have overcome these problems (Weaver and Turner 1994).

Emissions. Because natural gas is mostly methane, natural gas vehicles have lower exhaust emissions of non-methane hydrocarbons than gasoline vehicles, but higher emissions of methane. Controlled tests conducted by AQIRP on catalyst-equipped light-duty vehicles (U.S. 1992 and 1993 production models) showed that non-methane emissions of CNG vehicles were about one-tenth those of their counterpart gasoline-fueled vehicles, while methane emissions were ten times higher. CO and NO_x emissions varied significantly among the three pairs of test vehicles (passenger car, van, and pick-up truck). CO emissions were 20 to 80 percent lower with CNG compared to gasoline; NO_x emissions ranged from 80 percent lower with CNG (for the pick-up truck) to about the same levels for both CNG and gasoline (for the medium-duty van). Toxic air pollutant emissions were dramatically lower with CNG; formaldehyde emissions were reduced by 50 percent and acetaldehyde

emissions by 80 percent for CNG vehicles compared to their gasoline counterparts. Benzene and 1,3 butadiene are virtually nonexistent in CNG exhaust (AQIRP 1995c). Since the fuel system is sealed, there are no evaporative or running-loss emissions, and refueling emissions are negligible. Cold-start emissions from natural gas vehicles also are low, since cold-start enrichment is not required, and this reduces both non-methane hydrocarbon and carbon monoxide emissions. Natural gas vehicles are normally calibrated with somewhat leaner fuel-air ratios than gasoline vehicles, which also reduces CO emissions. Given equal energy efficiency, CO_2 emissions from natural gas vehicles will be lower than for gasoline vehicles, since natural gas has a lower carbon content per unit of energy. In addition, the high octane value for natural gas (RON of 120 or more) makes it possible to attain increased efficiency by increasing the compression ratio. Optimized heavy-duty natural gas engines can approach diesel efficiency levels. Emissions of nitrogen oxides from uncontrolled natural gas vehicles may be higher or lower than comparable gasoline vehicles, depending on the engine technology but are typically somewhat lower. Light-duty natural gas vehicles equipped with modern electronic fuel control systems and three-way catalytic converters have achieved NO_x emissions more than 75 percent below the stringent California ULEV standards (table 5.11).

In the past few years a number of heavy-duty engine manufacturers have developed diesel-derived lean-burn natural gas engines for use in emissions-critical applications such as urban transit buses and delivery trucks. These engines incorporate the low-NO_x technology used in stationary natural gas engines, and typically an oxidation catalyst as well. They are capable of achieving very low levels of NO_x, and particulate matter (less than 2.0 g/bhp-hr NO_x and 0.03 g/bhp-hr PM) with high efficiency, high power output, and (it is anticipated) long life. Five such engines—the Cummins L10 and Detroit Diesel Series 50 engines for transit buses and the Hercules 5.6l and 3.7l and Cummins 6B engines for schoolbuses and medium trucks—have been certified in California. A comparison of emissions from a two-stroke diesel engine and a Cummins L10 natural gas engine is shown in table 5.12. The emissions performance of DAF GKL1160 diesel engines converted to lean-burn naturally-aspirated CNG engines (with an oxidation catalyst) for use in city and regional bus services in the Netherlands is presented in table 5.13.

Fuel storage and refueling. Pipeline-quality natural gas is a mixture of several different gases. The primary constituent is methane (CH_4), which typically makes up 85 to 99 percent of the total. The rest is primarily ethane

Table 5.11   Emissions Performance of Chrysler Natural Gas Vehicles
(grams per mile)

	Mileage	NMOG	CO	NO_x	HCHO
LEV Std[a]	50,000	0.195	5.0	1.1	0.022
	120,000	0.280	7.3	1.5	0.032
ULEV Std[b]	50,000	0.050	2.2	0.4	0.009
	100,000	0.070	2.8	0.5	0.013
Chrysler B350 Ramvan (5,751–8,500 lbs)					
Gasoline	50,000	0.19	3.4	0.51	n.a.
CNG	50,000	0.031	2.3	0.05	0.002
(LEV)	120,000	0.040	3.1	0.05	0.003
Chrysler Minivan (3,751–5,750 lbs)					
Gasoline	50,000	0.20	1.2	0.19	n.a.
CNG	50,000	0.021	0.4	0.04	0.0002
(ULEV)	100,000	0.035	0.4	0.05	0.0002

n.a. = Not available
Notes: NMOG = Non-methane organic gases.
  HCHO = Formaldehyde
a. LEV standard for medium-duty vehicles: 5,751–8,500 lbs.
b. ULEV standard for light-duty trucks: 3,751–5,750 lbs.
Source: California Air Resources Board 1994

Table 5.12   Emissions from Diesel and Natural Gas Bus Engines in British Columbia, Canada
(grams per mile)

Pollutant	Current two-stroke engine diesel[a]	Natural gas[b]	Difference	Percent reduction
HC	0.79	0.02	0.77	97.5
CO	20.85	0.03	20.82	99.9
NO_x	20.36	10.26	10.10	49.6
PM	3.32	0.12	3.20	96.4

a.   A Detroit Diesel 6V-92TA two-stroke diesel engine.
b.   A Cummins L10 natural gas engine.
Source: B.C. Ministry of Energy, Mines, and Petroleum Resources 1994

Table 5.13   Emissions from Diesel and Natural Gas Bus Engines in the Netherlands
(g/kwh)

Bus type	NO_x	HC	CO	PM
CNG city bus	4.3	2.1	0.4	<0.05
CNG regional bus	2.9	3.1	2.5	<0.05
Diesel bus	14.0	1.2	4.0	0.55
EU 1996 standards	7.0	1.1	4.0	0.15

Source: IANGV/IGU 1994

(C_2H_6), with smaller amounts of propane, butane, and inert gases such as nitrogen, argon, and carbon dioxide. The mix of minor constituents varies considerably depending on the source and processing of the gas. The Society of Automotive Engineers (SAE) has recommended acceptable compositional limits for natural gas intended for introduction into the fuel container of natural gas vehicles and for use as an automotive fuel.

Refueling technology is one of the key elements in developing a successful market for natural gas vehicles which, on average, require refueling twice as frequently as gasoline vehicles. Refueling stations are categorized as either slow-fill or fast-fill. The fast-fill station operates in a manner comparable to gasoline stations with the vehicle refueling in 3-5 minutes. The slow-fill station accomplishes the refueling of one or more vehicles over a period of several hours, typically 12-14 hours. The cost of a slow-fill station is about U.S.$3000–4000 while a fast-fill station may cost between U.S.$100,000–250,000. A Vehicle Refueling Appliance (the home compressor) has been developed in the United States for the residential home market for refueling one passenger car overnight. The cost of the Vehicle Refueling Appliance is about U.S.$2,500 (IANGV/IGU 1994).

Natural gas under normal temperature and pressure creates significant problems with fuel storage aboard the vehicle. At present, natural gas is stored either as a gas (CNG) in high-pressure cylinders or as a cryogenic liquid (LNG) in an insulated tank. Both forms of storage are considerably heavier, more expensive, and bulkier than storage for an equivalent amount of gasoline or diesel. The costs of compressing or liquefying natural gas in order to store it are also substantial. The high-pressure cylinders needed for CNG weigh more and occupy more space than the vacuum-cryogenic tanks used for LNG, but the cost of the two storage systems is about the same. Nearly all the natural gas vehicles now in use rely on CNG rather than on LNG. This is due to the greater cost of liquefaction and the handling difficulties

involved with LNG. Use of LNG may be attractive in cases where large quantities are available at little or no incremental cost (as in LNG-importing countries), and for vehicles such as locomotives and long-haul trucks, which need to carry large amounts of fuel.

Figure 5.2 compares the weight of different CNG cylinders and fuel with a gasoline tank containing the same amount of energy. As this figure shows, the early plain steel CNG cylinders were far heavier than a gasoline tank. The additional weight resulted in a fuel economy penalty of 0.5 liters per 100 km for a CNG passenger car fitted with steel cylinders (Hutcheson 1995). With advances in cylinder technology, however, this weight disadvantage has been greatly reduced.

Fuel price and supply. After coal, natural gas is the most abundant fossil fuel. The ratio of proven gas reserves to annual production is double that of petroleum, and a larger proportion of world gas reserves than petroleum reserves is found outside the Middle East. Today, most major urban centers and many minor ones in industrial countries are served by a large network of gas pipelines. Other technologies for natural gas transportation

Figure 5.2 A Comparison of the Weight of On-Board Fuel and Storage Systems for CNG and Gasoline

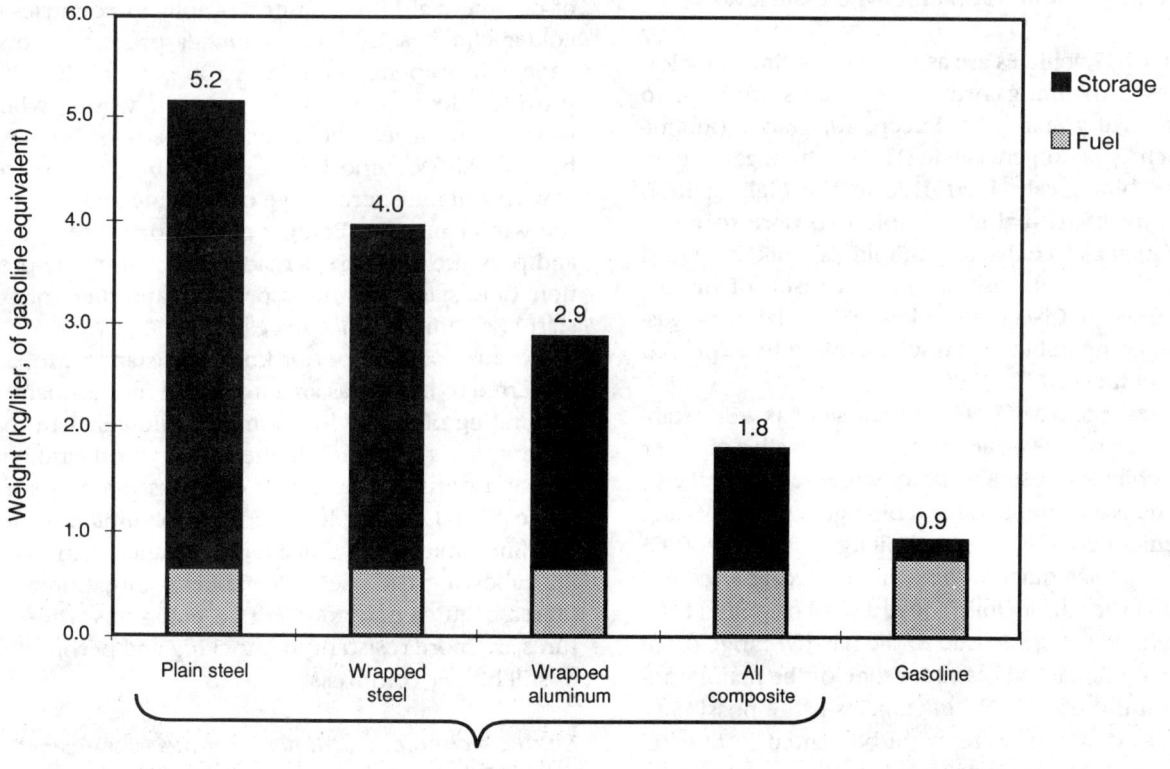

Source: Weaver and Chan 1995

and distribution include liquefaction and shipment in liquid form (LNG), and short-distance transport of CNG in large banks of cylinders. Japan, Korea, Taiwan (China), and many countries of Western Europe now import significant quantities of natural gas in the form of LNG.

Owing to transportation difficulties, the cost of natural gas varies greatly from country to country and even within countries. Where gas is available by pipeline from the field, its price is normally set by competition with residual fuel oil or coal as a burner fuel. The market-clearing price of gas under these conditions is typically about U.S.$3.00 per million BTU (equivalent to about 0.41 U.S. cents per gallon of diesel fuel equivalent). Compression costs for CNG can add another U.S.$0.50 to U.S.$2.00 per million BTU, depending on the size of the facility and the natural gas supply pressure.

The cost of LNG also varies considerably, depending on specific contract terms (there is no effective "spot" market for LNG). The cost of small-scale liquefaction of natural gas is about U.S.$2.00 per million BTU, making it uneconomic compared with CNG in most cases. Where low-cost remote gas is available, however, LNG production can be quite economic. Typical 1987 costs for LNG delivered to Japan were about U.S.$3.20 to U.S.$3.50 per million BTU. The costs of terminal receipt and transportation would probably add another U.S.$0.50 per million BTU, at the wholesale level.

Safety. CNG vehicles are as safe as gasoline vehicles. Compressed natural gas storage cylinders are built to rigorous quality standards. Except for trace contaminants such as hydrogen sulfide (H_2S), natural gas is nontoxic and biologically inert. Due to the high ignition temperature of natural gas, simple exposure to a hot surface (such as an exhaust manifold) is unlikely to lead to a fire. Overall, the risk of fire as a result of uncontrolled release of CNG is much lower than that from gasoline, and comparable to that which might be expected with diesel fuel (OECD 1995).

The safety record of LNG vehicles is not as well established as that of CNG, due to the much smaller number of LNG vehicles in use. As with compressed natural gas, vaporized LNG is non-toxic and biologically inert. Being a cryogenic fluid, LNG (with a boiling point of –160°C) tends to vaporize quickly when spilled. Experience has shown that the vapor cloud above a pool of spilled LNG is very difficult to ignite, due to the narrow range of inflammability of natural gas vapor. One of the major concerns with the use of LNG in vehicles is the possibility that excess vapor pressure might be vented from inactive vehicles left in an enclosed area such as a garage for long periods of time, possibly causing an explosion. LNG tanks are equipped with pressure relief valves, and newer technology fuel tanks guarantee fuel storage for

up to eight to ten days without pressure relief valves being activated. Another danger associated with the use of LNG is the possibility of cryogenic burns due to contact with spilled LNG during refueling or as the result of an accident. LNG nozzles, hoses, and dispenser, however, are equipped with valves to prevent excessive release of LNG in the event of an accident. The fact that LNG tanks are tough and designed to fail predictably reduces the likelihood of contact with spilled fuel during an accident (Hutcheson 1995).

Liquefied Petroleum Gas (LPG)

Liquefied petroleum gas, at present the most widespread of gaseous fuels, powers an estimated four millions vehicles in several countries, notably the Australia, Canada, Italy, Japan, Korea, Netherlands, New Zealand, Thailand, and the United States. As a fuel for spark-ignition engines, it has many of the same advantages as natural gas with the additional advantage of being easier to carry aboard the vehicle. Its major disadvantage is the limited supply, which rules out any large-scale conversion to LPG fuel.

LPG is typically a mixture of several gases in varying proportions. Major constituent gases are propane (C_3H_8) and butane (C_4H_{10}), with minor quantities of propene (C_3H_6), various butenes (C_4H_8), iso-butane, and small amounts of ethane (C_2H_6). The composition of commercial LPG is quite variable. In countries with colder climates, LPG has a higher proportion of propane and propene (as high as 100 percent) in order to provide adequate vapor pressure in winter, while in warmer countries LPG consists mostly of butane and butenes. LPG composition may also be varied seasonally with a higher percentage of propane and propene in the winter months. Being a gas at normal temperature and pressure LPG mixes readily with air in any proportion. Cold starting is not a problem and therefore cold-start enrichment is not necessary.

Because of its superior knock-resistance, propane is preferred to butane as an automotive fuel. Propane's octane rating of 112, while somewhat lower than that of natural gas, is still much higher than typical gasoline values, and permits the use of compression ratios in the range of 11-12:1. The lean combustion limit of propane-gasoline mixtures is considerably leaner than for gasoline, allowing the use of lean-burn calibrations, which increase efficiency and reduce emissions. These mixtures are more resistant to knocking and permit the use of still higher compression ratios.

Engine technology and performance. Engine technology for LPG vehicles is similar to that for natural gas vehicles, with the exception that LPG is not commonly used in dual-fuel diesel applications due to its relatively poor knock resistance (Hutcheson 1995). Both stoichiometric and lean-burn LPG engines have been developed with good results. Nearly all LPG vehicles currently in operation are aftermarket retrofits of existing gasoline vehicles, mostly using mechanical (as opposed to electronic) conversion systems. The costs of converting from gasoline to LPG are considerably less than those of converting to natural gas, due primarily to the lower cost of the fuel tanks. For a light-duty vehicle, conversion costs of U.S.$800–U.S.$1,500 are typical. As with natural gas, the cost of conversion for high-use vehicles can typically be recovered within a few years through lower fuel costs. Owing to the lack of strong industry support, research on LPG as a vehicle fuel has been limited in comparison to natural gas.

Emissions. LPG has many of the same emission characteristics as natural gas. On an energy basis LPG has a lower carbon content than gasoline or diesel fuel. When used in spark-ignition engines, LPG produces near-zero particulate emissions, very little CO and moderate HC emissions. Variations in the concentration of different hydrocarbons in LPG can affect the species composition and reactivity of HC exhaust emissions. As olefins (such as propene and butene) are much more reactive in contributing to ozone formation than paraffins (such as propane and the butanes), an increase in the olefin content of LPG is likely to result in increased ozone-forming potential of exhaust emissions. Due to the gas-tight seals required on the fuel system, evaporative emissions are negligible. LPG emissions during refueling are significant, and U.S. codes require a vapor vent on the tank to be opened to avoid overfilling. Technology to eliminate these emissions exists, however. Exhaust NMHC and CO emissions are lower with LPG than gasoline. The carbon dioxide emissions typically are also somewhat lower than those for gasoline due to the lower carbon-energy ratio and the higher octane quality of LPG. NO_x emissions are similar to those from gasoline vehicles, and can be effectively controlled using three-way catalysts. Overall, LPG provides less air quality benefits then CNG mainly because the hydrocarbon emissions are photochemically more reactive and emissions of carbon monoxide are higher.

Modern European dual-fueled LPG cars have achieved impressive results in reducing emissions. Average emissions and fuel consumption test results for five dual-fueled passenger cars fitted with closed-loop three-way catalysts and third generation LPG equipment are summarized in table 5.14. The tests were conducted over the ECE+EUDC cycle. Table 5.15 shows the limited emissions data available for LPG vehicles in California.

Modern spark-ignition LPG-fueled engines equipped with a three-way catalyst can easily meet (Euro 2 and 3) stringent heavy-duty emission standards. Lean burn engines in combination with an oxidation catalyst can also achieve very low emission results (Hutcheson 1995). The very low levels of particulate emissions with both stoichiometric and lean-burning LPG engines continue to be their strongest point, particularly as this is attainable with low NO_x emissions. With respect to CO_2 emissions and energy consumption, LPG-fueled heavy-duty vehicles typically consume 20-30 percent more energy. As LPG has a higher energy content per kilogram than diesel and a lower carbon mass fraction, the CO_2 emissions of LPG in heavy-duty use are roughly comparable to diesel. LPG also has an advantage over CNG in that it is stored at relatively low pressure in lighter tanks, which impose a lower energy penalty.

Fuel storage. LPG is stored on the vehicle as a liquid under pressure. LPG tanks, since they must contain an

Table 5.14   Comparison of Emissions and Fuel Consumption for Five Modern Dual-Fueled European Passenger Cars Operating on Gasoline and LPG

Emissions and fuel consumption	Gasoline	LPG
CO (g/km)	0.87	0.72
HC (g/km)	0.14	0.12
NO_x (g/km)	0.12	0.16
Fuel consumption (l/100km)	8.7	11.3
Energy consumption (MJ/km)	2.8	2.7

Source: Hutcheson 1995

Table 5.15   Pollutant Emissions from Light- and Heavy-Duty LPG Vehicles in California

Vehicle type	NO_x	NMHC	CO
Passenger car (g/mile)	0.2	0.15	1.0
Heavy-duty engine (g/bhp-hr)	2.8	0.5	23.2

Source: CARB 1991

internal pressure of 20–40 atmospheres, are generally cylindrical with rounded ends and are much stronger than the tanks used for storing gasoline or diesel fuel (albeit much less so than those used for CNG). Propane can be pumped from one tank to another like any liquid, but the need to maintain pressure requires a gastight seal. Except for the need for a standardized, gastight connection, LPG used as vehicle fuel can be dispensed in much the same way as gasoline or diesel fuel. So that some vapor space is always available for expansion, LPG tanks used in automotive service must never be filled to more than 80 percent capacity. Automatic fill limiters are incorporated in the tanks for this purpose.

Fuel price and supply. LPG is produced in the extraction of heavier liquids from natural gas and as a by-product in petroleum refining. LPG supply exceeds the demand in most petroleum-refining countries, so the price is low compared with other hydrocarbons. Wholesale prices for consumer-grade propane in the United States have ranged between 0.25 and 0.30 U.S. cents a gallon for several years, or about 30 percent less than the wholesale cost of diesel on an energy basis. Depending on the locale, however, the additional costs of storing and transporting LPG may more than offset this advantage.

Because the supply of LPG is limited (about 5-10 percent of the amount of petroleum produced and approximately 3 percent of the quantity of natural gas), and small in relation to other hydrocarbon fuels, any large-scale conversion of heavy-duty vehicles to LPG use would likely absorb the existing glut, causing prices to rise. For this reason LPG probably makes the most sense as a special fuel for use in vehicles—such as rickshaws, taxis, urban buses and delivery trucks—operating in especially pollution-sensitive areas. LPG is used in urban taxis in many Asian cities, and is also a popular fuel found in converted three-wheelers. Several hundred LPG-fueled city buses have been in use in Vienna, Austria.

Safety. LPG poses a greater safety risk than CNG. Unlike natural gas, LPG vapors are heavier than air, so that leaks from the fuel system tend to "pool" at ground level, where they might contact ignition sources. To some extent, the same considerations apply to conventional liquid fuels, although their volatility is somewhat lower. The flammability limits of LPG vapor in the air are also broader than those for natural gas. However, where appropriate ventilation systems and work practices are employed, LPG vehicles can be parked and maintained in enclosed premises without any problems. In addition, the risk of leaks from modern fuel systems is very small. Nevertheless, vehicles fitted with LPG systems may be subject to restrictions on parking in enclosed spaces and may be prohibited from using underground

communal parking facilities. Like natural gas, LPG is non-toxic. Also like natural gas, LPG is stored on the vehicle in sealed pressure vessels which are much stronger than typical gasoline fuel tanks. The probability of a tank rupturing and releasing fuel is thus less than for gasoline (Hutcheson 1995).

Methanol

Widely promoted in the United States as a "clean fuel," methanol has many desirable combustion and emissions characteristics, including good lean-combustion characteristics, low flame temperature (leading to low NO_x emissions), and low photochemical reactivity. Methanol is dispensed from fuel pumps in a manner similar to gasoline. The major drawback of methanol as a fuel is its cost and the volatility of pricing. While methanol prices have been highly volatile in the past, there is little prospect for it to become price-competitive with conventional fuels unless world oil prices increase greatly.

With an octane number of 112 and excellent lean combustion properties, methanol is a good fuel for lean-burn Otto-cycle engines. Its lean combustion limits are similar to those of natural gas, while its low energy density results in a low flame temperature compared with hydrocarbon fuels, and in lower NO_x emissions. To this can be added its low vapor pressure, a characteristic that contributes to a significant reduction in evaporative losses.

Engine technology and performance. The low energy density of methanol means that a large amount (roughly twice the mass of gasoline) is required to achieve the same power output. This intrinsic disadvantage is partially compensated by the high octane number of methanol and its charge cooling capability, so that 1.5 liters of methanol give the same mileage as one liter of gasoline in a dedicated use vehicle. The high heat of vaporization of methanol, combined with the large amounts required, makes it difficult to ensure complete vaporization and requires special attention to the design of intake manifolds and cold-start procedures. Otto-cycle engines using pure methanol (M100) become nearly impossible to start below 5°C without special pilot fuels or supplemental heating techniques.

The low-temperature starting and other problems with pure methanol have led the developers of light-duty methanol vehicles to specify an 85 percent methanol/15 percent gasoline blend (M85) for the current generation of methanol vehicles. The added gasoline increases the vapor pressure of the mixture, improving starting and making the headspace mixture too rich to burn. It also makes the flame luminous. Most of the emissions benefits of methanol (such as low evaporative emissions) are lost with the switch to M85, however. If the fuel mixture contains even a small percentage of methanol, the vapor pressure rises dramatically.

Flexible fuel vehicles—capable of running on any combination of gasoline and up to 85 percent methanol—have been developed, and fleets of these vehicles are being tested. The engines and emission control systems on these vehicles are similar to those for advanced-technology gasoline vehicles, and the overall energy efficiency and emission properties are also similar. The incremental cost of the flexible-fuel vehicle, compared with one designed for gasoline only, is estimated to be about U.S.$300-400 in large-scale production.

Heavy-duty engines also can be operated on methanol, using a variety of technical approaches. A number of heavy-duty methanol engines have been developed and are reported in the literature. The most promising approach is to inject the methanol in liquid form, as in a diesel engine. Engines using this approach can attain diesel-like efficiencies. Detroit Diesel Corporation in the United States certified its heavy-duty transit bus engine, the 6V-92TA, for methanol use in 1991, and a number of buses using this fuel are now in operation. This engine uses direct injection of methanol in liquid form.

Maintenance costs with methanol-fueled engines could be higher compared to gasoline-fueled engines. Methanol combustion products are fairly corrosive, and certain engine designs tested to date suggest that the engine life of these designs is likely to be considerably shorter. Other designs have experienced no significant added wear. Oil change intervals also may need to be shortened to counteract the increased corrosiveness of the combustion products.

Emissions. While the potential emissions benefits of neat methanol (M100) remain a matter of some conjecture, the use of M85 offers some advantages. As methanol contains no sulfur or complex organic compounds, it promises two air quality benefits over gasoline: lower ozone-forming potential in some areas and reduced emissions of a number of toxic air contaminants (especially benzene and other polycyclic aromatic hydrocarbons). Emissions of formaldehyde (a primary combustion product of methanol), however, can be significantly higher; more than five times the amount emitted by comparable gasoline-fueled vehicles. Formaldehyde is toxic and probably a carcinogenic. In confined spaces such as garages and tunnels, emissions from methanol-fueled vehicles could result in harmful concentration levels of formaldehyde. The difficult cold-starting characteristics of methanol also lead to high unburned fuel and CO emissions during cold starts. HC emissions from methanol engines are mostly unburned methanol and formaldehyde. Since methanol shows less photochemical reactivity than most hydrocarbons, it was long thought that methanol vehicles could help reduce urban ozone problems. More recent studies, however, have shown that the high reactivity of the formaldehyde offsets the low reactivity of methanol, so that net ozone benefits are small.

The use of M85 results in reduced emissions of formaldehyde and 1,3 butadiene, but acetaldehyde emissions (not significant with neat methanol) approach the levels associated with ethanol-gasoline blends (OECD 1995). Certification emissions of NMHC, CO, and NO_x are compared for gasoline and M85 flexible fuel vehicles (FFVs) in table 5.16, which also shows the applicable 1994 U.S. emission standards and transitional low emission vehicle (TLEV), low-emission vehicle (LEV), and ultra-low emission vehicle (ULEV) standards for flexible fuel vehicles. The use of M85 in FFV's resulted in significantly reduced NMHC and NO_x emissions compared to equivalent gasoline models. CO emission were similar for both M85 and gasoline-fueled vehicles.

AQIRP investigations have shown that, compared to gasoline, the use of M85 in earlier (pre-1990) prototype

Table 5.16 Standards and Certification Emissions for Production of M85 Vehicles Compared with Their Gasoline Counterparts

(g/mile)

Vehicle type	NMHC	CO	NO_x
Standard			
Current	0.390	0.4	7.0
TLEV	0.250	0.4	3.4
LEV	0.125	0.2	3.4
ULEV	0.075	0.2	1.7
Certification			
Gasoline			
Ford Taurus	0.170	0.11	2.3
Chrysler/Dodge Spirit	0.130	0.1	2.1
M85			
Ford Taurus	0.091	0.1	1.4
Chrysler/Dodge Spirit	0.040	0.3	1.1

Source: CARB 1994

flexible fuel and variable fuel vehicles (VFVs) reduced CO emissions (by 31 percent), increased NO_x emissions (by 23 percent), while exhaust NMHC emissions remained about the same. Exhaust benzene, 1,3 butadiene, and acetaldehyde emissions were lower with M85 than gasoline, while formaldehyde emissions were five times higher. Evaporative NMHC also increased significantly with M85. The use of M10 (a blend of 10 percent methanol and 90 percent gasoline) in the prototype FFVs and VFVs reduced all exhaust emissions but increased evaporative emissions (AQIRP 1993).

Heavy-duty methanol engines produce significantly lower NO_x and PM emissions than similar heavy-duty bus diesel engines, while NMHC, CO, and formaldehyde emissions, tend to be much higher. These emissions, however, have been controlled successfully by catalytic converters.

Safety. Methanol exhibits a number of safety and handling problems which have led to concerns over its possible widespread use. Unlike hydrocarbon fuels, methanol burns with a nearly non-luminous flame that is impossible to see in daylight. This has generated concern over effects on firefighters and passers-by in the event of a fire. Methanol's vapor pressure is also such that it can form a flammable mixture in the headspace of its fuel tank at normal ambient temperatures. The toxicity of methanol, and its lack of taste or odor, indicate that poisoning may be a much greater problem than with gasoline (Maggio and others 1991; EIA 1994; Hutcheson 1995).

Fuel price and supply. Methanol can be produced from natural gas, crude oil, coal, biomass or cellulose. At current prices the most economical feedstock for methanol production is natural gas, especially natural gas found in remote regions where it has no ready market. The current world market for methanol is for a commodity chemical rather than a fuel. World methanol production capacity is limited (about 7 billion gallons per annum) and projected to be tight at least through the 1990s. Methanol is used as feedstock in the production of MTBE, and the huge increase in MTBE demand for reformulated gasoline caused methanol prices to reach high levels in 1994. AQIRP analysis indicates that M85 is likely to be more costly than conventional gasoline in the short to medium term. Estimates range from U.S.$0.15 – 0.17 per liter in the short term. Transaction costs for methanol attaining a significant market penetration are likely toe high (AQIRP 1993). Methanol may be best used in specialized, ultra-low-emission applications such as urban buses and trucks. Any large-scale conversion of vehicles to methanol would require extensive new methanol production capacity if prices were not to rise significantly.

The price of methanol on the world market has fluctuated dramatically in the past decade, from about U.S.$0.06 per liter in the early 1980s to U.S.$0.16-0.18 in the late 1980s, to as much as U.S.$0.47 in 1994. The lower prices reflect oversupply; the higher prices reflect shortages. Estimates of the long-term supply price of methanol for the next decade range from U.S.$0.11 – 0.15 per liter (Wagner and Tatterson 1987; DiFiglio and Lawrence 1987; AQIRP 1993; Hutcheson 1995). This would be equal to U.S.$0.27 – 0.31 per liter on an energy-equivalent basis compared to a spot gasoline price of U.S.$0.18 – 0.20 per liter. In addition to new methanol supply capacity, any large-scale use of methanol as a vehicle fuel would require substantial investments in fuel storage, transportation, and dispensing facilities, which would further increase the delivered cost of the fuel.

Ethanol

Ethanol has attracted considerable attention as a motor fuel due to the success of the Brazilian Pro-alcohol program initiated in 1975 as a response to the global oil crisis of the 1970s. Despite the technical success of this program — around one-third of Brazil's 12 million cars are powered by ethanol fuel — the high cost of producing ethanol (compared with hydrocarbon fuels) has required large direct and indirect subsidies amounting to over U.S.$1.0 billion per annum.

As the next higher of the alcohols in molecular weight, ethanol resembles methanol in most combustion and physical properties, except that it is considerably cleaner, less toxic and less corrosive. In addition, ethanol has a higher volumetric energy content. Ethanol or grain alcohol can be produced by processing agricultural crops such as sugar cane or corn but it is more expensive to produce than methanol and requires large harvest of these crops and large amounts of energy for its production. This can lead to environmental problems, particularly soil degradation (OECD 1995).

Ethanol is most conveniently manufactured in one of the two forms: a 95 percent mixture with water, known as hydrous or hydrated ethanol, and a 99.5 percent mixture with benzene, known as anhydrous or absolute ethanol. Fuel grade ethanol, as produced in Brazil, is manufactured by distillation, and contains several parts water (by volume).

When ethanol is blended with ordinary gasoline in proportions up to 22 percent, the resulting mixture known as gasohol, may be burned in ordinary spark-ignition automobile engines. Pure ethanol is used extensively as a blendstock for gasoline in Brazil, South Africa, and the United States. While ethanol is completely miscible in gasoline, the presence of even a small amount of water can result in phase separation. Hence, only anhydrous ethanol is used in the gasohol mixture. Gasohol blends form azeotropes which cause a disproportionate increase in vapor pressure together with a reduction in front-end distillation temperature. This effect varies with ethanol concentration but is particularly significant at low ethanol concentrations up to around 10 percent. Such an increase in vapor pressure can cause hot drive-ability problems in vehicles. As a consequence, the base gasoline must be tailored to accept ethanol. Compared to the base gasoline, gasohol has a higher RON while the MON is about the same. Under low speed and acceleration conditions, road performance of gasohol is generally similar to or better than gasoline with the same octane quality. The high-speed, high-load performance of gasohol, however, tends to be inferior to that of an equivalent - octane quality gasoline (Hutcheson 1995; Weiss 1990).

Engine technology and performance. Anhydrous ethanol has about 65 percent of the heat energy content of an equal volume of gasoline. However, the physio-chemical and combustion characteristics of alcohol permit it to achieve a higher thermal efficiency in an internal combustion engine. Up to a proportion of about 20 percent anhydrous ethanol this increase in efficiency compensates almost fully for the slightly lower energy content of gasohol blend, so that there are no noticeable mileage penalties in substituting gasohol for gasoline. The fuel economy of vehicles using gasohol is dependent mostly on the engine type and driving conditions. Dynamometer test results and road trials on non-catalyst cars show no change in fuel economy for ethanol contents up to 5 percent volume. For closed-loop catalyst-equipped cars there is a noticeable reduction in fuel economy associated with gasohol (Hutcheson 1995).

A special engine is required to burn mixtures richer than 22 percent ethanol. "All-alcohol" cars bearing such engines normally burn hydrated ethanol and require special adaptations mainly to prevent corrosion. These adaptations costing about U.S.$500 per vehicle, include coating the insides of the fuel tank with tin, the fuel lines with copper and nickel, and the carburetor with zinc. The pistons also have to be strengthened to permit the use of higher compression ratios possible with the use of alcohol. Such engines cannot compensate for the lower energy content of a given volume of ethanol compared to gasohol, even though their higher compression ratios achieve even higher thermal efficiency than

those achieved with gasohol in an ordinary engine. About 1.25 litres of pure hydrated alcohol give the same mileage as one liter of gasoline—a significant fuel economy penalty. Ethanol-powered cars tend to be poor starters in cold weather (Homewood 1993; Weiss 1990). To ease this problem, flexible-fuel vehicles have been developed in the United States to operate on any blend of gasoline and ethanol up to 85 percent ethanol by volume.

Emissions. By blending 22 percent anhydrous ethanol with gasoline to produce gasohol, Brazil has been able to eliminate completely the requirement for lead as an octane enhancer. Emissions from a sample of non-catalyst Brazilian vehicles using ethanol and gasohol fuels are compared in table 5.17.

Emissions from ethanol-powered vehicles are not as well characterized as for other alternative fuels, but are believed to be high in unburnt ethanol, acetaldehyde (more than 12 times compared to gasoline vehicles), and formaldehyde. These emissions, however, can be effectively controlled with a catalytic converter. Ethanol-run vehicles generate 20 to 30 percent less CO and roughly 15 percent less NO_x compared to gasoline-fueled vehicles. As the vapor pressure of pure ethanol is much lower than gasoline, evaporative emissions from ethanol-powered vehicles are significantly lower. The low vapor pressure of ethanol, however, could cause cold-starting problems in colder climates and result in higher cold-start exhaust emissions. Emissions of benzene, 1,3 butadiene and particulate matter are also substantially lower for ethanol-powered vehicles (Homewood 1993; Pitstick 1993). Limited ozone modeling of speciated emissions from ethanol-fueled vehicle suggests that the ozone-forming potential of neat ethanol is less than that of gasoline and diesel, about the same as that of reformulated gasoline, and higher than that of LPG, methanol, and CNG.

Some of the limitations of pure ethanol such as cold-start problems and the need for dedicated OEM vehicles can be addressed by blending ethanol with gasoline to form E85 fuel (85 percent ethanol and 15 percent gasoline) for use in dedicated Variable Fuel Vehicles (VFVs)/Flexible Fuel Vehicles (FFVs). The gasoline component vaporizes more readily than ethanol, and makes cold starting possible. But ethanol blended with gasoline di-

Table 5.17 Average Emissions from Gasohol and Ethanol Light-Duty Vehicles in Brazil
(g/km)

Fuel type	CO	HC	NO_x
Gasohol	40.5	3.8	1.4
Ethanol	18.8	1.6	1.1

Source: Murgel and Szwarc 1989

minishes some of the inherent benefits of using ethanol in the first instance such as low volatility, no benzene in fuel, and lower CO emissions. FFV/VFVs running on ethanol-gasoline blends are likely to have higher evaporative emissions (with greater ozone-forming potential) than gasoline vehicles as well as increased emissions of CO and benzene (Pitstick 1993; OECD 1995).

A limited investigation by AQIRP to compare the exhaust emissions of catalyst-equipped FFV/VFVs operated on E85 fuel and gasoline showed that NO_x emissions were significantly reduced with E85, by as much as 50 percent. Total toxic emissions were two to three times greater with E85 compared to gasoline due to a large increase in acetaldehydes. Formaldehyde emissions with E85 also increased to twice the level of gasoline while benzene and 1,3 butadiene were greatly reduced. CO emissions also increased with E85. Energy specific fuel economy changed less than one percent while volumetric fuel economy was 25 percent lower with E85 (AQIRP 1995d).

Safety. In much the same manner as other fuels, ethanol presents a fire hazard if handled improperly. The explosion hazard of ethanol is rated as moderate when exposed to flame. Although ethanol is less volatile than gasoline, it is considered to be more explosive. Vapors that form above a pool of ethanol are potentially explosive. Repeated overexposure to ethanol will cause redness and irritation of the skin. The fuel is not considered to be particularly hazardous and inhalation of small amounts of ethanol vapor are not considered toxic. Excessive ingestion of ethanol is dangerous and will require medical care. As an intoxicating beverage, ethanol presents a special supervisory challenge. Supplies of ethanol must be carefully monitored (Maggio and others 1991).

Fuel prices and supply. Ethanol is produced primarily by fermentation of starch from grains or sugar from sugar cane. As a result the production of ethanol for fuel competes directly with food production in most countries. The resulting high price of ethanol (ranging from U.S.$0.26 – 0.41 per liter equivalent to U.S.$0.40 – 0.65 per liter of gasoline on an energy basis) has effectively ruled out its use as a motor fuel except where (as in Brazil and the United States) it is heavily subsidized. The Brazilian Pro-alcohol program has attracted worldwide attention as the most successful example of implementing an alternative fuel program. Despite the availability of a large and inexpensive biomass resource, this program has required massive government subsidies for its viability. To address a severe shortage of ethanol in 1990, a blend of methanol, ethanol, and gasoline was developed for use in dedicated alcohol vehicles (box 5.5). Because of earlier disruptions in supply of fuel alcohol and reductions in direct subsidies, the share of

ethanol-powered cars in new car sales in Brazil dropped to less than one percent in 1996.

Biodiesel

Vegetable oils considered as possible substitutes for diesel fuel can be produced from a variety of sources including rape (also known as canola or colza), sunflower, sesame, cotton, peanut, soya, coconut, and oil palm. Vegetable oils have been promoted as possible replacement for diesel fuel because of their good ignition quality. However, their high viscosity (about 20 times that of diesel), which results in poor fuel atomization, fuel injector blockage, ring stickings, and contamination of lubricating oil, makes them best used as blends with diesel fuels in mixtures of up to 50 percent. When used in blends, vegetable oils produce higher emissions of CO, HC, and PM as compared to pure diesel fuels (OECD 1995; IEA 1993). Biodiesel is produced by reacting vegetable or animal fats with methanol or ethanol (trans-esterification) to produce a lower viscosity fuel (oil ester) that is similar in physical characteristics to diesel, and can be used neat or blended with petroleum diesel. Care must be exercised to completely remove the glycerin residues produced as a by-product. Engines running on neat biodiesel or blended with petroleum diesel tend to have lower black smoke and CO emissions but higher NO_x and possibly higher emissions of particulate matter. These differences are not large, however. Other advantages of biodiesel include high cetane number, low viscosity, very low sulfur content, and the fact that it is a renewable resource. Disadvantages include high cost (U.S.$0.39 – 0.91 per liter before tax) and reduced energy density (resulting in lower engine power output). The effects of biodiesel on engine performance and emissions over a long period in actual service are not well documented.

Although the use of vegetable oils in diesel engines dates to 1900, the modern use of biodiesel fuel started in the early 1980s, when vegetable oils—soybean oil, sunflower oil, corn oil, cottonseed oil, and rapeseed oil—were first used as experimental substitutes for diesel fuel in farm tractors. Interest in biodiesel plunged along with the price of oil in 1983. Recently, interest in cleaner air and renewable energies has led to renewed private and public focus on biodiesel fuel. A small-scale program to convert palm oil to methyl ester for use in taxis has been successful in Malaysia (Ani, Lal, and Williams 1990).

Properties. The properties of biodiesel are compared with those of diesel fuel in table 5.18. The important properties include viscosity, which indicates the flow characteristics of the fuel; pour point, which describes the cold-flow characteristics of the fuel; alkaline content or acid number (soap content), which measures

Box 5.5 Brazil's 1990 Alcohol Crisis: The Search for Solutions

A one billion liter shortage of ethanol in 1990 - the result of a 50 percent drop in sugarcane prices over the past several years and the corresponding reduction in sugarcane acreage from 1987 to 1989—compelled Brazilian government agencies and private organizations to search for an emergency solution to guarantee fuel supply for the 4.2 million ethanol-fueled vehicle fleet. The initial recommendations were to:

- Reduce ethanol content in gasohol from 22 to 12 percent
- Add up to 5 percent gasoline to the hydrated ethanol fuel.

Because these recommendation fell short of satisfying the ethanol shortfall, the use of methanol as a supplementary alcohol was considered. After an intensive investigation, a new fuel blend of 60 percent hydrated ethanol (E), 33 percent methanol (M), and 7 percent gasoline (G), by volume, was identified as a suitable substitute for ethanol in Original Equipment Manufacture (OEM) vehicles designed to run on ethanol and gasohol mixtures. The addition of methanol tended to compensate for gasoline in the E60/M33/G7 blend: the methanol in the blend resulted in leaner air-to-fuel ratio and a low caloric value; the gasoline fraction in the blend fully compensated for this effect by providing a richer air-to-fuel ratio and a high caloric value. The volatility of this blend, however, was 80 percent higher than ethanol fuel in terms of Reid Vapor Pressure (RVP). Although a high RVP improves cold starts (less exhaust emissions and better fuel economy), the 80 percent rise in RVP resulted in a two- to threefold increase in evaporative emissions, particularly for pre- 1970 Brazilian vehicles. Fuel economy improved by 2 percent with this blend. Exhaust emissions from the new methanol/ethanol/gasoline (MEG) blend are compared below with ethanol and other other fuel blends evaluated as part of this investigation:

	Stoichiometric air/fuel ratio	Density (kg/l)	Relative emissions (base=100)			
			CO	HC	NO_x	Aldehydes
Hydrous ethanol(E100)	8.3	0.79	100	100	100	100
95% ethanol-5% gasoline (E95/G5)	n.a.	0.78	125	110	100	n.a
22% ethanol-78% gasoline (E22/G78)	8.9	0.77	175	122	138	36
12% ethanol-88% gasoline (E12/G88)	n.a.	0.76	282	134	106	n.a.
100% gasoline (G100)	14.8	0.74	350	140	115 (e)	10 (e)
60% ethanol-33% methanol-7% gasoline (E60/M33/G7)[a]	**8.2**	**0.79**	**98**	**90**	**112**	**54**

a. Emissions in grams per kilometer (FTP cycle) for this blend were 17 g/km CO; 2g/km HC; 1.1g/km NO_x, and 0.07 g/km aldehydes.
n.a. = Not available
(e) = Estimate

In summary, the ethanol-methanol gasoline blend helped to meet a critical fuel shortage without compromising the country's vehicle pollution control program (PROCONVE) or requiring special tuning, recalibration, or conversion of the dedicated alcohol engines. There have been no adverse effects reported from the use of this blended fuel.

Source: Szwarc and others 1991

the amounts of deposit-forming sodium and potassium in the fuel; sulfur and nitrogen contents, which affect the amount of PM and NO_x emissions formed; and iodine number, which indicates the fuel stability (level of saturation of the fuel). Biodiesel has many physical properties (such as the cetane number) quite similar to conventional diesel. It differs from ordinary diesel fuel in terms of its lower energy content (due to its higher oxygen content), which reduces engine power output; it has a higher density and a lower volatility compared with diesel, and a stronger affinity for water; and it contains no aromatics and only trace amounts of sulfur. Both the low sulfur content and the significant oxygen content should help to reduce soot formation and emissions from biodiesel. Low volatility and poor cold flow properties could constrain the use of biodiesel in colder climates. Being hydrophilic, biodiesel requires special care in storage. It is fully biodegradable – 98 percent of the spilled material is broken down within three weeks and the remainder within five weeks (Hutcheson 1995).

Use. Owing to the high price of conventional diesel fuel in Europe (U.S.$3 to U.S.$4 a gallon including taxes), a sizeable number of diesel vehicles in Austria, Belgium, France, and Italy run on biodiesel fuel made primarily from rape-seed oil (U.S. DOE 1994). In the

Table 5.18 Physical Properties of Biodiesel and Conventional Diesel Fuel

Property	Methyl ester (biodiesel)	Common diesel fuel
Viscosity @ 40°C	4.1 – 4.5	2.3 –2.8
Pour point, °C	–4	–29
Cloud point, °C	–6.7	–20
API gravity @ 15.6 °C	27.8	35.5
Net heat of combustion, MJ/kg	37.1–37.7	42.8
Cetane number	46–52	47–51
Oxydation stability, mg/1000ml	0.06	0.11
Carbon, % weight	77–78	86
Hydrogen, % weight	11–12	13
Oxygen, % weight	10–11	0
Sulfur, % weight	0.002 –0.03	0.15–0.2
Water, mg/kg	500	50
Nitrogen, ppm	29	0
Total acid no., mg KOH/g	0.14–0.45	n.a.
Flash point, °C	152–171	61–92
Iodine value	133	n.a.
Density at 15°C, (kg/m^3)	882–885	850–860
Distillation		
T10, °C	334	227
T50, °C	336	283
T95, °C	345	348

n.a. = Not available
Source: Reed 1994; Ziejewski, Goettler, and Pratt 1986; Alfuso and others 1993; Tritthart and Zelenka 1990; Perkins, Peterson, and Auld 1991, Scholl and Sorenson 1993; U.S. DOE 1994

United States biodiesel fuel has been used mostly in demonstration projects in which it is blended with conventional diesel fuel to reduce sulfur and particulate emissions from diesel vehicles. Some limited demonstration projects using neat biodiesel in diesel vehicles have also been initiated (U.S. DOE 1994; Reed 1994).

Engine emissions, performance, and durability. Results from laboratory studies (Alfuso and others 1993; Pagowski and others 1994; Scholl and Sorenson 1993; Tritthart and Zelenka 1990; Reed 1994; Perkins, Peterson, and Auld 1994; Ziejewski, Goettler, and Pratt 1986 generally agree that blended or neat biodiesel has the potential to reduce diesel CO emissions (although these are already low), smoke opacity, and HC emissions particularly polycyclic aromatic hydrocarbons. But these studies show an increase in NO_x emissions for biodiesel fuel compared with diesel fuel at normal engine conditions. The higher NO_x emissions from biodiesel-fueled engines are partly due to the higher cetane number of biodiesel, which causes a shorter ignition delay and higher peak cylinder pressure. Some of this increase may be due to the nitrogen content of the fuel. The reduction in smoke emissions is believed to be the result

of better combustion of the short-chain hydrocarbons found in biodiesel, as well as the effects of the oxygen content. Other data have shown that mixing oxygenates with diesel fuel helps reduce smoke.

Several European studies have also investigated the emissions performance of rape-seed methyl ester (RSME) as a neat fuel or as a blend with conventional diesel fuel up to 20 percent. As reported by Hutcheson (1995), RSME compared to conventional diesel fuel has the following emission characteristics:

- Lower HC and CO emissions.
- Consistently higher NO_x emissions.
- Generally lower particulate emissions particularly under high-speed and moderately high-load conditions but a dramatic increase under low-speed and light-load conditions (Cold ECE 15).

The increased NO_x emissions may be attributed to increased combustion temperatures due to better availability of oxygen within the combustion zone. This phenomena may also influence the formation of soot.

As for the HC emissions, research by Scholl and Sorenson (1993) shows a reduction in HC emissions when biodiesel is used. There is some concern, however, that the organic acids and oxygenated compounds in biodiesel may affect the response of the flame ionization detector, thus understating the actual HC emissions. Scholl and Sorenson also mention that the behavior of these compounds with respect to adsorption and desorption on the surfaces of the gas sampling system is unknown. Thus more studies are needed to understand the organic constituents in the exhaust from biodiesel-fueled engines before firm conclusions can be drawn regarding the effects on HC emissions.

There is controversy concerning the effect of biodiesel on particulate matter emissions. Alfuso and others (1993) report that some studies claim a reduction of particulate emissions when biodiesel is used, while some studies show an increase. Alfuso and others found that the particulate matter produced by biodiesel was higher than that produced by diesel fuel in light-duty transient driving (ECE 15 testing cycle). This effect may depend on the selection of the injection timing. According to an analysis of particulate composition reported by Hutcheson (1995), soot emissions (insolubles) were significantly reduced with RSME but the proportion of the emissions composed of fuel-derived hydrocarbons (fuel solubles) condensed on the soot was much higher (a five to six fold increase compared to diesel). This suggests that RSME may not burn to completion as readily as diesel fuel. Although the total PM emissions for diesel (0.275 g/mile) were significantly higher than RSME (0.224 g/mile), the soluble organic fraction (16 percent) in diesel emissions was substantially lower compared to RSME (51 percent). In addition, a general trend toward lower emissions of PAH and aldehydes has been found with RSME, although in cold FTP and ECE tests, an increase in these emissions was observed.

In general, no engine, ignition system, or fuel injector modification is necessary for diesel engines to operate on biodiesel. However, the solvent characteristics of the fuel require the replacement of the hoses and fuel lines that contact the fuel. Furthermore, the work of Scholl and Sorenson (1993) has shown that a 5 degree retardation of injection timing is required to lower NO_x emissions using biodiesel. Scholl's research also showed that the peak cylinder pressure and the peak rate of pressure rise were higher for biodiesel-fueled engines than for diesel-fueled engines. This suggests that biodiesel engine carrying higher stress than components engines using diesel, and this may affect their durability and reliability. These findings imply that some engine adjustments or modifications may be needed with the use of bio-diesel to optimize engine performance, emissions, and durability.

Research has shown that the combustion characteristics of biodiesel are comparable to diesel fuel (Scholl and Sorenson 1993; Alfuso and others 1993), but the sodium and potassium atoms in biodiesel fuel can form deposits in the combustion chamber. As these deposits build up, they can affect combustion characteristics, and influence engine performance and operability as well as increase exhaust emissions. Substantial deposit build-up in the combustion chamber can also degrade engine durability.

Another common finding on the use of biodiesel as a vehicle fuel is that biodiesel tends to dilute engine oil. More frequent oil changes may be required with biodiesel fuel. Prolonged engine operation with diluted engine oil may cause the oil to break down and the engine to seize. Other concerns regarding the use of biodiesel fuel are that the vehicle is likely to have problems operating in cold weather, due to the fuel's relatively high cloud and pour point (Perkins, Peterson, and Auld 1991). Although pour point depressant and cold-flow improvers may help alleviate this problem, such additives might affect the biodegradability of the fuel, which is an important advantage of biodiesel (U.S. DOE 1994).

Fuel cycle emissions and fuel cost. OECD/IEA (1994) includes a detailed evaluation of the economics of biofuels. The production of vegetable oil–based biodiesel fuel involves plowing the ground (tillage), planting the crops (rape, soybean, sunflower, corn), and fertilizing and harvesting the crops. These processes require the use of motorized farm equipment, which produce exhaust emissions and use fossil energy. In addition, methanol or ethanol are needed in the trans-esterification process, and the emissions emitted to produce these compounds should be taken into account. In a life-cycle study on rape-seed oil and its esters, the German Federal Environmental Agency has noted that other factors such as odor generated by fuel combustion and large emissions of nitrous oxide (N_2O) in the production cycle render the use of this fuel less attractive (OECD 1995). Therefore the overall energy consumption and full fuel cycle emissions for biodiesel fuel might be as high as those from conventional diesel fuel. Thus biodiesel fuel might not be as attractive as other alternative fuels, such as natural gas, as a substitute for conventional diesel fuel.

The cost of biodiesel fuel is one of the principal barriers making it less attractive as a substitute for diesel. The cost of vegetable oils is U.S.$2 – 3 a gallon. If the credit for glycerol, which is a by-product of the biodiesel trans-esterification process and a chemical feedstock for many industrial processes, is taken into account, the cost of converting vegetable oils to biodiesel is approximately U.S.$0.50 a gallon (Reed 1994). Thus the total cost for biodiesel fuel is U.S.$2.50 – 3.50 a gallon. This is substantially higher than the cost of conventional diesel, about U.S.$0.75 – 0.85 a gallon before tax. If waste vegetable oil is used, the cost of biodiesel could be reduced to about U.S.$1.50 a gallon (Reed 1994). Since the heating value for biodiesel is less than that for diesel, more fuel must be burned to provide the same work output as diesel. This adds further to the cost disadvantage of biodiesel. Considering these limitations and the availability of less greenhouse gas intensive alternatives at similar cost, the use of rape-seed oil or its derivatives cannot be justified from the standpoint of environmental protection.

Hydrogen

Although hydrogen has the potential to be the cleanest-burning motor fuel, it has many properties that make it extremely difficult to use in motor vehicles. Hydrogen's potential for reducing exhaust emissions stems from the absence of carbon atoms in its molecular structure. Because of the absence of carbon, the only pollutant produced in the course of hydrogen combustion is NO_x (of course, the lubricating oil may still contribute small amounts of HC, CO, and PM). Hydrogen combustion also produces no direct emissions of CO_2. Indirect CO_2 emissions depend on the nature of the energy source used to produce the hydrogen. In the long-term event of drastic measures to reduce CO_2 emissions (to help reduce the effects of global warming), the use of hydrogen fuel produced from renewable energy sources would be a possible solution. Mackenzie (1994) provides an excellent review of the prospects and potential of hydrogen vehicles.

Fuel supply. Hydrogen suffers from two major problems: production and storage. Hydrogen is not a fossil fuel, and is not found in significant quantities in nature. It therefore must be manufactured. Hydrogen can be produced by a number of methods, of which the most common are electrolysis of water, reforming natural gas, or partial oxidation and steam reforming of other fossil fuels. (It should be noted that hydrogen produced by reforming fossil fuels does not reduce greenhouse emissions because the CO_2 emitted in the reforming process is comparable to that which would have been emitted if the fossil fuels were used directly in a vehicle.) The most economical source of hydrogen is from reforming natural gas. Hydrogen is currently manufactured in limited quantities as an industrial chemical, as rocket fuel, and in larger quantities for use in petroleum refineries and chemical plants. Electrolysis is sometimes used to produce high-purity hydrogen for special purposes but cannot compete economically with natural gas reforming for large-scale production.

Advocates of a future "sustainable" hydrogen economy have proposed many ideas for renewable sources of hydrogen. These potential future sources include electrolysis of water using large-scale, cheap photoelectric or solar-electric systems, or cheap, abundant, and environmentally benign hydroelectric or nuclear-electric plants. In the nearer term, renewable hydrogen could be produced from biomass using any of several processes. It is unclear, however, why hydrogen would be produced from biomass instead of easier-to-handle fuels such as methanol, ethanol, or methane. Hydrogen also could be produced on a large scale through in-situ gasification of coal, but this would result in large emissions of CO_2, as well as localized environmental damage due to subsidence, possible groundwater contamination, and so on.

In order for hydrogen to become available as a motor fuel, significant investments would be needed in the infrastructure for the delivery, storage, and dispensing of the fuel. These investments are unlikely to occur as long as the costs of hydrogen production and use exceed those of other available motor fuels.

Fuel storage. Hydrogen can be stored on-board a vehicle as a compressed gas or as a liquid, or in chemical storage in the form of metal hydrides. Hydrogen also can be manufactured on-board the vehicle by reforming natural gas, methanol, or other fuels, or by the reaction of water with sponge iron. In the latter case the sponge iron is oxidized and must be regenerated periodically. Compressed hydrogen occupies roughly fourteen times the space of an equivalent amount of gasoline. Thus compressed gas storage would add significant bulk and weight to a vehicle—about three times the bulk and weight of an equivalent volume of compressed natural gas. Vehicle range and refueling frequency would be significant constraints.

Hydrogen can also be stored on-board a vehicle as a liquid, but this requires that it be cooled below its boiling point of –423°F (at 36°F above absolute zero, the lowest boiling point of any substance except helium). Even as a liquid, however, hydrogen would still occupy roughly four times the volume of an equivalent amount of gasoline. Furthermore, the energy required for refrigeration to liquefy the hydrogen could easily exceed the energy value of the fuel itself, resulting in an extremely inefficient system.

The third option for hydrogen storage is in the form of metal hydrides. Current metal hydride storage systems are heavy and bulky, but less so than the storage cylinders required for compressed hydrogen (Billings 1993). This is considered the leading technology for storing hydrogen for vehicular use. Research is under way in several countries to improve these characteristics, and some vehicle manufacturers in Germany and Japan are developing prototype vehicles.

Engine and vehicle technology. Aside from its low-emission characteristics, hydrogen is a poor fuel for internal combustion engines. This is due to its poor knock resistance compared with other gaseous fuels (such as natural gas or propane). Hydrogen-fueled spark-ignition engines are susceptible to detonation or knock, which limits the compression ratios that can be used and thus the efficiency attainable. Rotary engines have met with limited success in past engine research. Some research also has studied the possibility of using hydrogen in an ignition-assisted diesel engine, which would avoid the problem of knock.

The most promising approach to the use of hydrogen for vehicular propulsion is to react it with oxygen in a fuel cell to supply electric energy to a hybrid-electric vehicle. Fuel cells are extremely efficient compared with internal combustion engines, and hydrogen-air fuel cells produce no pollutant emissions. Thus this system would be similar to a battery-electric propulsion system, except that the fuel cell and fuel supply would require significantly less weight and bulk than a battery system for the same energy storage. The fuel could be supplied and stored as hydrogen or could be produced on-board by the reformation of methanol, natural gas, or other hydrocarbon fuels, or by the reaction of water with sponge iron. Producing hydrogen for the fuel cell by reforming hydrocarbon fuels may be especially attractive—both because of the increased ease of handling compared with pure hydrogen and because this approach could continue to use the fuel supply infrastructure already in place.

Research and development work on hydrogen-powered vehicles is progressing in several countries including Canada, Germany, Japan, and the United States. One of the most promising developments is a prototype 60-passenger hydrogen bus produced by Canada's Ballard Power Systems, a leader in fuel cell technology. The proton-exchange-membrane (PEM) fuel-cell bus has been put in service in Vancouver, Canada. It travels at normal operating speeds and can go more than 400 km on a single hydrogen charge. The Ballard prototype, however, costs three times as much to operate as a comparable diesel bus. Economies of scale and improved fuel cells might make it competitive with conventional diesel buses beyond 2000 (AASHTO 1996; MacKenzie 1994).

Emissions. Hydrogen has the potential to be the cleanest-burning motor fuel. With the virtual elimination of CO and HC exhaust emissions, only nitrogen oxide emissions would be present in any significant quantity in the vehicle exhausts; NO_x emissions from existing prototype hydrogen vehicles are similar to those from gasoline vehicles (Kukkonen and Shelef 1994).

Economics. Due to the high cost of production, lack of storage reserves, and the large quantities required for hydrogen to be used as a motor fuel, it is unlikely that hydrogen will be a cost-effective fuel in the near future. New production and storage techniques will be required before hydrogen fuel becomes economically—much less technologically—feasible.

Safety. There are numerous safety concerns using hydrogen as a motor fuel. These concerns are associated primarily with its extreme flammability. Although hydrogen is lighter than air, it mixes rapidly with air to create a combustible or explosive mixture. Hydrogen also is difficult to store as a gas because its molecules are quite small, which enables them to diffuse out through most common containment materials. Special materials for tanks and fuel lines are required to contain the fuel and to minimize leakage.

Electric and Hybrid-Electric Vehicles

Electric vehicles—in the form of trolleys used[7] for public transit—were a familiar sight in many cities in Latin America, Asia, and Eastern Europe. Although they are quiet and emit virtually no pollution, they have high operating costs, and very high capital costs compared with diesel-fueled buses. The flexibility of operations with trolley buses is also limited due to the need to follow overhead wires, which are vulnerable to accidents and sabotage. As a result they have been replaced by conventional diesel buses in most developing countries over the past 30 years. A few battery-electric buses have been built as demonstration projects, but these have extremely limited range and are suitable only for short "shuttle" routes. Because of the disadvantages of electric buses, the use of emission-controlled diesel or alternative-fuel buses is frequently a better and more cost-effective solution to the problem of diesel bus emissions in urban areas.

The major focus at present is on the development of battery-powered electric passenger cars and light vans

---

7. Worldwide there are over 370 trolleybus systems. Western nations (including the American hemisphere), operate 73 systems totaling some 5,400 vehicles, the largest system is in São Paulo, Brazil, with 480 trolley buses. Eastern countries, including China and the successor states of the former Soviet Union, account for about 300 systems with an estimated 25,000 vehicles. The largest of these systems is in Moscow with an estimated 2,000 vehicles. All of Western Europe has 52 systems with 2,200 trolleybuses. The United States has five systems, (Boston, Dayton, Philadelphia, San Francisco, and Seattle) with a total of 727 vehicles. Canada has two systems in Vancouver and Edmonton with 355 trolleybuses. Trolleybuses cost between 50-100 percent more than diesel buses of comparable size and are economical to operate only in cities with steep streets and substantial height differences (AASHTO 1995).

for commuter and pickup-and-delivery services in urban areas. With improvements in electric motor and battery technology, a number of vehicle designs having acceptable performance characteristics have been developed. A brief review of electric and hybrid electric vehicle technology is presented in appendix 5.2. The resources expended on electric vehicle development have increased greatly since 1990, when the California Air Resources Board (CARB) adopted new emission regulations requiring a minimum of two percent of vehicle sales in California (about 20,000 cars) to be zero emission vehicles (ZEVs) beginning in 1998. This percentage was to rise 5 percent in 2001 and 10 percent in 2003[8]. The current state-of-the-art in battery technology still limits battery-electric vehicles to short ranges—on the order of 100-150 kilometers under the most favorable conditions. Under less favorable conditions the achievable range may be much less. Factors contributing to reduced range include use of heaters or air conditioners, use of lights for night driving, and rain, snow, wind, or other factors that increase rolling resistance or air drag.

Despite their limitations, electric vehicles are potentially attractive in some kinds of vehicle service—especially those, such as commuting and urban pickup and delivery, where short range is not a problem. Of course, the main advantage of electric vehicles is the absence of pollutant emissions at the point of use, although the incremental emissions from the power plant to produce the electricity may be significant. In addition, electric vehicles have a potential efficiency advantage in applications—such as buses—that involve repeated stopping and starting. This is because of the potential for regenerative braking. By using the motors as generators to decelerate, it is possible to recover for later use much of the kinetic energy of the vehicle, rather than dissipating this energy as heat as is done in a traditional friction brake.

Battery technology is now the subject of intensive research and development aimed at increasing energy and power density/efficiency, increasing useful life, reducing manufacturing costs, and finding substitutes for environmentally undesirable materials such as lead and cadmium. Other on-board energy storage devices such as high-speed flywheels and ultracapacitors are also undergoing intensive research and development and may ultimately provide a better solution than electrochemical batteries (OTA 1995; Mackenzie 1994).

Another promising electric-vehicle technology is the hybrid-electric vehicle, which would combine many of the advantages of the battery-electric vehicle (regenerative braking, quiet operation), with much longer range and greater energy efficiency overall. These vehicles could also be designed to have very low emissions, possibly less than those of the power plant supplying the battery-electric vehicle. The motor in a hybrid-electric vehicle would be designed to give the best emissions and fuel economy at one or two specific operating points and would then always operate at these points. The motors themselves might be advanced internal combustion engine designs or could include such other technologies such as gas turbines and Stirling engines. The motor would turn a generator and would be sized to supply the average electric power required by the vehicle, while an on-board electric energy storage system would supply surge power for acceleration and would accept the power returned by regenerative braking. In addition to avoiding the broad range of engine operating conditions experienced by current vehicle engines, this arrangement would avoid engine transients, a major control problem and a significant source of pollutant emissions (Sperling 1995; OTA 1995).

Electric vehicles are very expensive compared to other clean-vehicle technologies. Low-production models such as the Chrysler TEVan sell for more than U.S.$100,000 each. General Motor's EV1, the first electric car in modern times to have been specifically designed by a major car maker for the mass market, is priced at U.S.$35,000. Estimates of the costs of battery-electric vehicles in volume production are heavily influenced by assumptions regarding battery cost and efficiency. Estimates of the incremental cost (compared with a gasoline-fueled vehicle meeting current U.S. emission standards) range from U.S.$3,000 to more than U.S.$20,000 per vehicle. To these costs need to be added the costs of the electric-charging infrastructure, and the incremental capital costs of new power plants. OTA (1995), Sperling (1995), Mackenzie (1994), and IEA (1993) provide detailed assessments of the technology and economics of electric vehicles. Barring a crisis that could accelerate the development schedules, it might be 2020 or 2025 before advanced electric vehicles produced for the mass market have fully penetrated the new vehicle fleet, and it would be another 10 to 15 years before penetration of the entire fleet takes place (OTA 1995). Despite these limitations, electric vehicles have become popular in some European countries and are beginning to penetrate the motorcycle/moped market in Asia.

In France, buyers of electric cars qualify for a FF15,000 (U.S.$2,500) government grant to offset the additional cost compared to conventional cars. France has operated the world's biggest trial program for elec-

---

8.   In January 1996, CARB decided to suspend this regulation as the technology to produce battery-powered ZEV's is not adequately advanced in terms of vehicle range, cost or performance for the industry to achieve these targets. Despite rescinding its 1998 deadline, CARB remains committed to imposing targets for the number of ZEVs and has obtained a commitment from the motor industry to sell about 2000 electric vehicles in Los Angeles and Sacramento (which have the worst smog records) between 1998 and 2000.

tric cars, involving Peugot-Citröen. A major demonstration project has been in progress in the port city of La Rochelle. Under an agreement involving the city, the French electricity authority and the automaker, 50 battery-powered Citroen Ax and Peugeot 106 cars have been rented to private users from FF900 to FF1000 (U.S.$150 to $110) a month, including insurance, maintenance and rental of nickel-cadmiums batteries. In addition to the availability of public recharge bays with "quick charge" dispensers, participants with their own garages may plug into a recharger overnight; this allows them to drive 75 to 100 km for about 8–9 francs (U.S.$1.50) of electricity (AASHTO 1994). A new experiment in Tours will test the demand for specifically-designed urban electric cars, which can be rented and returned after short periods, similar to taxis. Potential backing for electric cars in France has been influenced by the cheap nuclear-generated power (Simonian 1996).

Despite their limited range, there are about 250,000 electric vehicles in France; compared to just 3,000 such vehicles in the United States. Sweden has also experimented with electric cars for use by municipal organizations under the auspices of Newtek, a Swedish State agency. In Finland the postal service has more than 40 electric vans in daily use. In England, a bus fleet powered by electricity and guided by a satellite navigation system has been launched in London. The 12-seater buses use an electric drive system developed and supplied by Wavedriver, a joint venture between the Technology Partnership, a research company, and PowerGen, the UK power generator. The batteries, which can run for 50 miles, can be re-charged in less than an hour. The Oxford County Council (U.K.) has also successfully operated four 18-seater, battery-powered electric buses to provide inner city shuttle service over a two-year period. Electric vehicle (EV) demonstration projects have been sponsored by other jurisdictions in U.K., such as Camden, Conventry, and Ipswich.

Germany's Ministry of Research and Development in association with the automotive industry and battery manufactures is supporting a four-year demonstration program on the Baltic Sea Island of Ruegen to test a fleet of 37 new private cars, 20 vans, and 3 minibuses retrofitted with electric motors. The experiment includes both new nickel-cadmium and conventional batteries. The demonstration program, besides assessing the technical performance of the latest generation of battery-powered vehicles, expects to determine the degree of driver acceptance and accommodation to the operational characteristics of electrical vehicles. In Japan, the National Environment Agency has offered to subsidize 50 percent of the cost of electric vehicles (mini vans and delivery vehicles) purchased by local governments. The objective is to raise environmental awareness and encourage local businesses to purchase electric vehi-

cles. The electric delivery van has a top speed of 85 km/hr and can travel nearly a hundred miles after being charged overnight (AASHTO 1993).

In Nepal, electric three-wheelers (Safa Tempos[9]) have been introduced in Kathmandu as a replacement for conventional diesel-fueled tempos under a joint program supported by the national government, the United States Agency for International Development (USAID), and the private sector. The program is managed by Nepal Electric Vehicle Industry (NEVI) and has received strong government support in the form of reduced import duties and other tax incentives (box 5.6).

Factors Influencing the Large-Scale Use of Alternative Fuels

Cost

Large scale introduction of alternative transport fuels depends on the cost of production and the additional costs of fuel storage, distribution, and end use. Production costs are a function of the abundance or scarcity of the resources from which the fuel is produced, as well as the technology available for extracting those resources. Gasoline and diesel substitutes made from heavy oils, natural gas or biomass require relatively minor changes to existing distribution and end-use systems, whereas CNG, LPG, and alcohol fuels require larger modifications.

Alternative fuel costs (inclusive of production, distribution, and end-use) as estimated by OECD and based on 1987 prices and technology are shown in table 5.19. According to the OECD's International Energy Agency (IEA), CNG and LPG could be economically competitive with conventional gasoline. Methanol and synthetic gasoline made from natural gas could become marginally competitive under optimistic assumptions about gas prices. Methanol from coal or biomass and ethanol from biomass were estimated to cost at least twice as much as gasoline (IEA 1990).

A study by the World Bank (Moreno and Bailey 1989) found that at crude oil prices of U.S.$10 per barrel or lower (in 1988 prices) alternative fuels were generally uncompetitive. With crude oil cost between U.S.$10 and U.S.$20 per barrel, custom-built, high-mileage LPG vehicles and retrofitted, slow-fill CNG vehicles (mostly captive fleets of urban buses, taxis, and delivery vehi-

---

9.   The Safa Tempo (electric three-wheeler) runs on a 72-volt direct current (D.C.) motor (8.5 hp @3,800 rpm), powered by a pack of 12 (6-volt, 220 amp. hr) or 36 (2-volt, 150 amp. hr) batteries. The maximum speed is set at 40 km/hr and range per charge of batteries is 60 kilometers. The daily duty range (with exchange of batteries) is 150 kilometers. The average power consumption is 150-200 watt-hr per kilometer. The maximum pay load is 600 kg, equivalent to about 10 passengers plus driver.

Box 5.6 Electric Vehicle Program for Kathmandu, Nepal

Kathmandu suffers from severe air pollution due to vehicle emissions and provides ideal circumstances for the introduction of zero emission electric vehicles. With a population of approximately 1.5 million people, the city occupies an area roughly 12 kilometers wide. Thus distances traveled are quite short. Speeds seldom exceed 40 kilometers per hour and are generally below 30 kilometers per hour. Inexpensive, non-polluting hydroelectric power is available. And the total number of motor vehicles including motorcycles is under 100,000. Because they are on the streets all day, the 3,500 three-wheelers providing public transportation account for 25 percent of the road traffic and, because they are poorly maintained, more than their share of the air pollution. Thus, at a relatively small cost and in a short time, a dramatic change can be made in Kathmandu's air quality.

Global Resources Institute initiated the first phase of the electric vehicle program for Kathmandu in September 1993 with the conversion of a diesel three-wheeler to electric power. The Vikram three-wheeler was chosen because of its reputation as the worst polluter in Kathmandu. Following extensive tests on the initial vehicle to optimize the drive system, seven new three-wheelers were built and the first converted vehicle was retrofitted with the new drive system. In August 1995, these eight vehicles were placed into service as a six-month demonstration project with a company providing public transportation. Ongoing work is focused on developing a sustainable electric vehicle industry in Nepal.

The Electric Vehicle Program for Kathmandu has gone well. Since its inception in September 1993, the Safa Tempo ("safa" means clean and "tempo" is the local generic name for a three-wheeler) program has become widely recognized and supported by the public. It has been featured in the press and on local and worldwide television and radio. In July 1995, the government reduced import duties on components for local assembly of electric three-wheelers from 60 percent to 5 percent and on fully-assembled electric vehicles of all types from as much as 150 percent to 10 percent.

During the first six months of service the demonstration fleet of Safa Tempos carried over 200,000 people and traveled 175,000 kilometers. Within four months of operation on a new route, the fleet reached 95 percent of its target revenue figures to make it more profitable than diesel three-wheelers. Public response has been overwhelmingly favorable – the only criticism being the presence of too few Safa Tempos. Two companies have formed to assemble, manufacture, import, and service electric vehicles in Nepal. In addition several transport companies plan to operate fleets of 30 to 50 electric three-wheelers. Additional companies have expressed interest in manufacturing components and supporting services for the electric vehicles. Banks in Nepal and Bangladesh have indicated interest in providing loans to the electric vehicle industry.

Other cities in South and East Asia have become aware of the developments in Kathmandu and have asked for guidance to promote electric vehicle programs to reduce air pollution.

Source: Moulton 1996

Table 5.19 Costs of Substitute Fuels, 1987
(1987 U.S. dollars)

Fuel	Cost per barrel of gasoline energy equivalent
Crude oil (assumed price)	$18
Conventional gasoline	$27
Compressed natural gas	$20-46
Very heavy oil products	$21-34
Methanol (from gas)	$30-67
Synthetic gasoline (from gas)	$43-61
Diesel (from gas)	$69
Methanol (from coal)	$63-109
Methanol (from biomass)	$64-126
Ethanol (from biomass)	$66-101

Source: IEA 1990

cles with high annual mileage but restricted range) become competitive. Between U.S.$20 and U.S.$30 per barrel, fast-fill CNG and propane-fueled low-mileage vehicles would be competitive. Methanol from natural gas becomes competitive above U.S.$50 per barrel. Synthetic gasoline and diesel do not become competitive until the price of crude oil reaches U.S.$70 per barrel. In view of the high cost of fuel transport in tube trailers, CNG as a motor fuel becomes competitive at the crude oil prices noted above only if filling stations are located near a natural gas pipeline or distribution network.

Table 5.20 Comparison of Truck Operating Costs Using Alternative Fuels
(constant 1987 U.S. dollars)

Cost[a]	Current diesel	Reformulated diesel	Methane (M-100)	CNG	LNG	LPG
Tractor price	$70,000	$71,850	$74,990	$86,320	$73,200	$73,550
Fuel (annual cost)	$25,200	$25,800	$46,400	$26,400	$29,200	$24,600
Fuel (cents/mile)	25.2	25.8	46.4	26.4	29.2	24.6
Engine maintenance (cents/mile)	2.0	2.3	2.4	2.1	2.1	1.3
New facilities (cents/mile)	Base (0)	Base (0)	1.8	3.5	2.5	1.3
Net increase (cents/mile)	Base (0)	1.2	24.2	7.6	7.1	1.5
Total increase in cost	Base[b]	1%	24%	8%	7%	2%

a. Costs are based on a drayage operation, excluding training and dispatching. Two new fueling and maintenance bays are provided for every twenty-five alternative-fuel trucks in the fleet (each traveling 100,000 miles annually). Applicable depreciation schedules for tractors and facilities are used in the cost analysis. Fuel costs are based on U.S.$1.26 per gallon of ordinary diesel and five miles per gallon for each tractor.
b. The base cost is about U.S.$1.00 per mile.
Source: Battelle and Ganett Fleming 1990

A 1990 study for the American Trucking Associations compared vehicle operating costs for heavy-duty trucks using alternative fuels (table 5.20). The study concluded that for U.S. conditions the use of alternative fuels would increase truck operating costs by 1 to 24 percent, compared with diesel costs. New fueling facilities were considered a significant (although not a major) contributor to the higher cost of alternative fuels (Battelle and Ganett Fleming 1990). An investigation of alternative fuel options for urban buses in Santiago, Chile, concluded that LPG-, methanol- and gasoline-fueled buses were unlikely to gain a substantial share of the urban transport market. Well-maintained diesel-fueled buses had a strong competitive advantage while CNG-fueled buses could become competitive (appendix 5.3).

Compared to gasoline or diesel, the cost of biofuels (bioethanol and rape-seed methyl ester) is 3-4 times higher. As the energy and emissions benefits from the use of biofuels are relatively small, widespread adoption of biofuels does not appear to be justified. Electric and hybrid-electric cars, even under the most favorable scenarios, are expected to cost U.S.$5,000-20,000 more than comparable conventionally-fueled cars, on a life-cycle cost basis. The additional cost of electric vehicle technology, however, has a substantial payoff in near-zero emissions and increased fuel economy (OTA 1995).

End-Use Considerations

Alternative fuels require changes in distribution, marketing, and end-use systems. Regardless of the economics, inadequate fuel supply or unreliable distribution systems could inhibit consumer acceptance of alternative transportation fuels. Experience with ethanol in Brazil and CNG in New Zealand suggests that the main factors influencing large-scale introduction of CNG and alcohol fuels are price competitiveness, avail-

ability, cost of feedstock (sugarcane for ethanol, natural gas for CNG), fuel safety and quality standards, the reliability of distribution systems, and the technical quality of vehicles (driveability, durability, safety). Brazil's experience with ethanol (box 5.7) and New Zealand's experience with CNG (box 5.8) clearly show that it is possible to develop a large market for alternative fuels within a reasonable time frame if the financial incentives are favorable and efforts are made to overcome uncertainty on the part of industry and consumers (Sathaye and others 1989). In both instances substantial subsidies were provided to fuel producers and to private motorists to induce the transition to alternative fuels (Moreno and Bailey 1989).

Substantial market penetration by alternative fuel vehicles may not occur if the refueling infrastructure is inadequate, as is often the case. The early evidence from New Zealand suggests that lack of fueling stations was a significant barrier to vehicle conversions to CNG and LPG. A rapid buildup of the retail network for a new fuel to a level approximately 15 percent of the retail gasoline network is likely to allay consumer concerns regarding fuel availability (Kurani and Sperling 1993). Alternative fuel vehicles, in addition, face disadvantages in terms of range, weight, and vehicle space relative to gasoline and diesel vehicles, as demonstrated by the British Columbia (Canada) experience (table 5.21).

Many of the disadvantages of alternative-fuel vehicles—including incremental vehicle costs—could be minimized with original equipment manufacture (OEM) vehicles, provided that there is sufficient market demand to permit mass production (5,000-10,000 vehicles of a particular make and model are required for a normal production run). Production of alternative-fuel OEM vehicles, however, is influenced significantly by the ability of conventional gasoline- and diesel-fueled

Box 5.7 Ethanol in Brazil

Ethanol as a neat fuel or as a fuel additive, has been used extensively in Brazil, and also in the United States, Sweden, Kenya, South-Africa and Zimbabwe. The Brazilian experience has been extensively researched and shows the critical importance of consistent government support for the promotion of alternative fuel technologies.

Brazil has used small amounts of ethanol for decades as an enhancer in gasoline and to provide a by-product market for the sugar industry. In the late 1970s, rising oil prices coupled with high interest rates and a crash in the world sugar market created a foreign debt-service crisis. This compelled the Brazilian government to look at options to reduce petroleum imports.

During 1975-79, the government-sponsored *Pro-alcohol* program increased the ethanol percentage in gasohol to 20 percent (and after 1984 to 22 percent). From 1979 to 1986, Pro-alcohol promoted dedicated ethanol vehicles and in 1985, ethanol vehicles comprised 95 percent of new car sales. From 1986 onwards, however, alcohol production could not keep pace with the demand. This reflected mainly a lag between sugarcane and fuel alcohol prices. Brazil produced 11.7 million cubic meters and consumed 12.5 million cubic meters of fuel alcohol in 1988. In the first phase of the Pro-alcohol program, the government provided up to 75 percent subsidies for ethanol producer investments and assured a 6 percent return on investment. At this stage, however, car manufacturers were unwilling to produce ethanol-only vehicles. During the second phase consumer incentives were introduced for vehicle purchases, and the pump price of ethanol was guaranteed to be no more than 65 percent that of gasoline. The car industry was encouraged by the government's commitment and started producing ethanol vehicles. Despite the poor redesign and inadequate performance of alcohol-powered vehicles initially, the consumer take-up was massive.

The world sugar market began improving in the late 1980s, and the government increased the ethanol price from 54 percent of the gasoline price toward the 65 percent limit. After 1989, this ratio was further raised to 75 percent. In addition, credit subsidies for distilleries were suspended. The combination of vigorous demand and a production level which repeatedly fell short of the projections and commitments made by distilleries led to the depletion of strategic inventories in 1988 and to disruptions in fuel alcohol supply in the following years. In 1990 a new fuel blend, referred to as MEG (methanol-ethanol-gasoline) was introduced in São Paulo and some cities in the State of Minas Gerais, made up of a volumetric blend of about 60 percent hydrated alcohol, 33 percent methanol, and 7 percent gasoline, to substitute for neat ethanol in alcohol-powered vehicles (box 5.5). The disruption in alcohol fuel supply produced a rapid fall in purchases of ethanol vehicles. The government later regained public confidence by restoring incentives but Pro-alcohol's credibility was irreparably damaged. By 1996 the share of ethanol cars had dropped to less than one percent of new car sales.

The opportunity cost of ethanol production in Brazil was claimed to be about $30-35/toe. This would require sugar prices below U.S.$100 per ton. Sugar prices often have exceeded U.S.$200 per ton and independent analyses have put the Brazilian ethanol cost at nearer U.S.$70/toe. At U.S.$0.43 per liter this exceeds the net-of-tax price of gasoline in most countries by a large margin. The economic subsidy attributed to the alcohol fuel program in 1995 was about U.S.$1.9 billion (alcohol fuel consumption of 12.3 billion liters and an estimated subsidy of U.S.$0.15 per liter).

Although an outstanding technical success, the Pro-alcohol program became an economic liability when the world energy prices plunged in mid-1980s. Oil prices would have to return to U.S.$25-30 a barrel for the program to cover even the operating costs making it a costly insurance policy against an unlikely contingency. Pro-alcohol offered extremely generous investment incentives and price guarantees to expand sugarcane production and distillery capacity. These arrangements, however, placed a heavy implicit tax on the Brazilian motorist (compared to the price of imported gasoline), and encouraged the production of alcohol-fueled cars well past any possible economic justification. With ethanol production costs at U.S.$30-40 per barrel (159 liters) and no immediate risk of a hike in world oil prices, it is quite likely that production of ethanol cars may cease by year 2000 with ethanol becoming simply a gasoline additive. Until then, there remains of fleet of about 4.5 million ethanol-fueled cars in Brazil that will continue to require a steady and reliable supply of fuel alcohol.

Source: Michaelis 1991; Weiss 1990; 1993; Petrobras 1996

---

vehicles to meet increasingly strict emission standards at reasonable cost.

Life-Cycle Emissions

End-use emissions constitute only a part of the life-cycle emissions associated with the production and use of motor vehicles. Automotive life-cycle emissions may be divided into three major stages: production of the motor vehicle, production of the fuel to run the vehicle, and the use of the vehicle on the road.

The energy use and emissions related to the manufacture of comparable vehicles designed to run on different fuels do vary significantly, with at most a 10 percent variation among various fuels. A greater variation occurs among vehicles of different size and between vehicles with different specifications within the same model type and range. The life cycle emissions from fuel production and end-use have received considerable attention in recent years with the emergence of a variety of alternative fuels for which the end-use emissions are considerably lower or approach zero as in the case of electric vehicles. For electric and hydrogen-fueled vehicles, the emissions are heavily dependent on the source of electricity. The emissions from

Box 5.8 Compressed Natural Gas in New Zealand

The New Zealand government launched a CNG program in 1979, aiming to convert 150,000 vehicles to CNG by the end of 1985. By 1986, 110,000 vehicles (11 percent of all cars and light trucks) had been converted, and New Zealand had 400 filling stations dispensing CNG. In 1979, CNG sold for NZ$0.29 per liter compared to NZ$0.42 per liter for premium gasoline. The price advantage for CNG peaked in 1984 when CNG cost about 40 percent the price of gasoline. Following deregulation of fuel prices in 1986, gasoline prices fell more rapidly in real terms than CNG prices, significantly narrowing the price difference between the two fuels. Incentives were provided in the form of vehicle conversion grants. The cost of conversion at the time was about NZ$1,500 (U.S.$750 in 1984 prices), of which the government provided NZ$150. Initial take-up was poor, but it was improved in late 1980 by an increase in the grant to NZ$200 and increased subsidies to fueling stations, along with tax benefits to consumers. The average payback period for car conversion was about two years. Motorists, however, were reluctant to convert to the new technology and were concerned about fuel availability. Some poor conversions contributed to consumer doubts. A large increase in government support in 1983 included a low-interest loan for conversions.

The increase in conversions was halted in 1985 when a new administration reduced support for the program. The conversion rate fell from 2,400 a month in 1984 to 150 a month in 1987. By 1988, sales of conversion kits had slumped to levels below the earliest years of the program.

Source: Michaelis 1991; Kurani and Sperling 1993

Table 5.21 Alternative Fuel Vehicles: Refueling Infrastructure Costs and Operational Characteristics
(Canadian dollars)

Fuel type	Infrastructure costs	Vehicle retrofit cost	Incremental OEM vehicle cost; current and future mass production	Vehicle range (with equal cargo space)[a]	Weight of fuel and tank (equivalent 75 liter capacity)[a]
Natural gas (CNG)	$300,000 (quick fill) $3,500 (5-hr fill)	$2,500-3,000	$3,000-4000 $1,000 (mass)	20%	5 times higher (299 kg)
Propane (LNG)	$40,000-80,000 $50,000 (new pump)	$1,500-2,000 No conversion available	$2,000 $600 (mass) $2,000	70%	1.7 times higher (102 kg) 2.2 times higher
Methanol	$10,000 (station retrofit)		$200 (mass)	60%	(131 kg)
Gasohol	No additional infra. needed	No adjustment needed	Special vehicle not needed	Slightly lower	No difference
Ethanol	Same as methanol	No conversion available	Same as methanol	67%	1.65 times higher (99 kg)

a. Compared to an equivalent gasoline-powered car.
Source: British Columbia, Ministry of Energy, Mines, and Petroleum Resources 1994

the production and distribution of fossil-derived alternative fuels (e.g, LPG or CNG) are similar to those for gasoline although there may be major differences with respect to specific pollutants e.g., low SO_2 emissions but high methane (CH_4) emissions from the production of CNG (Lewis and Gover 1996; OECD 1993; OECD 1995).

Emissions from vehicle use comprise the largest part of life-cycle emissions, and depend on many factors, including engine technology, emissions control technology, the type of trip-making (congested urban or free-flowing motorway), the driving style (aggressive or steady), and the level and quality of vehicle maintenance. Estimated life-cycle emissions for gasoline-fueled light-duty vehicles are shown in table 5.22. Vehicle use is the largest lifetime contributor to CO_2, CO, and NO_x emissions, fuel production the greatest contributor to hydrocarbon emissions, while vehicle production is dominant in SO_2 and PM emissions. The relative share of vehicle use to fuel production and distribution emissions in the gasoline life-cycle depends mainly on the lifetime distance traveled by the vehicle.

Estimated life-cycle emissions from cars for a variety of fuels is presented in table 5.23. While there is a large variation among various fuel types with respect to the stage (vehicle production, fuel production and distribution, vehicle use) at which emissions occur, the range in aggregate life-cycle emissions is much smaller. By changing fuels, it is possible to transfer the source of emissions from the vehicle-use stage to the fuel production/distribution stage but it is more difficult to reduce the overall magnitude of emissions (Lewis and Gover 1996). This is

Table 5.22 Aggregate Life-Cycle Emissions for Gasoline-Fueled Cars with Respect to Fuel Production, Vehicle Production, and In-Service Use
(grams per kilometer)

	CO_2	CO	NO_x	HC^a	SO_2	PM	N_2O
Fuel production	47.0	0.061	0.174	0.388	0.185	0.011	0.0066
Vehicle production	54.5	0.021	0.160	0.105	0.493	0.016	0.0007
Vehicle use	186.3	3.371	0.224	0.299	0.020	0.005	0.0538
Total	287.8	3.453	0.558	0.792	0.699	0.032	0.0611
Percent of total emissions							
Fuel production	16.3	1.8	31.1	49.1	26.5	35.7	10.9
Vehicle production	18.9	0.6	28.7	13.2	70.6	48.8	1.2
Vehicle use	64.7	97.6	40.2	37.7	2.9	15.5	88.0

a. Hydrocarbon figures include VOC emissions from paint and adhesive operations.
Note: Vehicle utilization based on a vehicle lifetime of 12 years at 9,500 miles per annum.
Source: Lewis and Gover 1996

Table 5.23 Aggregate Life-Cycle Emissions from Cars for Conventional and Alternative Fuels
(grams per kilometer)

	CO_2	CO	NO_x	HC	SO_2	PM	N_2O
Gasoline	287.8	3.453	0.558	0.792	0.699	0.032	0.0611
Diesel	227.1	0.489	0.981	0.384	0.702	0.131	0.005
LPG	239.0	3.889	0.482	0.443	0.649	0.027	0.505
CNG	242.0	0.863	0.457	1.137	0.575	0.022	0.815
Methanol	233.7	3.292	0.729	0.914	0.549	0.023	0.038
Biomethanol	292.0	3.419	0.784	0.597	0.646	0.039	0.046
Electricity	228.1	0.068	0.520	0.451	1.005	0.040	0.008

Note: Vehicle utilization based on a vehicle lifetime of 12 years at 9,500 miles per annum.
Source: Lewis and Gover 1996

of particular relevance to greenhouse gas emissions. By transferring the source of emissions from vehicle use to fuel production, as in the case of electric vehicles, local pollutants such as CO, HC, and NO_x may be eliminated entirely but there may be a considerable increase in CO_2, SO_2, and PM emissions (e.g., from coal-fired power generation plants).

Conclusions

Although alternative fuels (including electricity) power less than two percent of the global motor vehicle fleet, they have the potential to reduce urban air pollution when used in vehicles with dedicated engines and optimized emission control systems. Among the alternative fuels, the greatest emission reductions are obtained with hydrogen, followed by natural gas and LPG. Electric vehicles have zero emissions at the point of use, although emissions of NO_x and SO_2 associated with

power generation for electric vehicle use can exceed those from conventional gasoline and diesel vehicles.

Commercial vehicle fleets with high annual mileage (taxis, urban buses, pick-ups, and delivery vehicles) are likely to remain the main niche market for alternative fuels, with CNG capturing a major share of the market in developing countries. Inconveniences associated with gaseous fuel and electric vehicles—slow refueling, bulky fuel storage, reduced range—tend to limit their appeal in the consumer vehicle market. Increasing consumer familiarity with the technology and the establishment of more extensive fuel dispensing networks should make alternative-fuel vehicles attractive to consumers.

In developing countries after-market conversions to LPG or CNG use will remain the primary mode of alternative fuel use. The depth of market penetration will depend on retail fuel prices, inclusive of taxes. The fuel price differential should allow recovery of retrofit costs in three to four years. A potential market for alternative fuels in developing countries could be as additives or feedstocks for reformulated gasolines. MTBE and ETBE (produced from methanol and ethanol), are the preferred additives because they require no change in vehicle technology. With reasonable fiscal incentives, appropriate fuel pricing, and technical support, electric two-and three-wheelers could substitute for their gasoline-fueled counterparts in many highly congested urban areas in the developing world.

References

AASHTO (American Association of State Highway and Transportation Officials). 1993. "Major Electric Vehicle Test on Baltic Island" and "Electric Vehicle Subsidy Programs: U.S., Japan, and France". *International Transportation Observer* (Summer 1993), AASHTO and International Center of the Academy for State and Local Government (ASLG), Washington, D.C.

____. 1994. "La Rochelle (France) Starts Major EV Demo". *International Transportation Observer* (Spring 1994), Washington, D.C.

____. 1995. "Trolleybuses - A Trolleybus Renaissance?". *International Transportation Observer* (Spring 1995), Washington, D.C.

____. 1996. "Energy Alternatives - the Bus that Sheds Water" *International Transportation Observer* (Spring 1996), Washington, D.C.

ACEA/EUROPIA (European Automobile Manufacturers Association/European Petroleum Industry Association). 1995a. *European Programme on Emissions, Fuels, and Engine Technologies Report*, Brussels.

____. 1995b. "Fuel Additives, Deposit Formation, and Emissions: An Evaluation of Existing Literature." Report of European Commission — Industry Working Group (February 2, 1995), European Commission, Brussels.

____. 1994. "Effects of Fuel Qualities and Related Vehicle Technologies on European Vehicle Emissions: An Evaluation of Existing Literature and Proprietary Data", Report of European Commission—Industry Working Group (February 24, 1994), European Commission, Brussels.

Aguila, J., N. Chan, and J. Courtis. 1991. "Proposed Regulations for California Phase 2 Reformulated Gasoline." Technical Support Document, California Air Resources Board, Sacramento, California.

Alfuso, S., M. Aurlemma, G. Police, and M.V. Prati. 1993. "The Effect of Methyl-Ester of Rapeseed Oil on Combustion and Emissions of DI Diesel Engines." *SAE Paper* 932801. SAE International, Warrendale, Pennsylvania.

Andrews, G.E., and L.A. Charalambous. 1991. "An Organic Diesel Fuel Additive for the Reduction of Particulate Emissions." *SAE Paper* 912334. SAE International, Warrendale, Pennsylvania.

Ani F., M. Lal, and A. Williams. 1990. "The Combustion Characteristics of Palm Oil and Palm Oil Ester." Paper presented at the Third International Conference on Small Engines and Their Fuels for Use in Small Areas (July 17), University of Reading, U.K.

AQIRP (Air Quality Improvement Research Program). 1990. "Inertial Mass Exhaust Emissions Results from Reformulated Gasolines." Technical Bulletin 1, Auto/Oil Coordinating Research Council Inc., Atlanta, Georgia.

____. 1991a. "Emissions Results of Oxygenated Gasolines and Changes in RVP." Technical Bulletin 6, Auto/Oil Coordinating Research Council Inc., Atlanta, Georgia.

____. 1991b. "Exhaust Emissions of Toxic Air Pollutants Using Reformulated Gasolines." Technical Bulletin 5, Auto/Oil Coordinating Research Council Inc., Atlanta, Georgia.

____. 1991c. "Mass Exhaust Emissions Results from Reformulated Gasolines in Older Vehicles." Technical Bulletin 4, Auto/Oil Coordinating Research Council Inc., Atlanta, Georgia.

____. 1992a. "Comparison of Effects of MTBE and TAME on Exhaust and Evaporate Emissions, Air Toxics, and Reactivity." Technical Bulletin 9, Auto/Oil Coordinating Research Council Inc., Atlanta, Georgia.

____. 1992b. "Effects of Fuel Sulfur on Mass Exhaust Emissions, Air Toxics, and Reactivity." Technical Bulletin 8, Auto/Oil Coordinating Research Council Inc., Atlanta, Georgia.

____. 1992c. "Estimated Costs of Modifying Gasoline Properties." Economics Bulletin 2, Auto/Oil Coordinating Research Council Inc., Atlanta, Georgia.

____. 1993. *Auto/Oil Air Quality Improvement Research Program Phase I Final Report*. Auto/Oil Coordinating Research Council Inc., Atlanta, Georgia.

____. 1995a. "Effects of Gasoline T50, T90, and Sulfur on Exhaust Emissions of Current and Future Vehicles", Technical Bulletin 18, Auto/Oil Coordinating Research Council Inc., Atlanta, Georgia.

____. 1995b. "Gasoline Reformulation and Vehicle Technology Effects on Exhaust Emissions", Technical Bulletin 19, Auto/Oil Coordinating Research Council Inc., Atlanta, Georgia.

____. 1995c. "Exhaust Emissions of Compressed Natural Gas Vehicles Compared with Gasoline Vehicles." Technical Bulletin 15, Auto/Oil Coordinating Research Council Inc., Atlanta, Georgia.

_____. 1995d. "Exhaust Emissions of E85 Ethanol Fuel and Gasoline Flexible/Variable Fuel Vehicles." Technical Bulletin 16, Auto/Oil Coordinating Research Council Inc., Atlanta, Georgia.

ARCO Chemical Europe. 1993. "European Clean Gasoline." Maidenhead, Berk., U.K.

Armstrong, J.D. and Wilbraham, K. 1995. "Fuels - Their Properties and Respective International Standards", Paper Presented at the Seminar on Environmentally Friendly Petroleum Fuels (Nov. 18-19, 1995), Directorate General of Oil, Ministry of Petroleum and Natural Resources, Islamabad, Pakistan.

British Columbia, Ministry of Energy, Mines, and Petroleum Resources. 1994. "Cleaner Fuels for Cleaner Air: The Role of Alternative Transportation Fuels in British Columbia." Victoria, British Columbia, Canada.

Barakat, M.A. 1995. "Availability of Improved Petroleum Products as a Function of Quality". Paper Presented at the Seminar on Environmentally Friendly Petroleum Fuels (Nov. 18-19, 1995), Directorate General of Oil, Ministry of Petroleum and Natural Resources, Islamabad, Pakistan.

Battelle and Gannett Fleming. 1990. "Effects of Alternative Fuels on the U.S. Trucking Industry." American Trucking Association Research Foundation, Alexandria, Virginia.

Billings, R.E. 1993. "The Hydrogen World View." Project Energy 1993, Conference Proceedings, International Academy of Science, Independence, Montana.

Braddock, J.N. 1988. "Factors Influencing the Composition and Quantity of Passenger Car Refueling Emissions - Part II." *SAE Paper* 880712, SAE International, Warrendale, Pennsylvania.

Camarsa, M. and M. Hublin. 1995. "Preliminary Results - Diesel Fuel", European Programme on Emissions, Fuels, and Engines (EPEFE), Brussels.

CARB (California Air Resources Board). 1991. "Technical Support Document for Proposed Reactivity Adjustment Factors for Transitional Low-Emission Vehicles." Mobile Source Division, Research Division, El Monte, California.

_____. 1994. "Zero Emission Vehicle Update." Draft Technical document for the Low-emission Vehicle and Zero Emission Vehicle Workshop, March 25.

CONCAWE (Conservation of Clean Air and Water in Europe). 1987. "Diesel Fuel Quality and its Relationship with Emissions for Diesel Engines." *Report 10/87.* The Hague, Netherlands.

_____. 1989a. "Economic Consequences of Limiting Benzene/Aromatics in Gasoline." The Hague.

_____. 1989b. "Costs to Reduce the Sulfur Content of Diesel Fuel." *Report 10/89.* The Hague.

_____. 1992. "Motor Vehicle Emission Regulations and Fuel Specifications - 1992 Update." *Report 2/92.* Brussels.

_____. 1994. "Motor Vehicle Emission Regulations and Fuel Specifications - 1994 Update." *Report 4/94.* Brussels.

_____. 1995. "Motor Vehicle Emission Regulations and Fuel Specifications in Europe and the United State - 1995 Update." *Report 5/95.* Brussels.

COST302. 1988. *Prospects for Electric Vehicles in Europe.* F. Fabre and A. Klose, eds. Commission of the European Communities, Luxembourg.

DiFiglio, C., and M.F. Lawrence. 1987. "Economic and Security Issues of Methanol Supply." In METHANOL—Promise and Problems. *SAE Paper* 872062. SAE International, Warrendale, Pennsylvania.

Draper, W.M., J. Phillips, and H.W. Zeller. 1988. "Impact of a Barium Fuel Additive on the Mutagenicity and Polycyclic Aromatic Hydrocarbon Content of Diesel Exhaust Particulate Emissions." *SAE Paper* 881651. SAE International, Warrendale, Pennsylvania.

Eager, J. 1995. "The Potential of High Quality Fuels from Consumer's Point of View", Paper Presented at the Seminar on Environmentally Friendly Petroleum Fuels (Nov. 18-19, 1995), Directorate General of Oil, Ministry of Petroleum and Natural Resources, Islamabad, Pakistan.

EIA (Energy Information Administration). 1994. *Alternatives to Traditional Transportation Fuels—An Overview.* U.S. Department of Energy, Washington, D.C.

Furey, R.L., and D.R. Monroe. 1981. "Fuel Sulfur Effects on the Performance of Automotive Three-Way Catalysts During Vehicle Emissions Tests." *SAE Paper* 811228. SAE International, Warrendale, Pennsylvania.

Lewis, C.A. and M.P. Gover. 1996. "Life-Cycle Analysis of Motor Fuel Emissions", Estimation of Pollutant Emissions from Transport, *COST 319*, Interim Report and Proceedings Workshop Nov 27-28, 1995, CEC/DGVII, European Commission, Brussels.

Hochhauser, A.M., J.D. Benson, V. Burns, R.A. Grose, W.S. Koehl, L.J. Painter, B.H. Rippon, R.M. Reuter, and J.A. Rutherford. 1991. "The Effect of Aromatics, MTBE, Olefins, and T90 on Mass Exhaust Emissions from Current and Other Vehicles." The Auto/Oil Air Quality Improvement Research Program, Atlanta, Georgia.

Holman C. 1990. *Pollution from Diesel Vehicles.* Friends of the Earth Ltd., U.K.

Homewood, B. 1993. "Will Brazil's Cars Go to the Wagon?" *New Scientist*, Jan 9, 1993 (pp 22–23).

Hutcheson R.C., and C. van Paasen. 1990. "Diesel Fuel Quality into the Next Century." Paper presented at the Institute of Petroleum Conference on the European Auto Diesel Challenge. Shell International Petroleum Company, London.

Hutcheson, R.C. 1995. "Alternative Fuels in the Automotive Market" *CONCAWE Report No. 2/95.* Brussels.

IANGV/IGU (International Association of Natural Gas Vehicles/International Gas Union). 1994. "Task Force Report, Milan 1994". Natural Gas Vehicle Task Force, Auckland, New Zealand.

IEA (International Energy Agency). 1990. *Substitute Fuels for Road Transport: A Technology Assessment.* Organization for Economic Cooperation and Development, Paris.

_____. 1993. *Electric Vehicles: Technology, Performance, and Potential,* Organization for Economic Cooperation and Development, Paris.

Jamal, M.Q. 1995. "Impact of Lead Phase-Out", Paper Presented at the Seminar on Environmentally Friendly Petroleum Fuels (Nov. 18-19, 1995), Directorate General of Oil, Ministry of Petroleum and Natural Resources, Islamabad, Pakistan.

Kukkonen, C.A., and M. Shelef. 1994. "Hydrogen as an Alternative Automotive Fuel: 1993 Update." *SAE Paper* 940766. SAE International, Warrendale, Pennsylvania.

Kurani, S.K. and D. Sperling. 1993. "Fuel Availability and Diesel Fuel Vehicles in New Zealand", TRB Reprint Paper No. 930992, Institute of Transportation Studies, University of California, Davis.

Lang, C. J., and F. H. Palmer. 1989. "The Use of Oxygenates in Motor Gasolines." In K. Owen ed., *Gasoline and Diesel Fuel Additives.* John Wiley and Sons, New York.

Mackenzie, J.J. 1994. *The Keys to the Car: Electric and Hydrogen Vehicles for the 21st Century.* World Research Institute, Washington, D.C.

Maggio, M.E., T.H. Maze, and K.M. Waggoner. 1991. *Alternative Fuels: What You Need to Know.* American Public Works Association, Chicago.

McArragher, J.S., W.E. Betts, J. Brandt, D. Kiessling, G.F. Marchesi, K. Owen, J.K. Pearson, K.P. Schug, and D.G. Snelgrove. 1988. "Evaporative Emissions from Modern European Vehicles and their Control." *SAE Paper* 880315, SAE International, Warrendale, Pennsylvania.

Michaelis, L. 1991. "Air Pollution from Motor Vehicles." Draft position Paper Prepared for the World Bank, Washington, D.C.

_____. 1995. "The Abatement of Air Pollution from Motor Vehicles, The Role of Alternative Fuels", *Journal of Transport Economics and Policy,* Volume XXIX, No.1 (pp 71-84).

Moreno, R. Jr., and D.G. Fallen Bailey. 1989. *Alternative Transport Fuels from Natural Gas.* World Bank Technical Paper 96. Washington, D.C.

Moulton, P. 1996. "The Electric Vehicle Transportation Program for Kathmandu-A Brief Status Report", Global Resources Institute, Eugene, Oregon.

Murgel, E.M., and A. Szwarc. 1989. "Condições de Tráfego e a Emissão de Poluentes." *Ambiente* 3 (1), São Paolo, Brazil.

NRDC and CAPE 21. 1994. "Four in '94: Assessing National Actions to Implement Agenda 21: A Country-by-Country Report." National Resources Defense Council, Washington, DC.

OECD (Organization for Economic Development and Cooperation). 1988. *Transport and Environment.* Paris.

_____. 1993. *Choosing an Alternative Trasnportation Fuel, Air Pollution and Greenhouse Gas Impacts,* Paris.

_____. 1995. *Motor Vehicle Pollution Reduction Strategies Beyond 2010.* Paris.

OTA (Office of Technology Assessment, U.S. Congress). 1990. *Replacing Gasoline: Alternative Fuels for Light-Duty Vehicles,* U.S. Government Printing Office, Washington, D.C.

_____. 1995. *Advanced Automotive Technology: Visions of a Super-Efficient Family Car,* OTA-ETI-638, U.S. Government Printing Office, Washington, D.C.

Owen K., and T. Coley. 1990. *Automobile Fuels Handbook.* SAE International, Warrendale, Pennsylvania.

Pagowski, Z., S. Oleksy, B. Wislicki, J. Szczecinski, and B. Zdrodowska. 1994. "An Investigation of Polish Biofuel Used in High-Speed Diesel Car Engines." *3rd Poster Proceedings,* INRETS, Arcueil, France.

Perkins, L.A., Peterson, C.L., and Auld, D.L. 1991. "Durability Testing of Transesterified Winter Rape Oil (Brassica Napus L.) as Fuel in Small Bore, Multi-Cylinder, DI, CI Engines." *SAE Paper* 911764. SAE International, Warrendale, Pennsylvania.

Petrobras (Brazilian Oil Company). 1996. "Personal Communications." Rio de Janeiro, Brazil

Piracha, M. A. 1993. "Fuel Quality in Pakistan - Product Specifications and Typical Values." Personal Communication, Pakistan Refinery Limited, Karachi.

Post, R. 1993. "Flywheel Technology", in *Advanced Components for Electric and Hybrid Electric Vehicles: Workshop Proceedings.* NIST Special Publication 860, National Institute of Standards and Technology, Gaithersburg, Maryland.

Reed, T.B. 1994. "An Overview of The Current Status of Biodiesel." Department of Chemical Engineering, Colorado School of Mines, Golden, Colorado.

Sathaye J., B. Atkinson, and S. Meyers. 1989. "Promoting Alternative Transportation Fuels: The Role of Government in New Zealand, Brazil, and Canada." *Energy:* Vol. 14 (10): 575–84. Pergamon Press.

Scholl, K.W., and Sorenson, S.C. 1993. "Combustion of Soybean Oil Methyl Ester in a Direct Injection Diesel Engine." *SAE Paper* 930934. SAE International, Warrendale, Pennsylvania.

Schwartz, J., H. Pitcher, R. Levin, B. Ostro, and A.L. Nichols. 1985. "Costs and Benefits of Reducing Lead in Gasoline: Final Regulatory Impact Analysis." Report EPA-230-05-85-006. U.S. EPA, Washington, D.C.

Seisler, J., G. Sperling, C.S. Weaver, and S.H. Turner. 1993. "Alternative Fuel Vehicles", *in Energy Law and Transactions.* Oakland, California: Matthew Bender and Co.

Shell. 1988. "Diesel." Shell Science and Technology, Shell International Petroleum Company, London.

———. 1989. "Gasoline." Shell Science and Technology, Shell International Petroleum Company, London.

Simonain, H. 1996. "Smog Clears over Carmakers", *Financial Times* (January 11, 1996), London.

Smith, A.K., M. Dowling, W.J. Fowler, and M.G. Taylor. 1991. "Additive Approaches to Reduced Diesel Emissions", *SAE Paper 912327.* SAE International, Warrendale, Pennsylvania.

Sperling, D., ed. 1989. *Alternative Transportation Fuels: Environmental and Energy Solutions.* Westport, Conn.: Quorom Books.

Sperling, D. 1995. *Future Drive: Electric Vehicles and Sustainable Transportation,* Island Press, Washington, D.C.

Szwarc, A., G. Murgel Branco, E.L. Farah, and W. Costa. 1991. "Alcohol Crisis in Brazil: The Search for Alternatives." Paper Presented at the Ninth International Symposium on Alcohol Fuels, November 12-15, Florence, Italy.

Transnet. 1990. *Energy, Transport, and the Environment.* Innovation in Transport, London.

Tritthart, P., and Zelenka, P. 1990. "Vegetable Oils and Alcohols: Additive Fuels for Diesel Engines." *SAE Paper 905112.* SAE International, Warrendale, Pennsylvania.

U.S. DOE (Department of Energy). 1994. "Biofuels Update 2 (1)." National Alternative Fuels Hotline, Arlington, Virginia.

U.S. EPA (Environmental Protection Agency). 1987. "Draft Regulatory Impact Analysis: Control of Gasoline Volatility and Evaporative Hydrocarbon Emissions From New Motor Vehicles." Office of Mobile Sources, Washington, D.C.

———. 1988. "Mobile Source Emission Standards Summary." Office of Air and Radiation, Washington, D.C.

———. 1993. "Motor Vehicle-Related Air Toxic Study," Public Review Draft, Washington, D.C.

Wagner, T.O., and D.F. Tatterson. 1987. "Comparative Economics of Methanol and Gasoline." *SAE Paper 871579.* SAE International, Warrendale, Pennsylvania.

Wall, J.C., and S.K. Hoekman. 1984. "Fuel Composition Effects on Heavy-Duty Diesel Particulate Emissions." *SAE Paper 841364.* SAE International, Warrendale, Pennsylvania.

Walsh, M.P. 1996b. "Car Lines." Issue 96–4, 3105 N. Dinwiddie Street, Arlington, Virginia.

Weaver, C.S. 1986. "The Effects of Low-Lead and Unleaded Fuels on Gasoline Engines." *SAE Paper 860090.* SAE International, Warrendale, Pennsylvania.

———. 1989. *Natural Gas Vehicles: A Review of the State of the Art.* Natural Gas Vehicle Coalition. Arlington, Virginia.

Weaver, C.S. 1995. "Low-Lead as a Transition Measure", Informal Technical Note, Engine, Fuel, and Emissions Engineering, Inc., Sacramento, California. (mimeo)

Weaver, C.S., and R. Klausmeier. 1988. "Heavy-Duty Diesel Vehicle Inspection and Maintenance Study." (4 volumes). Report to the California Air Resources Board. Radian Corporation, Sacramento, California.

Weaver, C.S. and S.H. Turner. 1994. "Dual Fuel Natural Gas/Diesel Engines: Technology, Performance, and Emissions", Report to the Gas Research Institute, Engine, Fuel, and Emissions Engineering, Inc., Sacramento, California.

Weaver, C.S. and S.H. Turner. 1995. "Natural Gas Vehicles: State of the Art, 1995," Report to the Natural Gas Vehicle Coalition." Engine, Fuel, and Emissions Engineering, Inc., Sacramento, California.

Weaver, C.S. and L.M. Chan. 1995. "Company Archives: Multiple Sources." Engine, Fuel and Emissions Engineering, Inc., Sacramento, California.

World Bank. 1981. "Alternative Fuels for Use in Internal Combustion Engines." Energy Department Paper 4. Washington, D.C.

World Bank. 1995. "Argentina - Managing Environmental Pollution: Issues and Options", Volume II- Technical Report (14070-AR), Environment and Urban Development Division, Country Dept. I, The World Bank, Washington, D.C.

Ziejewski, M., H. Goettler, and G. L. Pratt. 1986. "Comparative Analysis of the Long-Term Performance of a Diesel Engine on Vegetable Oil Based Alternate Fuels." *SAE Paper 860301.* SAE International, Warrendale, Pennsylvania.

Appendix 5.1

International Use of Lead in Gasoline

Worldwide, leaded gasoline has been estimated to account for three quarters or more of atmospheric lead emissions. (Nriagu and Pacyna 1988). Octel Ltd. (88 percent owned by Great Lakes Chemical and 12 percent owned by Chevron) is the main producer of lead gasoline additives worldwide. The other producers are in Germany—less than 4,000 tons of lead per year, and in Russia—about 5,000 tons per year. Ethyl Corporation was, until March 1994, the world's other major producer of lead additives. Ethyl has ceased production of tetra-ethyl lead (TEL) and has entered into an agreement with Octel for its supply, so that Ethyl and Octel will remain the the two main suppliers of TEL. Ethyl supplies about a third of the world market for TEL; this accounts for about 70 percent of Ethyl's earnings (Ottenstein and Tsuei 1994).

The total annual sales of lead additives are in the order of U.S.$1 billion (TEL sells for U.S.$7 per kg, equivalent to U.S.$0.023 per g Pb). The TEL market is estimated to be decreasing at about 7 percent annually. Great Lakes and Ethyl have priced ahead of TEL volume declines to maintain earnings. Thus annual price increases of 7 percent or more can be expected in coming years (Ottenstein and Tsuei 1994).

Annual world use of lead in gasoline additives was estimated to be about 70,000 tons in 1993. An estimated one half of this was used in the former Soviet Union, Eastern Europe and the Far East. The rest was used, in approximately equal amounts, in Western Europe, Africa, the Middle East, and the Americas (Thomas 1995). A number of countries have effectively eliminated the use of leaded gasoline. In Japan, reduction of lead in gasoline began in the 1970s, and leaded gasoline has now been eliminated (OECD 1993). In Canada, the use of leaded gasoline was banned in 1990, except in farm, marine and commercial transport and aviation, in which lead content is limited to 0.026 g/liter (OECD 1993). Brazil uses no leaded gasoline, with its vehicle fleet operating entirely on fuel alcohol and gasohol. Colombia eliminated leaded gasoline in 1991. As of 1994, leaded gasoline accounted for about one percent of total U.S. automotive gasoline use. The addition of lead to gasoline has been prohibited in the United States since January 1, 1995. Austria banned leaded gasoline in 1993 (NRDC and CAPE 1994). In Sweden and South Korea, use of leaded gasoline was eliminated in 1994. Slovakia and Thailand banned leaded gasoline in 1995.

A number of other countries are in the process of phasing out leaded gasoline. In Mexico, as of 1993, all new cars must have catalytic converters, and thus must use unleaded gasoline (Driscoll and others 1992). In Australia, the government is urging all motorists to switch to unleaded gasoline, and the concentrations of lead in gasoline are being reduced (Australian CEPA 1994). In the European Union, the lead content of gasoline is limited to 0.15g/l, and all new cars are required to have catalytic converters, which require use of unleaded gasoline. Unleaded gasoline accounted for 62 percent of gasoline sales in the European Union as a whole, in 1994. A full ban on the use of leaded gasoline in the European Union will become effective in 2000. China also plans to ban production of leaded gasoline by 2000 (Walsh 1996b).

Table A5.1.1 shows the available data on worldwide use of leaded gasoline. The reliability of some of the data, regarding both concentrations of lead in gasoline and the leaded fraction of gasoline sold, is low. In many cases, the maximum concentration and not the average is reported. Overall, however, the table shows that leaded gasoline continues to be used heavily in Africa, Central and South America, Asia and Eastern Europe; in many countries, unleaded gasoline remains unavailable.

This appendix was prepared by Ms. Valerie M. Thomas, Researcher Staff, Princeton University and is based on her paper published in the *Annual Review of Energy and Environment* 20:301-24.

Not listed in Table A5.1.1 is aviation gasoline, used by aircraft with piston engines (nor jet engines). Aviation gasoline is typically leaded, but there is little information on the lead content of aviation gasoline in most countries. In Switzerland, lead in aviation gasoline is limited to 0.56 g/l (OECD1993). In the United States, lead in aviation gasoline is typically 0.5-0.8g/l. About 1.2 billion liters of aviation gasoline are used in the United States annually, for a total lead emission of about 840 tons per year. This is about eight times the amount of lead contained in the highway gasoline still used in the US in 1993. Many piston aircraft do not have hard-ened exhaust valve seats, and so may risk valve seat recession if lead is not used. However, in tests of automobile engines, the risk of valve seat recession appears to have been exaggerated, and even in the most strenuous tests, valve seat recession is prevented at lead concentrations of 0.05 g/l, an order of magnitude less than is apparently used in aviation gasolines (Thomas 1995).

Table A5.1.1 includes gasoline used for boat engines. While in most countries this is a minor market, in small island nations it could represent a significant fraction of gasoline consumption. Leaded gasoline is used for car

Table A5.1.1 Estimated World Use of Leaded Gasoline, 1993

Region/countries	Motor gasoline consumption 10^9 l/yr	Lead content of leaded gasoline, g/l	Total added lead, tons/yr	Leaded gasoline share percent (1993)*
North America				
Canada	32.0	–	–	0
Mexico	26.0	0.07	1,300	1
United States	430.0	0.026	100	70
Total	488.0	–	1,400	
Central & South America and Caribbean				
Antigua	0.02	–	–	0
Argentina	4.50	0.10	600	70
Brazil	17.00	–	–	0
Chile	1.60	0.42	660	99
Colombia	6.40	–	–	0
Ecuador	1.50	0.84	1,200	>95
Peru	1.20	0.84	920	91
Puerto Rico	2.90	0.13	380	–
Surinam	0.06	–	–	0
Trinidad & Tobago	0.36	0.40	140	100
Venezuela	7.70	0.37	2,600	90
Virgin Islands	0.23	1.12	260	–
Other [a]	8.00	0.40	3,000	
Total	51.47		9,760	
Western Europe				
Austria	3.5	–	–	0
Belgium	3.7	0.15	240	43
Denmark	2.2	0.15	79	24
Finland	2.7	0.15	120	30 (1992)
France	27.0	0.15	2,400	59
Germany	42.0	0.15	699	11
Greece	3.3	0.15	380	–
Ireland	1.2	0.15	110	62
Italy	19.0	0.15	2,200	76
Luxemborg	5.2	0.15	240	31
Netherlands	4.7	0.15	180	25
Norway	2.4	0.15	140	40
Portugal	1.9	0.40	600	79
Spain	11.0	0.15	1,600	94
Sweden	5.7	0.15	9	1
Switzerland	5.0	0.15	260	35 (1992)
United Kingdom	33.0	0.15	2,300	47
Other [b]	0.2	0.15	30	
Total	173.7		11,587	

* Unless indicated otherwise

Table A5.1.1 (Continued)

Region/countries	Motor gasoline consumption 10^9 l/yr	Lead content of leaded gasoline, g/l	Total added lead, tons/yr	Leaded gasoline share percent (1993)*
Eastern Europe[c] and Former Soviet Union				
Former Czechoslovakia	2.3	0.15	330	91 (1991)
Former Soviet Union	100.0	0.20	10,000	60 (1994)
Former Yogoslavia	3.6	0.50	1,800	98
Hungary	2.1	0.15	240	75
Poland	3.7	0.15	490	88
Romania	2.8	0.60	1,700	100
Turkey	4.4	0.15	650	90
Total	118.9		15,210	
Middle East				
Iran	8.1	0.19	1,500	100
Iraq	4.4	0.40	1,700	100
Israel	2.1	0.15	310	100
Kuwait	1.2	0.53	620	100
Qatar	0.4	0.40	75	47 (1992)
Saudi Arabia	9.3	0.40	3,700	100
Syria	1.5	0.24	360	100
United Arab Emirates	1.3	0.40	530	100
Other	4.6	0.40	1,900	
Total	28.5		8,995	
Africa[d]				
Algeria	2.6	0.60	1,700	100
Egypt	2.0	0.35	700	100
Libya	2.0	0.80	1,600	100
Nigeria	6.4	0.66	4,200	100
South Africa [e]	6.4	0.40	2,600	100
Other[f]	6.4	0.40	2,600	100
Total	25.8		13,400	
Far East and Oceania				
Australia	17.0	0.40	2,700	55 (1994)
China	30.0	0.70	900	41 (1994)
Hong Kong	0.4	0.15	32	32
India[g]	4.8	0.70	700	95 (1994)
Japan	44.0	-	-	0
Malaysia	3.8	0.15	300	46 (1994)
New Zealand	2.8	0.40	640	57 (1994)
Singapore	0.6	0.12	35	42 (1994)
Sri Lanka	0.2	0.20	50	100
Taiwan (China)	4.9	0.10	70	41 (1994)
Thailand	3.7	0.15	440	58 (1994)
Republic of Korea	3.8	-	-	0
Other[h]	15.0	0.40	3,167	
Total	114.0			
World	1,000.3		60,352	

* Unless otherwise indicated.

a. Honduras 100% of gasoline is leaded, 0.84g Pb/l; unleaded gasoline introduced in Barbados on June 1, 1994.

b. Iceland-the share of leaded gasoline is 70% (1993), lead content: 0.15g Pb/l.

c. Bulgaria-unleaded gasoline available; 0.15g Pb/l (1993) in leaded gasoline.

d. No unleaded gasoline available in Sub-Saharan Africa (1993); lead content as follows: Benin 0.84g Pb/l, Botswana 0.44g Pb/l, Ethiopia 0.76 g Pb/l, Kenya 0.40g Pb/l, Namibia 0.40g Pb/l, Nigeria 0.66g Pb/l, Malawi 0.53g Pb/l, South Africa 0.49g Pb/l, Uganda 0.80g Pb/l, and Zimbabwe 0.84g Pb/l.

e. Unleaded gasoline expected to be introduced by 1996 in South Africa.

f. Morocco – 0.3% of gasoline unleaded (1991), 0.5g Pb/l in leaded gasoline; Tunisia-no unleaded (1991), 0.5g Pb/l.

g. India-lead content to be reduced to 0.5g Pb/l by 1996.

h. Pakistan-no unleaded gasoline; 0.42 - 0.82 g Pb/l to be reduced to 0.35g Pb/l in 1996; Philippines-unleaded gasoline available in Metro Manila, Cebu, and Davao; 0.15g Pb/l in leaded gasoline with a market share of 90 percent (1994); Myanmar-no unleaded gasoline available, 0.56g Pb/l; Indonesia-the only ASEAN country not using unleaded gasoline; average lead content in gasoline, 0.15g Pb/l.

Source: Shah 1994; Thomas 1995; *Oil and Gas Journal* 1995; Walsh 1996b

racing, although unleaded racing gasoline is available in Canada (Mally 1993).

While in many countries the refineries may not be capable of producing unleaded gasoline with sufficient octane, this is not always the case. For example, in Venezuela and in Trinidad and Tobago (both oil producers), unleaded gasoline is made for export, while leaded gasoline is produced for the domestic market (Garip-Bertuol 1994). Oman has the capacity to produce unleaded gasoline, but, due to lack of demand, does not yet do so (MEED 1994).

In Table A5.1.1 the former Soviet Union is shown as the largest user of lead gasoline additives. While some lead additives are manufactured in Russia, they are mostly purchased from Octel, making the former Soviet Union one of Octel's major markets for lead additives (Great Lakes Chemical 1992). In a number of countries, use of leaded gasoline has been reduced in major cities. For example, while the lead content of leaded gasoline in most of Greece is 0.4g/l, in Athens it is 0.15 g/l (OECD 1993). Only unleaded gasoline is reported to be sold in Moscow and St. Petersburg. Lower concentrations of lead are used in Caracas than in the rest of Venezuela (Garip-Bertuol 1994). In Mexico City, use of leaded gasoline has been eliminated for the most part, but its use will continue in the rest of Mexico. India introduced unleaded gasoline in its four largest metropolitan cities (Bombay, Calcutta, Delhi and Madras) on April 1, 1995. By the end of 1998, all commercial gasoline sold in Beijing, Tianzing, Shanghai, Guangzhou and Shenzhen in China will be unleaded (Walsh 1996b).

A number of other countries have introduced unleaded gasoline, although there is no available information on its market share. These include Chile (1991), Israel (1991), Peru (1992), Morocco (1994) and the Philippines (1994). The Philippines also reduced the lead content of its leaded gasoline from 0.6g/l to 0.15g/l.

Other countries, including South Africa, India, Saudi Arabia and Bahrain plan to introduce unleaded gasoline in the near future (1995-96). Kuwait plans to have completely eliminated the use of leaded fuel by 1997 (Jansen 1994), and Argentina well before that deadline (Lloyd's list 1993). Ecuador will stop production of leaded gasoline in 1996 (Inter Press Service 1993), while Taiwan (China) will eliminate its use by 1999 (CENS 1992).

References

CENS (China Economic News Service). 1992. "Taiwan: Government to Advance Mandatory Unleaded Gas by Two Years." (December 15).

Driscoll W. Muschak J. Garfias J. Rothenberg SJ. 1992. "Reducing Lead in Gasoline: Mexico's Experience." *Environmental Science and Technology 26(9): 1702-05.*

Garip-Bertuol P. 1994. "Leaded Gasoline: Waiting for a Choice." Venectomy Monthly (April).

Great Lakes Chemical Corporation. 1992. "1991 Annual Report." West Lafyette, Indiana.

Inter Press Service. 1993. Ecuador: "Alarming Levels of Lead Poisoning in Quito." (September 27).

Janssen N. 1994. "Cars Must Switch for Unleaded Kuwait by '97." Moneyclips (March 10).

Lloyd's List 1993. "Argentina: Petrol Plan." YPF (August 21), London.

Natural Resources Defense Council (NRDC) and Campaign for Action to Protect Earth (CAPE 21). 1994. "Four in '94: Assessing National Actions to Implement Agenda 21: A Country-by-Country Report." NRDC, Washington, DC.

Oil & Gas Journal. 1995. "Asian Product Specifications Tighten by 2000." 93(32): 63-65.

OECD (Organization for Economic Cooperation and Development). 1993. *Risk Reduction Monograph No. 1: Lead.* Environmental Directorate, OECD, Paris.

Ottenstein R and Tsuei A. 1994. "Ethyl Corporation." CS First Boston Corporation, New York.

Nriagu, J.O. and J.M. Pacyna. 1988. "Quantitative Assessments of Worldwide Contamination of Air, Water and Soils by Trace Metals." *Nature 333:134-139.*

Malloy G. 1993. "Firm Scores with Unleaded Race Fuel." *Toronto Star. Oct. 9, p.L13.*

MEED (Middle East Economic Digest). 1994. "Oman; Oil Refining Company Denies Report of Unleaded Petrol Production." (August 1).

Shah, J. 1994. "Fact Sheet for Unleaded, Low-leaded Gasoline and Diesel." Asia Technical Department, World Bank, Washington, D.C. (mimeo)

Thomas, V.M. 1995. "The Elimination of Lead in Gasoline." *Annual Review of Energy and Environment 20:301-24.*

Walsh, M.P. 1996b. "Car Lines." Issue 96-4, 3105 N. Dinwiddie Street, Arlington, Virginia, U.S.A.

Appendix 5.2
Electric and Hybrid-Electric Vehicles

The distinguishing feature of an electric vehicle (EV) is that it is propelled by one or more electric motors, rather than by an internal-combustion engine. Electric vehicles can be separated into three groups, based on the how and where the electricity is produced:

- Vehicles relying on *continuous electric supply* from an off-board generation system. These include trolley buses supplied by overhead wires, as well as most electric rail transportation systems. Because of their dependence on continuous electric supply, these vehicles are suitable only for very limited niches, and they will not be discussed further in this appendix.
- Vehicles relying on *stored electricity* from an off-board generation system. These include battery-electric vehicles, as well as vehicles using other energy storage media such as flywheels. These are frequently referred to as zero-emission vehicles (ZEVs), but this is true only if emissions from the off-board power generation system are ignored.
- Vehicles relying on *on-board electric generation* to supply their needs. These include series electric hybrids, in which a small engine drives an electric generator while the wheels are powered exclusively by an electric motor, parallel electric hybrids, in which both the engine and the electric motor can drive the wheels, and fuel-cell electric vehicles.

The major advantage of electric propulsion is that it allows the two processes of propelling the vehicle and converting the chemical energy in the fuel into useful work to be separated. If the electrical energy to drive the vehicle is produced off-board, as in a battery-electric vehicle or a trollybus supplied by overhead wires, there is no pollution at the point of use. Of course, pollution may still be created at the power plant, but pollution control for a single, centralized power plant is often easier and cheaper than for a multitude of vehicles. Even where the electric power is generated on-board the vehicle, as in a hybrid or fuel-cell vehicle, the pollution produced is typically much less than would be produced by a vehicle with a conventional internal-combustion engine (ICV).

Because electric motors are capable of generating high torque at low speed, and can operate efficiently over a greater range of speeds than internal-combustion engines, the drivetrain of an EV can be simpler than that of an ICV. It can also be more efficient. The engine in an ICV must be designed to meet the peak power demand for acceleration, which is far more than required for normal driving. Figure A5.2.1 shows the cruising power required by a typical large passenger car as a function of speed and grade. Under straight-and-level conditions, an IC engine capable of producing more than 100 kW will typically be producing 10 kW or less. As a result, the engine's efficiency will typically be less than half of its efficiency under optimum conditions. The need to accommodate large, rapid variations in speed and load also impairs IC engine efficiency and emissions. While electric motors are also less efficient under light load than at full power, the reduction in efficiency is much less.

Electric vehicles can also save energy in stop-and-go driving through *regenerative braking*. In this technique, the electric motor is used as a generator, converting the kinetic energy of the vehicle's motion back into electric energy, rather than dissipating it as heat in the brakes. Provided there is sufficient on-board storage, regenerative braking can recover 50 to 80 percent of the kinetic energy of the vehicle for later use, while saving wear on the mechanical brakes. This is especially valuable for transit buses and other vehicles that stop and start frequently.

Figure A5.2.1 Vehicle Cruise Propulsive Power Required as a Function of Speed and Road Gradient

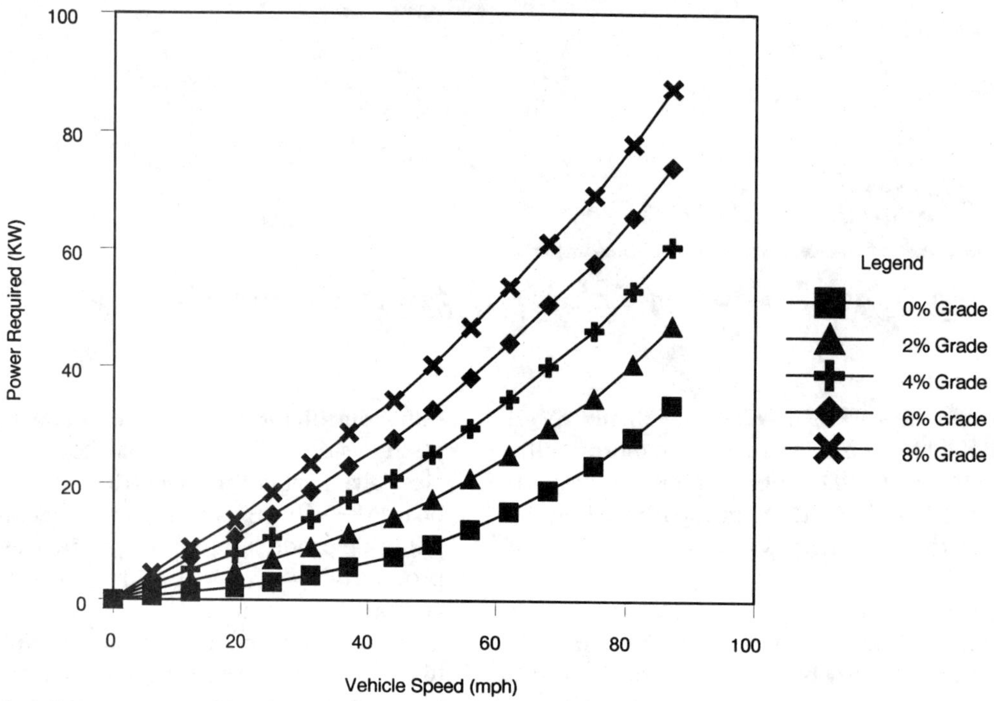

Source: Weaver and Chan 1995

Electric Propulsion Technology

Electric motors used in EV's are of three main types: DC motors, AC induction motors, and AC permanent-magnet synchronous motors. Table A5.2.1 compares the characteristics of these motor types. Most electric road vehicles produced up to this point have used DC motors, because this type of motor permits the use of a simpler and less expensive control system. Compared to AC motors producing the same power, DC motors are heavier and less efficient. The requirement for periodic servicing of the motor brushes is also a disadvantage. AC induction and permanent-magnet synchronous motors are lighter and more efficient, but can operate only at a rotational speed that is close to (induction motors) or equal to (synchronous motors) the frequency of the AC power source. For automotive use, they require a sophisticated and expensive control system to synthesize variable-frequency AC power. With recent developments in power semiconductor technology, the costs of variable-frequency AC motor controllers have been reduced considerably, and many new EV designs now incorporate AC propulsion. Permanent-magnet synchronous motors offer the highest power density and efficiency of the three motor types, but the exotic materials required for their magnets make them fairly expensive.

Energy Requirements of Electric Vehicles

Much confusion exists regarding the energy requirements of electric vehicles, and many published statements regarding EV energy requirements and range are half-truths, at best. EVs require energy for propulsion, to operate essential auxiliary systems such as headlights and windshield wipers, and for climate control. With the possible exception of a fuel-fired heater for climate control, all of this energy must be provided by the on-board storage or generation system.

Propulsion - Energy required for propulsion depends on the characteristics of the propulsion system; the mass, aerodynamic characteristics, and rolling resistance of the vehicle; the driving cycle; and the environmental conditions. The need to minimize on-board battery storage has led to considerable research on reducing propulsion energy requirements. Propulsive energy is required to overcome three main energy losses: aerodynamic drag, rolling resistance, and kinetic energy in the brakes when a vehicle slows down or stops. Kinetic energy is proportional to the vehicle mass and to the square of the speed, so that four times as much energy is required to accelerate a vehicle to 100 km/hr as to 50 km/hr. When a vehicle is slowed or stopped using conventional friction brakes, this energy is dissipated as

Table A5.2.1 Characteristics of Electric Motors for EV Applications

	DC Brushed	*AC Induction*	*AC Synchronous*
Efficiency (Max)	80-85%	85-90%	90-95%
Weight	—	+	++
Motor Cost	—	+	—
Controller Cost	++	—	—

Note: (-) = low,
(+) = moderate, and
(++) = high.
Source: Institute of Applied Energy, Japan 1992

heat. The use of regenerative braking makes it possible to recover much of the vehicle's kinetic energy once again as electricity, but electrical losses in the motor/generator and electric storage medium mean that some energy is still lost. The amount of this loss depends heavily on the driving pattern—jerky, start-and-stop driving wastes much more energy than gradual acceleration and deceleration.

Aerodynamic drag increases as the cube of velocity. At highway speeds, aerodynamic drag accounts for most of the energy loss from the vehicle, but this effect is negligible in city traffic. Headwinds and crosswinds can increase drag forces considerably. Aerodynamic drag is also proportional to the product of the *drag coefficient* C_D and the frontal area of the vehicle. It can be minimized by reducing the frontal area - making the vehicle thinner and flatter - and by designing the vehicle to have as low a C_D as possible, consistent with its intended use. This generally involves streamlining, eliminating protrusions and other disturbances in the airstream, and designing the rear of the vehicle to create a minimum of turbulence in its wake. Typical C_D values for vehicles are from 0.4 to 0.7 for cars, and up to 1.5 for trucks, but some advanced designs such as the GM Impact prototype vehicle have C_D values as low as 0.18 (Bartholomew 1993).

Rolling resistance is due mostly to energy lost in deformation of the tires in their contact with the road surface (and in deformation of the road surface itself, if it is other than clean, dry pavement). Friction in bearings and other moving parts of the vehicle contributes as well. Energy lost in the tires can be reduced by increasing the air pressure and reducing the cross-section of the tires, as well as by internal changes in tire construction. These changes may conflict with other tire-design goals such as safety and handling, however. Rolling resistance is lowest on clean, smooth, dry pavement. Rough pavement rain, snow, dirt or gravel on the road improperly inflated tires or improper tire alignment re-

sult in rolling resistance coefficients many times higher than the minimum.

The electrical energy required for propulsion is also affected by the efficiency of the propulsion system. Electrical losses in the motor controller and in the motor itself mean that this system is less than perfectly efficient. The combined efficiency of the electric controller and motor(s) in converting electrical energy from the battery into kinetic energy of the vehicle is typically in the range of 80 to 90 percent. Losses tend to increase with the *maximum* power capacity of the propulsion system, so that a large electric motor (allowing rapid acceleration) is likely to be less efficient under normal cruise conditions than a smaller and less powerful one. This is one of several reasons that EVs tend to be underpowered compared to ICVs.

Overall propulsive energy requirements for electric vehicles are usually quoted for idealized conditions of clean, dry pavement, no crosswinds or headwinds, and a moderate driving cycle. Under these conditions, propulsive energy requirements for a typical light-duty passenger vehicle are around 0.25 kWh per kilometer (measured at the output from the battery). Consumption levels for a van are about double this. Advanced energy-efficient small cars such as the GM Impact have achieved propulsive energy consumption levels around 0.1 kWh/km under these ideal conditions. Adverse weather and road conditions can increase this consumption by factors of two or more, however.

Auxiliary loads - In addition to propulsive power, vehicles require electrical or mechanical power for essential safety-related functions such as lights, windshield wipers, window defoggers, and - in large vehicles - power steering. These loads can amount to a further 0.3 to 0.7 kW for lights and other auxiliaries, while power steering may consume 0.2 to 1.5 kW (Bartholomew 1993). These essential auxiliary loads can easily amount to 5 or 10 percent of the propulsive energy requirement.

Climate control - Climate control (cabin heating and cooling) is a critical issue for electric vehicles. Depending on climatic conditions, adequate heating may be essential for vehicle safety (keeping windows defogged, etc.). Air-conditioning, although not critical to safety, may be essential for consumer acceptance. Adequate heating requires about 6 kW of thermal energy, which in ICVs is supplied from the engine waste heat. Hybrid and fuel-cell vehicles could also use waste heat for cabin comfort, but vehicles relying on off-board generation need to supply this either from the electricity stored on board or from a separate fuel-fired heater. The GM Impact uses an electric heat pump for both heating and cooling, depending on the ambient conditions. When air-conditioning is in use, the compressor consumes between 1.5 and 6 kW (Bartholomew 1993). At 60 km/hr, 3 kW for climate control would be equivalent to 0.05 kWh/km. Depending on the driving speed and ambient conditions, it is clear that the energy consumed by the climate control system could easily equal or exceed the energy required to propel the vehicle.

Electric energy consumption - All of the preceding discussion has referred to the demand for energy *supplied by the battery* (or other on-board electric supply system). For vehicles using off-board generation, additional losses will be incurred in electric transmission, in converting from AC line power to DC, voltage drops in the battery (or other storage system) during charging and discharging, and self-discharge (the gradual loss of charge experienced by batteries as they stand unused). For typical battery-electric systems, these losses amount to 30 percent to 50 percent of the electricity supplied[1], so that powerplant emissions per kWh of electricity actually provided to the propulsion system will be correspondingly greater.

Battery Technology

Battery technology is presently the main obstacle to widespread use of electric vehicles. If one takes 0.25 kWh/km as a *realistic* estimate of the stored energy requirement for a passenger vehicle under typical (non-ideal) operating conditions, attaining a minimally acceptable range of 200 km would require storage of 50

kWh of electrical energy. Present lead-acid batteries can store about 25 Wh per kg, so that storing 50 kWh would require about two tons of batteries. This is clearly impractical. For this reason, present EVs using lead-acid batteries have significantly shorter range than most consumers consider acceptable, especially under non-ideal (but realistic) operating conditions such as rain, night-driving, use of air-conditioning, etc. Major efforts to develop better batteries are under way around the world. Table A5.2.2 summarizes the mid-term and long-term goals of the U.S. Advanced Battery Consortium (ABC). These goals have not yet been attained. Even when the mid-term goals are attained, the batteries required to achieve 200 km range under realistic conditions would weigh 625 kg, would cost U.S.$7,500, would have a maximum power output of 94 kW, and a life of 5 years or 600 cycles. It is unclear whether a vehicle acceptable to consumers could be based on such batteries. U.S. ABC's far more ambitious long-term goals, if met, could result in commercially-acceptable electric vehicles, but the technological feasibility of meeting these goals is questionable. Table A5.2.3 summarizes the specific energy levels already achieved and the development goals for a number of battery technologies. As this table shows, only the lithium and aluminum-air batteries show significant promise of meeting the 200 Wh/kg goal of the U.S. ABC.

Other Energy Storage Media

Other energy storage media besides batteries could conceivably be used in an EV. Two of the most promising non-battery storage media are advanced flywheel systems and ultracapacitors. However, neither of these technologies is presently mature enough to offer a viable alternative to batteries as a primary energy storage medium. They may see their major applications in providing surge power and storage for regenerative braking in hybrid and fuel-cell electric vehicles.

Flywheel energy storage - A flywheel stores energy in the form of rotational kinetic energy, rather than the electrochemical potential energy of a battery. An electric motor/generator system combined with the flywheel serves to convert this rotational kinetic energy into electricity, and vice-versa. The use of high-strength materials allows the flywheel to rotate at extremely high speed, and thus to store a great deal of energy for its size and weight. Advantages of flywheels include high charge/discharge efficiency, high specific energy, and high specific power. Values quoted by a flywheel developer are 100 to 150 Whr/kg specific energy, and 5-10 kW/kg for specific power (Post 1993). The latter value is orders of magnitude higher than present electro-

---

1. See, for instance, G.D. Whitehead and A.S. Keller, "Performance Testing of the Vehma G Van Electric Vehicle", SAE Paper No. 910242. Between 42 and 49 percent of total AC electric power supplied was lost between the AC input to the charger and the DC output from the batteries (Whitehead and Keller 1991).

Table A5.2.2 Goals of the U.S. Advanced Battery Coalition

	Mid term	*Long term*
Specific energy Wh/kg (C/3 discharge rate)	80 (100 desired)	200
Energy density Wh/l (C/3 discharge rate)	135	300
Specific power W/kg (80% DOD/30 sec)	150 (200 Desired)	400
Power density W/l	250	600
Life (years)	5	10
Cycle life (80% DOD cycles)	600	1000
Ultimate price (US$/kWh)	<$150	$50
Operating environment	–30 to 65˚C	–40 to 85˚C
Recharge time	<6 hours	3 to 6 hours
Continuous discharge in one Hour (No failure)	75% (of rated energy capacity)	75% (of rated energy capacity)
Power & capacity degradation (Percent of rated spec)	20%	20%
Efficiency (C/3 discharge, 6 hour charge)	75%	80%
Self-discharge	<15% in 48 hours	<15% per month
Maintenance	No maintenance Service by qualified personnel only	No maintenance Service by qualified personnel only
Thermal loss (For high temp batteries)	3.2 W/kWh 15% of capacity, 48-hour period	3.2 W/kWh 15% of capacity, 48-hour period
Abuse resistance	Tolerant Minimized by on-board controls	Tolerant Minimized by on-board controls

Note: Wh/kg = Watt hour per kilogram; Wh/l = Watt hours per liter; C/3 = One third of capacity; DOD = Depth of discharge.
Source: Jemerson and others 1991

chemical batteries, making flywheels especially suitable for surge power applications. Charge/discharge efficiency, at 90 to 95 percent, is also considerably higher comparable to electrochemical batteries.

The major drawbacks of flywheel technology are an energy density little better than advanced battery systems, and very high costs. The projected cost of a modular 1 kW flywheel assembly is more than U.S.$400 after the 10,000th unit has been produced (Post 1993). To be economically viable, these costs would need to be reduced to U.S.$50 to U.S.$100 per kW at the very highest.

Ultracapacitors - Ultracapacitors are very high-capacity electronic charge storage devices. Because of their all-electronic design, they are extremely simple, and capable of generating very high specific power. Specific power in excess of 4 kW/kg has been achieved. These characteristics may make them useful for surge power applications in hybrid and fuel-cell ve-

hicles. Specific energy is presently 10-25 Wh/kg, which is no better than present lead-acid batteries. Thus, barring a significant breakthrough, these devices are unlikely to be useful for primary energy storage in EV applications.

Fuel cells. A fuel cell converts chemical energy directly into electrical energy, (same as a battery) but it differs from both a rechargeable (or secondary) storage battery and from a heat engine. While fuel cells and batteries are both electrochemical devices, the main difference is that in a battery, the electricity-producing reactants are regenerated in the battery by the recharging process, whereas in a fuel cell, the electricity-producing reactants are continually supplied from an external source such as air and a hydrogen storage tank.

A fuel cell system consists of the fuel-cell stack itself, which produces the electricity; a container to store the hydrogen or hydrogen-containing compound; an air-compressor to provide pressurized oxy-

Table A5.2.3 Specific Energies Achieved and Development Goals for Different Battery Technologies
(Wh/kg)

Type of battery	Specific energy			
	Achieved		*Goal*	
Lead-acid batteries				
Sealed	35	Battery	40-45	Battery
Vented	40	Battery	45-55	Battery
Alkaline batteries				
Ni/Cd	35	Battery	55-60	Battery
Ni/Fe	55	Battery	55-60	Battery
Ni/MH	55	Cell	70-80	Battery
Ni/Zn	70	Battery	70-80	Battery
Flow through batteries				
Zn/Br	70	Battery	70-80	Battery
High temp. batteries				
Na/S	110	Battery	120-130	Battery
Na/NiCl$_2$	80	Battery	100-120	Battery
LiAl/FeS	95	Module	100-120	Battery
Ambient temp. lithium batteries				
Li/MnO$_2$	100	Cell	160-180	Module
Li/LiCoO$_2$	115	Cell	170-190	Module
Metal-air batteries				
Fe/Air	70	Cell	100	Battery
Al/Air	250	Cell	250	Module

Source: OECD/IEA 1993

gen to the fuel cell; a cooling system to maintain the proper operating temperature; and a water management system to keep the fuel-cell membrane saturated and also to prevent product water from accumulating at the cathode. It will also have a reformer to convert the methanol into hydrogen and CO_2, if the vehicle stores hydrogen in the form of methanol (OECD/IEA 1993).

There are four types of fuel cells that could be used in motor vehicles: phosphoric acid, alkaline, solid oxide, and proton-exchange membrane. Phosphoric-acid fuel cells use a corrosive electrolyte, are too large and heavy to be used in light-duty motor vehicles, but may be satisfactory for use in heavy-duty vehicles. Alkaline fuel cells perform very well, but the electrolyte is intolerant of CO_2 and the system must be supplied with either bottled oxygen or air scrubbed of CO_2. Solid-oxide fuel cells appear promising with respect to performance, but it will take time before they become commercially available. On cold-start, they require a relatively long warm-up period to reach their operating temperature, although a battery could provide the energy required at start up.

Proton-exchange membrane (PEM) fuel cells offer relatively good performance and are able to provide a substantial amount of power at ambient temperatures. They contain no corrosive fluids, are relatively simple in construction, have a long life, and are potentially inex-

pensive to manufacture. Most electric vehicle research, development and demonstration programs are focusing on PEM fuel cells. It is believed that PEM fuel cells, which are expected to be commercially available within a few years are best suited for use in highway vehicles in the short term.

Several challenges face the development of PEM fuels cells; a detailed discussion of developments in fuel-cell technology in presented in OECD/IEA (1993):

- Improving the performance and reducing the cost of the membrane, without compromising its mechanical properties or making it susceptible to impurities in the gas stream.
- Finding a simple and effective way to keep the membrane moist but at the same time not allowing product water to build up at the cathode.
- Reducing the size and energy consumption of the air-compression system—by using a variable-speed "smart" air-compressor programmed to operate at the optimal efficiency point depending on the load.
- Reducing the weight, bulk, and manufacturing cost of the stack plates and assembly.
- Improving gas diffusion at low current densities and increasing the activity and active area of the catalyst.

Hybrid- and Fuel-Cell Electric Vehicles

A hybrid-electric vehicle combines electric propulsion with an on-board energy conversion system. Views of the relative roles of battery storage and on-board power appropriate for a hybrid vehicle vary. Two polar cases can be defined, type A and type C, to match the nomenclature used by the California Air Resources Board. A type A hybrid is essentially a battery-electric vehicle, with an on-board battery charger to extend its driving range. The expectation is that this vehicle would operate most of the time on battery power, with the batteries recharged from the electric utility grid. The on-board engine would be used only when the battery charge was near exhaustion - primarily on extra-urban trips. In this view, which appears to be that of the California Air Resources Board, a hybrid vehicle is essentially a way to make battery-electrics acceptable to the public by extending their driving range.

A type C hybrid, on the other hand, would not rely on the batteries for primary energy storage to any great extent. Instead, the engine would run continuously, at a more-or-less constant load, and the batteries, ultracapacitor bank, flywheel system, or other energy storage medium would function as an accumulator - smoothing out the load on the prime mover, and accepting the energy recovered by regenerative braking. This would allow the prime mover to be sized for the *average* power demand over the driving cycle (with some reserve), rather than the *peak* power demand, as it must be in a mechanical drive system without energy storage. Since the prime mover would operate at 50-100 percent of its designed power output most of the time, its average efficiency over the driving cycle would be much greater than that of present engines (for which the average load over the driving cycle is typically less than 10 percent of maximum power rating). The higher efficiency and the decoupling of engine and wheels should result in much lower emissions than a conventional vehicle. Decoupling the engine from the wheels (and thus from the need to respond to rapid changes in speed and power output) would also make possible the use of new, lower-emitting and possibly more efficient prime movers such as stirling engines, ultra-lean burn IC engines, and gas turbines. To eliminate evaporative emissions, these vehicles would probably be fueled by natural gas or neat methanol (M100). Ultimately, the primary energy conversion could be accomplished by a fuel cell, if future R&D results in fuel cell systems inexpensive enough for automotive use.

Unfortunately, none of these advanced electric vehicle technologies would qualify for ZEV status, and it is thus uncertain whether they will be developed fully. Even though emissions from each of these technologies would be comparable to or less than the powerplant emissions due to a battery-electric vehicle, the emissions from the vehicle itself would not be mathematically zero. In the Stirling and gas-turbine engines, non-zero emissions would result from catalytic combustion of natural gas or M100, while in the case of fuel cells, reforming of natural gas or M100 to produce the hydrogen required for the fuel cell would generate small but measurable amounts of NO_x and CO.

Table A5.2.4 shows the calculation of relative emissions from battery-electric, Stirling, gas-turbine, and fuel-cell technologies in heavy-duty engines. The first four columns in the table show the emissions from each technology per kilowatt-hour of electric power output. These are compared with projected average emissions from electric generation in the northeast United States. The fifth column shows the estimated loss in going from the electricity source to the motor input on the vehicle. In the case of the on-board generation technologies, these losses would be negligible. For the battery-electric alternative, it is assumed that cumulative losses due to AC-DC conversion and electrical losses in the battery would amount to 40 percent. Columns 6 through 9 show the emissions produced by the energy conversion system per bhp-hr of work output by the electric propulsion system. These values are comparable both with each other and with the emissions (in g/bhp-hr) from conventional engines over the transient test cycle.

As shown in table A5.2.4, NO_x and SO_x emissions from the power-plants supplying the battery-electric vehicle are much higher than the corresponding emissions from the on-board energy conversion systems. Since the energy-efficiency, range, and payload of these systems would also be better than those of a battery-electric, they clearly represent a superior alternative.

Electric Vehicle Costs and Availability

In-use electric vehicles around the world are mostly deployed in demonstration programs (IEA 1993). This excludes specialized, low-speed vehicles such as golf carts, milk delivery vans, and forklifts. Due to commercialization efforts in Japan, California, and elsewhere, this number of EVs is projected to increase to several hundred thousand by 2000, but the realism of these projections is open to question. Table A5.2.5 summarizes the characteristics and prices of some electric vehicles that are currently available. As this table indicates, the prices of present EVs are extremely high compared to those of ICVs.

The high costs of present EVs are due in part to the very low production volumes involved. These costs would presumably be reduced by mass production. On the other hand, the features and range required to make

Table A5.2.4 Relative Emissions from Battery-Electric and Hybrid-Electric Vehicles

	Emissions (g/kWh elec.)				Battery loss	Emissions (g/bhp-hr mech)			
	NO_x	CO	NMHC	SO_x		NO_x	CO	NMHC	SO_x
Hybrid vehicles: on-board generation									
Natural gas fuel cell									
Phosphoric acid	0.066	n.a.	n.a.	0.00	0%	0.049	—	—	0.00
Solid oxide	0.156	n.a.	n.a.	0.00	0%	0.116	—	—	0.00
Natural gas stirling engines									
MTI stirling	0.097	0.425	0.0005	0.00	0%	0.073	0.317	0.0004	0.00
DDC-STM	0.201	2.856	0.038	0.00	0%	0.150	2.131	0.028	0.00
Natural gas turbine (24 KW)	0.046	0.03	0.025	0.00	0%	0.034	0.022	0.019	0.00
Battery-electric vehicles: fixed power plants									
Future Northeast power plants (NESCAUM 1992)									
Avg. 1995	1.9	0.15	0.15	5.1	40%	2.387	0.187	0.187	6.32
Avg. 2000	0.4	0.15	0.013	3.9	40%	0.497	0.187	0.016	4.85
Avg. 2015	0.6	0.15	0.013	1.5	40%	0.746	0.187	0.016	1.87

n.a. Not available
— Not applicable
Source: Hirschenhofer 1992; Weaver and Chan 1992; MacKay 1992; Tennis 1992

Table A5.2.5 Examples of Electric Vehicles Available in 1993

Model	Manufacturer	Type of vehicle	Battery	Motor	Range	Top speed	Price
G-Van	Conceptor, Canada	Full-sized van	Pb-Acid	DC	100 km[1]	83 km/h	US$50,000
J5/C25	Peugeot/Citroen, France	Full-sized van	Pb-Acid	DC	75 km[1]	90 km/h	not available
Hi-Jet	Daihatsu, Japan	Micro-Van	Pb-Acid	DC	90 km[2]	75 km/h	Y 2,170,000
Panda	Fiat, Italy	Two-seater car	Pb-Acid	DC	70 km[2]	70 km/h	L 30,000,000
Volta	SEER, France	Mini-van	Pb-Acid	DC	80 km[2]	73 km/h	FF 125,000
Mini-el City	CityCorn, Denmark	Commuter car	Pb-Acid	DC	40 km[2]	35 km/h	KDr 40,000
Town	Toyota Motor Co., Japan	Van	Pb-Acid	AC	160 km[2]	110 km/h	Y 8,000,000
Libero	Mitsubishi, Japan	Four-seater station wagon	Ni/Cd or Pb-Acid	AC	250 km[3] 166 km[4]	130 km/h	Y 10,000,000
Elcat	Elcat, Finland	Van	Pb-Acid	DC	100 km[5] 60-70 km[1]	72 km/h	not available

Note:
1. City driving
2. Constant speed of 40 km/h
3. Constant speed of 40 km/h using Ni/Cd battery
4. Constant speed of 40 km/h using Pb/Acid battery
5. Speed of 50 km/h
Source: Institute of Applied Energy, Japan 1993

EVs acceptable to consumers are likely to increase costs significantly, due to the increased battery capacity and other added requirements. Estimates of the future price differential between EVs and ICVs range from little or no difference (California Air Resources Board 1994) to U.S.$21,000 per vehicle (Austin and Lyons 1994). Some U.S. auto manufacturers have estimated even higher cost differentials.

In addition to the difference in vehicle cost, it is important to take into account the differences in vehicle performance between feasible EVs and present ICVs. Consumers place a significant value on vehicle characteristics such as range and acceleration. Henderson and Rusin (1994) have calculated that -based on U.S. consumers' demonstrated willingness-to-pay for these characteristics in ICVs, the value provided to consumers by EVs would be between U.S.$6,700 and U.S.$7,700 less per vehicle. In other words, EVs would have to be priced between U.S.$6,700 and U.S.$7,700 *lower* than comparable ICVs in order to be attractive to a typical consumer.

It can be assumed that the *initial* purchasers of EVs would be consumers who do not value range and acceleration highly, so that this difference in values would be much less for those individuals. For EVs to make a significant contribution to air quality or energy savings, however, it would be necessary for them to be attractive to the typical consumer as well.

References

Austin, T.C. and J.M. Lyons. 1994. "Cost Effectiveness of the California Low Emission Vehicle Standards", *SAE Paper* No. 940471, SAE International, Warrendale, Pennsylvania.

Bartholomew, R.W. 1993. "Opening Remarks for the Parallel Session on Energy Conversion Systems" in *Advanced Components for Electric and Hybrid Electric Vehicles: Workshop Proceedings*, NIST Special Publication 860, National Institute of Standards and Technology, Gaithersburg, Maryland.

Bortone, C. 1992. "Outcomes of the Workshop on Electric Vehicles and Advanced Batteries" in *The Urban Electric Vehicle - Policy Options, Technology Trends, and Market Prospects*, Organization for Economic Co-Operation and Development (OECD), Paris, France.

CARB. 1994. "Zero Emission Vehicle Update," Draft Technical Document for the Low-emission Vehicle and Zero Emission Vehicle Workshop (March 25, 1994), California Air Resources Board, Los Angeles, California.

Fukino, Masato, Namio Irie, and Hideo Ito. 1992. "Development of an Electric Concept Vehicle with a Super Quick Charging System" *SAE Paper* No. 920442, SAE International, Warrendale, Pennsylvania.

Henderson, T.P. and M. Rusin. 1994. *Electric Vehicles: Their Technical and Economic Status*, Research Study No. 073, American Petroleum Institute, Washington, D.C.

Hirschenhofer, J.J. 1992. "Commercialization of Fuel Cell Technology", *Mechanical Engineering*.

Institute of Applied Energy, Japan. 1993.

Jemerson, F.E., M.A. Dzieciuch, and D.R. Smith. 1991. "United States Advanced Battery Consortium", Annual Automotive Technology Development Contractor's Coordination Meeting, Dearborn, Michigan (October, 1991).

MacKay, R. 1992. "Hybrid Vehicle Gas Turbines," *SAE Paper* No. 930044, SAE International, Warrendale, Pennsylvania.

OECD/IEA. 1993. *Electric Vehicles: Technology, Performance, and Potential*, Organization for Economic Co-operation and Development and International Energy Agency, Paris.

Post, R. 1993. "Flywheel Technology" in *Advanced Components for Electric and Hybrid Electric Vehicles: Workshop Proceedings*, NIST Special Publication 860, National Institute of Standards and Technology, Gaithersburg, Maryland.

Tennis, M.W. 1992. "Impact of Battery-Powered Electric Vehicles on Air Quality in the Northeast States," Report to NESCAUM, Boston.

Weaver, C.S. and L.M. Chan. 1992. "Technical Memorandum under GRI Contract." Engine, Fuel, and Emissions Engineering, Inc., Sacramento, California.

Weaver, C.S. and L.M. Chan. 1995. "Company Archives" Multiple Sources. Engine, Fuel and Emissions Engineering, Inc., Sacramento, California.

Whitehead, G.D. and A.S. Keller. 1991. "Performance Testing of the Vehma G Van Electric Vehicle." *SAE Paper* No. 910242, SAE International, Warrendale, Pennsylvania.

Appendix 5.3

Alternative Fuel Options for Urban Buses in Santiago, Chile: A Case Study

In 1989 the National Commission for Energy in Chile organized a pilot program with the objective of verifying under local conditions the conversion costs, energy efficiency, and impact on air pollution of using "cleaner" fuels in public transport buses in Santiago (CNE 1988).

The experiments were carried out over six months on thirteen buses—eleven representative of the regular fleet and two new vehicles. As originally designed, the program called for the use of properly maintained buses complying with factory specifications before conversion to alternative fuels. These requirements were impossible to meet due to the poor condition of the fleet, so a conditioning stage was introduced to bring the engines up to "first repair" specifications. Important conclusions were reached before the pilot program even started. First, the existing engines did not operate properly as diesels and were not suited for direct conversion to alternative fuels. Second, dramatic improvements in performance and emissions could be obtained with standard engine overhaul and maintenance.

Following a conditioning stage, six buses were modified to run on LPG and methanol, and two new buses (one gasoline-fueled, another CNG-fueled) provided by private owners were added to the experiment. Four conventional diesel-fueled buses, after overhaul, were used as controls under various bus maintenance regimes. A fifth diesel-fueled bus (one year old) was allowed to operate under its owner's rules, and an old diesel bus representing actual operating and maintenance conditions was added to the ex-periment. For the emissions test, the gasoline-fueled bus and a diesel bus were tested with and without a catalytic converter. Characteristics of the buses are summarized below:

Converted to liquefied petroleum gas (LPG)

- LPG-1; spark-ignited (converted diesel engine)
- LPG-2; spark-ignited (new spark ignition V-8 Otto-cycle engine)
- LPG-3; spark-ignited (factory-converted)

Converted to methanol

- Methanol-1; spark-ignited (converted diesel engine)
- Methanol-2; spark-ignited (new ethanol-adapted engine)
- Methanol-3; reconditioned engine (diesel engine using methanol with additives to facilitate ignition).

Gasoline

- Gasoline; spark-ignited (new Otto-cycle engine with an oxidation catalytic converter; unleaded gasoline)

Compressed natural gas (CNG)

- CNG; spark-ignited (new Otto-cycle engine)

Conventional diesel (reconditioned)

- Diesel-1; fitted with a ceramic filter (trap) and a re-generation system
- Diesel-2; with manufacturer-recommended mainte-nance
- Diesel-3; with average Santiago maintenance
- Diesel-4; without maintenance
- Diesel-5; reference bus (base case)

This appendix was contributed by Mr. Juan Escudero, University of Chile, Santiago. He was formerly the Executive Secretary of the Special Commission for the Decontamination of the Santiago Metropolitan Region.

Table A5.3.1 Emissions of Buses with Alternative Fuels, Santiago, Chile
(grams per kilometer)

Bus I.D./fuel type	CADEBUS (Santiago driving cycle)[a]				
	Particulate matter	CO	NO_x	VOC	Aldehydes
LPG-1	0.076	31.543	11.491	2.626	0.0127
LPG-2	0.061	166.253	3.727	6.264	0.0068
LPG-3	0.047	115.294	8.949	3.617	0.0050
METHANOL-1	0.164	116.438	4.746	11.866	0.0600
METHANOL-2	0.033	84.354	6.059	6.494	0.0471
METHANOL-3	0.154	16.598	4.150	7.227	0.0328
GASOLINE-1A[b]	0.195	2.568	3.705	0.842	0.0003
GASOLINE-1	0.304	23.423	3.857	15.341	—
CNG-1	0.043	28.976	7.937	5.136	0.0033
DIESEL-1A[b]	0.385	1.796	3.821	0.640	—
DIESEL-1	1.621	3.656	4.205	1.291	0.0029
DIESEL-2	0.705	2.136	4.632	1.422	0.0016
DIESEL-3	0.598	2.556	4.522	1.447	0.0022
DIESEL-4	1.518	4.606	7.779	2.295	—
DIESEL-5	0.341	1.833	5.362	0.885	0.0028
DIESEL-6	3.050	9.660	7.000	1.730	—

— Not applicable
a. All tests performed on a chassis dynamometer.
b. With a catalytic converter.
Source: Escudero 1991; CNE 1988

Table A5.3.2 Economics of Alternative Fuel Options for Urban Buses in Santiago, Chile

Bus I.D./fuel type	Engine conversion (U.S.$)	Fuel system conversion (U.S.$)	Fuel efficiency (km/l)	Fuel cost (U.S.$/km)	Total incremental cost (U.S.$/year)[a]
LPG-1	4,489	1,253	1.93	0.149	6,234
LPG-2	3,132	1,009	1.64	0.174	7,275
LPG-3	b/	1,427	1.65	0.173	8,273
METHANOL-1	4,495	303	0.81	0.183	8,627
METHANOL-2	b	303	1.16	0.128	4,084
METHANOL-3	450	303	1.30	0.114	4,386
GASOLINE-1	4,784 + b/	—	2.11	0.169	7,415
CNG-1	3,973 + b/	2,385	3.30[c]	0.080	2,437
DIESEL-1	1,288 (trap)	—	4.09	0.076	2,015
DIESEL-2	130	—	4.56	0.069	*
DIESEL-3	130	—	4.34	0.072	*
DIESEL-4	130	—	3.35	0.093	*
DIESEL-5 (Base case)[d]	130	—	4.58	0.068	0

— Not applicable
a. Compared to base case (DIESEL-5), and includes engine conversion/replacement, fuel and maintenance costs.
b. New engine costs in Chile were: LPG-3: U.S.$13,011; METHANOL-1: U.S.$7,250; GASOLINE-1: U.S.$6,894; CNG-1: U.S.$9,715. Engine replacement cost for LPG-2 was U.S.$6,097; for all other buses U.S.$7,250.
c. CNG in km/m^3.
d. Optimized diesel.
* not significantly different from base case
Source: Escudero 1991; CNE 1990

Old diesel

- Diesel-6; poorly maintained Santiago bus

The bus emission characteristics of alternative fuel options are summarized in table 5.3.1. All alternative fuel options were effective in reducing particulate emissions, with the highest reductions achieved with LPG, CNG, and methanol (with additives to facilitate compression ignition). All fuel alternatives increased CO emissions compared with diesel, to the point that catalytic converters were required (compare GASOLINE-1A with GASOLINE-1). In terms of reduced diesel emissions, the importance of maintenance is paramount; buses with overhauled and well-maintained diesel engines had three to seven times lower emissions than the typical Santiago bus.

An economic evaluation of alternative fuel options for Santiago buses (table 5.3.2) confirmed that diesel-fueled buses had a strong competitive advantage, while the CNG-fueled buses could become competitive. This evaluation demonstrated that LPG-, methanol-, and gas-oline-fueled buses were unlikely to gain a substantial market share in the Santiago metropolitan region. In light of the results of this pilot program, bus emissions control policy in Chile was targeted at maintenance improvement programs for diesel-fueled buses and conversion of public transport buses to natural gas.

References

CNE (Comisión Nacional de Energía). 1988. "Contaminación Atmosférica: Análisis de Alternativas de uso de Petroleo Diesel en Vehículos de Locomoción Colective de Santiago." Santiago, Chile.

_____. 1990. "Informe Final: Resultados, Conclusiones y Recomendaciones," Programa Piloto: Uso Combustibles Alternativos en Vehiculos de Locomocion Colectiva Urbana en Santiago, Santiago.

Escudero, J. 1991. "Notes on Air Pollution Issues in Santiago Metropolitan Region" Letter #910467, dated May 14. Comisión Especial de Descontaminación de la Region Metropolitana. Santiago, Chile.

Abbreviations and Conversion Factors

Abbreviations

AASHTO	American Association of State Highway and Transportation Officials
ABC	Advanced Battery Consortium
ACEA	European Automobile Manufacturers' Association
ACVEN	Advisory Council on Vehicle Emissions and Noise
AKI	Anti Knock Index
AQIRP	Air Quality Research Improvement Program
ASLG	Academy for State and Local Government
ASM	Acceleration Simulation Mode
ASTM	American Society for Testing and Materials
ATAC	Australian Transport Advisory Council
C$	Canadian Dollar
°C	Degrees Celsius
CAPE	Campaign for Action to Protect the Earth
CARB	California Air Resources Board
CETESB	Companhia de Tecnologia de Saneamento Ambiental, São Paulo, Brazil
CH_4	Methane
CNAIC	China National Automotive Industry Corporation
CNG	Compressed Natural Gas
CO	Carbon Monoxide
CO_2	Carbon Dioxide
CONAMA	Brazilian National Environmental Board
CONCAWE	Oil Companies' European Organization for Environmental Protection and Health
CONTA	Comision Nacional del Transporte Automotor
CSEPA	China State Environmental Protection Administration
CVS	Constant Volume Sampling
CVT	Continuously Variable Transmission
DC	Engine Displacement (cubic centimeters)
DI	Direct Injection
DIPE	Di-Isopropyl Ether
EC	European Community
ECE	Economic Commission for Europe
ECMT	European Conference of Ministers of Transport
ECS	Emission Control System
EEC	European Economic Community

EGR	Exhaust Gas Recirculation
EHC	Electrically Heated Catalyst
EHN	Ethyl Hexyl Nitrate
EOI	End of Ignition
EPEFE	European Programme on Emissions, Fuels, and Engine Technologies
ETBE	Ethyl Tertiary Buryl Ether
EU	European Union
EUDC	Extra Urban Driving Cycle
EUROPIA	European Petroleum Industry Association
EV	Electric Vehicle
FBP	Final Boiling Point
FTP	Federal Test Procedure

g/bhp-hr	grams per brake horse power per hour
g/km	grams per kilometer
g/l	grams per liter
g/mile	grams per mile
GUT	Graz University of Technology, Austria
GVRD	Greater Vancouver Regional District
GVW	Gross Vehicle Weight

HC	Hydrocarbons
HDV	Heavy-Duty Vehicle
HSU	Hartridge Smoke Unit

IBP	Initial Boiling Point
ICV	Vehicle with an Internal Combustion Engine
I/M	Inspection and Maintenance
IDI	Indirect Injection
IEA	International Energy Agency
IFP	Institut Francais du Petrole
IGRP	Indonesian German Research Project
ILEV	Inherently Low-Emission Vehicle
INRETS	Institut National de Recherche sur les Transports et Leur Securité, France
IR	Infra-red
ITRI	Industrial Technology Research Institute, Taiwan (China)
IVHS	Intelligent Vehicle-Highway Systems

kWh	kilowatt-hour
kPa	kilo Pascal

IAPAC	Injection Assisté par Air Comprimé
LDV	Light-Duty Vehicle
LEV	Low-Emission Vehicle
LHDV	Light Heavy-Duty Vehicle
LHV	Lower Heating Value
LNG	Liquefied Natural Gas
LPG	Liquefied Petroleum Gas

MBT	Minimum for Best Torque
MEG	Methane/Ethanol/Gasoline
MHDV	Medium Heavy-Duty Vehicle
MMT	Methylcyclopentadienyl Manganese Tricarbonyl
MON	Motor Octane Number

mpg	miles per gallon
msl	mean sea level
MTBE	Methyl Tertiary-Butyl Ether
MVMA	Motor Vehicle Manufacturer's Association
n.a.	not available
N.A.	Naturally Aspirated
NAFTA	North American Free Trade Agreement
NDIR	Non-Dispersive Infra-Red
NEVI	Nepal Electric Vehicle Industry
N_2O	Nitrous Oxide
NGV	Natural Gas Vehicle
NMHC	Non-methane Hydrocarbons
NMOG	Non-Methane Organic Gas
NO	Nitric Oxide
NO_2	Nitrogen Dioxide
NO_x	Nitrogen Oxides
NR	Not Regulated
NRDC	Natural Resources Defense Council
NZ$	New Zealand Dollar
O_3	Ozone
OBD	On-Board Diagnostic
OCP	Orbital Combustion Process
OECD	Organization for Economic Cooperation and Development
OFPE	l'Office Federal de la Protection de l'Environnement, Switzerland
OTA	Office of Technology Assessment, U.S. Congress
PAH	Polycyclic and Nitro-Polycyclic Aromatic Compounds
Pb	Lead
PCV	Positive Crankcase Ventilation
PEM	Proton-Exchange Membrane
PM	Particulate Matter
PNA	Polynuclear Aromatic Content
PROCONVE	Brazilian Vehicle Emissions Control Program
psi	pounds per square inch
ppm	parts per million
R&D	Research and Development
RFG	Reformulated Gasoline
RM	Reference Mass
rpm	revolutions per minute
RVP	Reid Vapor Pressure
RON	Research Octane Number
SAE	Society of Automotive Engineers
SHED	Sealed Housing for Evaporative Determinations
SIP	State Implementation Plans
SMD	Sauter Mean Diameter
SOF	Soluble Organic Fraction
SOI	Start of Ignition
SO_2	Sulfur Dioxide
SO_x	Sulfur Oxides
SOF	Soluble Organic Fraction

SPM	Suspended Particulate Matter
SULEV	Super Ultra Low-Emission Vehicle
SVRP	Smoking Vehicle Reporting Program
TAME	Tertiary Amyl Methyl Ether
TBA	Tertiary Butyl Alcohol
TC	Turbo-charged
TCA	Turbo-charged After-cooled
TDC	Top-Dead-Center
TEL	Tetra-Ethyl Lead
THC	Total Hydrocarbons
TLEV	Transitional Low-Emission Vehicle
TML	Tetra-Methyl Lead
TPM	Total Particulate Matter
TSP	Total Suspended Particulates
ULEV	Ultra Low-Emission Vehicle
UN	United Nations
UNCHS	United Nations Center for Human Settlement
UNDP	United Nations Development Programme
UNEP	United Nations Environment Programme
UNIDO	United Nations Industrial Development Organization
UNOCAL	Union of Oil Companies of California
U.S. EPA	United States Environmental Protection Agency
U.S.$	United States Dollar
VCO	Valve-Covers-Orifice
VDT	Vehicle Distance Traveled
VGT	Variable Geometry Turbochargers
VKT	Vehicles Kilometers of Travel
VOC	Volatile Organic Compounds
WHO	World Health Organization
WRI	World Resources Institute
ZEV	Zero Emission Vehicle

Conversion Factors

1 gallon	=	3.785 liters
1 mile	=	1.609 kilometers
1 psi	=	6.893 kPa
1 bhp	=	0.745 kW
1 lb	=	0.373 kg

Country Index

Distributors of World Bank Publications

Prices and credit terms vary from country to country. Consult your local distributor before placing an order.

ALBANIA
Adrion Ltd.
Perlat Rexhepi Str.
Pall. 9, Shk. 1, Ap. 4
Tirana
Tel: (42) 274 19; 221 72
Fax: (42) 274 19

ARGENTINA
Oficina del Libro Internacional
Av. Cordoba 1877
1120 Buenos Aires
Tel: (1) 815-8156
Fax: (1) 815-8354

AUSTRALIA, FIJI, PAPUA NEW GUINEA, SOLOMON ISLANDS, VANUATU, AND WESTERN SAMOA
D.A. Information Services
648 Whitehorse Road
Mitcham 3132
Victoria
Tel: (61) 3 9210 7777
Fax: (61) 3 9210 7788
http://www.dadirect.com.au

AUSTRIA
Gerold and Co.
Graben 31
A-1011 Wien
Tel: (1) 533-50-14-0
Fax: (1) 512-47-31-29
http://www.gerold.co.at/online
E-mail: buch@gerold.telecom.at

BANGLADESH
Micro Industries Development
  Assistance Society (MIDAS)
House 5, Road 16
Dhanmondi R/Area
Dhaka 1209
Tel: (2) 326427
Fax: (2) 811188

BELGIUM
Jean De Lannoy
Av. du Roi 202
1060 Brussels
Tel: (2) 538-5169
Fax: (2) 538-0841

BRAZIL
Publicações Tecnicas
Internacionais Ltda.
Rua Peixoto Gomide, 209
01409 Sao Paulo, SP.
Tel: (11) 259-6644
Fax: (11) 258-6990

CANADA
Renouf Publishing Co. Ltd.
1294 Algoma Road
Ottawa, Ontario K1B 3W8
Tel: 613-741-4333
Fax: 613-741-5439
http://fox.nstn.ca/~renouf
E-mail: renouf@fox.nstn.ca

CHINA
China Financial & Economic
Publishing House
8, Da Fo Si Dong Jie
Beijing
Tel: (1) 333-8257
Fax: (1) 401-7365

COLOMBIA
Infoenlace Ltda.
Apartado Aereo 34270
Bogotá D.E.
Tel: (1) 285-2798
Fax: (1) 285-2798

COTE D'IVOIRE
Centre d'Edition et de Diffusion
  Africaines (CEDA)
04 B.P. 541
Abidjan 04 Plateau
Tel: 225-24-6510
Fax: 225-25-0567

CYPRUS
Center of Applied Research
Cyprus College
6, Diogenes Street, Engomi
P.O. Box 2006
Nicosia
Tel: 244-1730
Fax: 246-2051

CZECH REPUBLIC
National Information Center
prodejna, Konviktska 5
CS – 113 57 Prague 1
Tel: (2) 2422-9433
Fax: (2) 2422-1484
http://www.nis.cz/

DENMARK
SamfundsLitteratur
Rosenoerns Allé 11
DK-1970 Frederiksberg C
Tel: (31)-351942
Fax: (31)-357822

EGYPT, ARAB REPUBLIC OF
Al Ahram
Al Galaa Street
Cairo
Tel: (2) 578-6083
Fax: (2) 578-6833

The Middle East Observer
41, Sherif Street
Cairo
Tel: (2) 393-9732
Fax: (2) 393-9732

FINLAND
Akateeminen Kirjakauppa
P.O. Box 23
FIN-00371 Helsinki
Tel: (0) 12141
Fax: (0) 121-4441
URL: http://booknet.cultnet.fi/aka/

FRANCE
World Bank Publications
66, avenue d'Iéna
75116 Paris
Tel: (1) 40-69-30-56/57
Fax: (1) 40-69-30-68

GERMANY
UNO-Verlag
Poppelsdorfer Allee 55
53115 Bonn
Tel: (228) 212940
Fax: (228) 217492

GREECE
Papasotiriou S.A.
35, Stournara Str.
106 82 Athens
Tel: (1) 364-1826
Fax: (1) 364-8254

HONG KONG, MACAO
Asia 2000 Ltd.
Sales & Circulation Department
Seabird House, unit 1101-02
22-28 Wyndham Street, Central
Hong Kong
Tel: 852 2530-1409
Fax: 852 2526-1107
http://www.sales@asia2000.com.hk

HUNGARY
Foundation for Market Economy
Dombovari Ut 17-19
H-1117 Budapest
Tel: 36 1 204 2951 or
36 1 204 2948
Fax: 36 1 204 2953

INDIA
Allied Publishers Ltd.
751 Mount Road
Madras - 600 002
Tel: (44) 852-3938
Fax: (44) 852-0649

INDONESIA
Pt. Indira Limited
Jalan Borobudur 20
P.O. Box 181
Jakarta 10320
Tel: (21) 390-4290
Fax: (21) 421-4289

IRAN
Kowkab Publishers
P.O. Box 19575-511
Tehran
Tel: (21) 258-3723
Fax: 98 (21) 258-3723

Ketab Sara Co. Publishers
Khaled Eslamboli Ave.,
6th Street
Kusheh Delafrooz No. 8
Tehran
Tel: 8717819 or 8716104
Fax: 8862479
E-mail: ketab-sara@neda.net.ir

IRELAND
Government Supplies Agency
Oifig an tSoláthair
4-5 Harcourt Road
Dublin 2
Tel: (1) 461-3111
Fax: (1) 475-2670

ISRAEL
Yozmot Literature Ltd.
P.O. Box 56055
Tel Aviv 61560
Tel: (3) 5285-397
Fax: (3) 5285-397

R.O.Y. International
PO Box 13056
Tel Aviv 61130
Tel: (3) 5461423
Fax: (3) 5461442

Palestinian Authority/Middle East
Index Information Services
P.O.B. 19502 Jerusalem
Tel: (2) 271219

ITALY
Licosa Commissionaria Sansoni
SPA
Via Duca Di Calabria, 1/1
Casella Postale 552
50125 Firenze
Tel: (55) 645-415
Fax: (55) 641-257

JAMAICA
Ian Randle Publishers Ltd.
206 Old Hope Road
Kingston 6
Tel: 809-927-2085
Fax: 809-977-0243

JAPAN
Eastern Book Service
Hongo 3-Chome,
  Bunkyo-ku 113
Tokyo
Tel: (03) 3818-0861
Fax: (03) 3818-0864
http://www.bekkoame.or.jp/~svt-ebs

KENYA
Africa Book Service (E.A.) Ltd.
Quaran House, Mfangano Street
P.O. Box 45245
Nairobi
Tel: (2) 23641
Fax: (2) 330272

KOREA, REPUBLIC OF
Daejon Trading Co. Ltd.
P.O. Box 34
Yeoeida, Seoul
Tel: (2) 785-1631/4
Fax: (2) 784-0315

MALAYSIA
University of Malaya Cooperative
  Bookshop, Limited
P.O. Box 1127
Jalan Pantai Baru
59700 Kuala Lumpur
Tel: (3) 756-5000
Fax: (3) 755-4424

MEXICO
INFOTEC
Apartado Postal 22-860
14060 Tlalpan,
Mexico D.F.
Tel: (5) 606-0011
Fax: (5) 624-2822

NETHERLANDS
De Lindeboom/InOr-Publikaties
P.O. Box 202
7480 AE Haaksbergen
Tel: (53) 574-0004
Fax: (53) 572-9296

NEW ZEALAND
EBSCO NZ Ltd.
Private Mail Bag 99914
New Market
Auckland
Tel: (9) 524-8119
Fax: (9) 524-8067

NIGERIA
University Press Limited
Three Crowns Building Jericho
Private Mail Bag 5095
Ibadan
Tel: (22) 41-1356
Fax: (22) 41-2056

NORWAY
Narvesen Information Center
Book Department
P.O. Box 6125 Etterstad
N-0602 Oslo 6
Tel: (22) 57-3300
Fax: (22) 68-1901

PAKISTAN
Mirza Book Agency
65, Shahrah-e-Quaid-e-Azam
P.O. Box No. 729
Lahore 54000
Tel: (42) 7353601
Fax: (42) 7585283

Oxford University Press
5 Bangalore Town
Sharae Faisal
PO Box 13033
Karachi-75350
Tel: (21) 446307
Fax: (21) 454-7640
E-mail: oup@oup.hi.erum.com.pk

PERU
Editorial Desarrollo SA
Apartado 3824
Lima 1
Tel: (14) 285380
Fax: (14) 286628

PHILIPPINES
International Booksource Center
Inc.
Suite 720, Cityland 10
Condominium Tower 2
H.V dela Costa, corner
Valero St.
Makati, Metro Manila
Tel: (2) 817-9676
Fax: (2) 817-1741

POLAND
International Publishing Service
Ul. Piekna 31/37
00-577 Warzawa
Tel: (2) 628-6089
Fax: (2) 621-7255

PORTUGAL
Livraria Portugal
Rua Do Carmo 70-74
1200 Lisbon
Tel: (1) 347-4982
Fax: (1) 347-0264

ROMANIA
Compani De Librarii Bucuresti
S.A.
Str. Lipscani no. 26, sector 3
Bucharest
Tel: (1) 613 9645
Fax: (1) 312 4000

RUSSIAN FEDERATION
Isdatelstvo <Ves Mir>
9a, Kolpachniy Pereulok
Moscow 101831
Tel: (95) 917 87 49
Fax: (95) 917 92 59

SAUDI ARABIA, QATAR
Jarir Book Store
P.O. Box 3196
Riyadh 11471
Tel: (1) 477-3140
Fax: (1) 477-2940

**SINGAPORE, TAIWAN,
MYANMAR, BRUNEI**
Asahgate Publishing Asia
  Pacific Pte. Ltd.
41 Kallang Pudding Road #04-03
Golden Wheel Building
Singapore 349316
Tel: (65) 741-5166
Fax: (65) 742-9356

SLOVAK REPUBLIC
Slovart G.T.G. Ltd.
Krupinska 4
PO Box 152
852 99 Bratislava 5
Tel: (7) 839472
Fax: (7) 839485

SOUTH AFRICA, BOTSWANA
For single titles:
Oxford University Press
  Southern Africa
P.O. Box 1141
Cape Town 8000
Tel: (21) 45-7266
Fax: (21) 45-7265

For subscription orders:
International Subscription Service
P.O. Box 41095
Craighall
Johannesburg 2024
Tel: (11) 880-1448
Fax: (11) 880-6248

SPAIN
Mundi-Prensa Libros, S.A.
Castello 37
28001 Madrid
Tel: (1) 431-3399
Fax: (1) 575-3998
http://www.tsai.es/mprensa

Mundi-Prensa Barcelona
Consell de Cent, 391
08009 Barcelona
Tel: (3) 488-3009
Fax: (3) 487-7659

SRI LANKA, THE MALDIVES
Lake House Bookshop
P.O. Box 244
100, Sir Chittampalam A.
  Gardiner Mawatha
Colombo 2
Tel: (1) 32105
Fax: (1) 432104

SWEDEN
Fritzes Customer Service
Regeringsgaton 12
S-106 47 Stockholm
Tel: (8) 690 90 90
Fax: (8) 21 47 77

Wennergren-Williams AB
P. O. Box 1305
S-171 25 Solna
Tel: (8) 705-97-50
Fax: (8) 27-00-71

SWITZERLAND
Librairie Payot
Service Institutionnel
Côtes-de-Montbenon 30
1002 Lausanne
Tel: (021)-341-3229
Fax: (021)-341-3235

Van Diermen Editions Techniques
Ch. de Lacuez 41
CH1807 Blonay
Tel: (021) 943 2673
Fax: (021) 943 3605

TANZANIA
Oxford University Press
Maktaba Street
PO Box 5299
Dar es Salaam
Tel: (51) 29209
Fax (51) 46822

THAILAND
Central Books Distribution
306 Silom Road
Bangkok
Tel: (2) 235-5400
Fax: (2) 237-8321

TRINIDAD & TOBAGO, JAMAICA
Systematics Studies Unit
#9 Watts Street
Curepe
Trinidad, West Indies
Tel: 809-662-5654
Fax: 809-662-5654

UGANDA
Gustro Ltd.
Madhvani Building
PO Box 9997
Plot 16/4 Jinja Rd.
Kampala
Tel/Fax: (41) 254763

UNITED KINGDOM
Microinfo Ltd.
P.O. Box 3
Alton, Hampshire GU34 2PG
England
Tel: (1420) 86848
Fax: (1420) 89889

ZAMBIA
University Bookshop
Great East Road Campus
P.O. Box 32379
Lusaka
Tel: (1) 213221 Ext. 482

ZIMBABWE
Longman Zimbabwe (Pte.)Ltd.
Tourle Road, Ardbennie
P.O. Box ST125
Southerton
Harare
Tel: (4) 6216617
Fax: (4) 621670